Lecture Notes in Computer Science 4613

Commenced Publication in 1973
Founding and Former Series Editors:
Gerhard Goos, Juris Hartmanis, and Jan van Leeuwen

Franco P. Preparata   Qizhi Fang (Eds.)

Frontiers
in Algorithmics

First Annual International Workshop, FAW 2007
Lanzhou, China, August 1-3, 2007
Proceedings

 Springer

Volume Editors

Franco P. Preparata
Brown University, Department of Computer Science
115 Waterman St., Providence, RI 02912-1910, USA
E-mail: franco@cs.brown.edu

Qizhi Fang
Ocean University of China, Department of Mathematics
23 Hong Kong Eastern Road, Qingdao 266071, China
E-mail: qfang@ouc.edu.cn

Library of Congress Control Number: 2007931053

CR Subject Classification (1998): F.2.2, G.2.1-2, G.1.2, G.1.6, C.2.2, I.3.5, E.1

LNCS Sublibrary: SL 1 – Theoretical Computer Science and General Issues

ISSN   0302-9743
ISBN-10   3-540-73813-4 Springer Berlin Heidelberg New York
ISBN-13   978-3-540-73813-8 Springer Berlin Heidelberg New York

Springer is a part of Springer Science+Business Media

springer.com

© Springer-Verlag Berlin Heidelberg 2007
Printed in Germany

Typesetting: Camera-ready by author, data conversion by Scientific Publishing Services, Chennai, India
Printed on acid-free paper   SPIN: 12097180   06/3180   5 4 3 2 1 0

Preface

FAW 2007, the 1st International "Frontiers in Algorithmics Workshop" took place in Lanzhou, China, August 1–3, 2007. The FAW symposium aims to provide a focused forum on current trends in research on algorithms, discrete structures, and their applications, and to bring together international experts at the research frontiers in those areas so as to exchange ideas and to present significant new results. In response to the Call for Papers, a total of 141 papers were submitted from 16 countries and regions, of which 33 were accepted. These papers were selected for nine special focus tracks in the areas of bioinformatics, discrete structures, geometric information processing and communication, games and incentive analysis, graph algorithms, Internet algorithms and protocols, parameterized algorithms, design and analysis of heuristics, approximate and online algorithms, and algorithms in medical applications.

We would like to thank the Conference General Chair, Maocheng Cai and Hao Li, and Advising Committee Chair, Danny Chen, for their leadership, advice and help on crucial matters concerning the conference. We would like to thank the International Program Committee and the external referees for spending their valuable time and effort in the review process. It was a wonderful experience to work with them.

Finally, we would like to thank the Organizing Committee, led by Lian Li and Xiaotie Deng, for their contribution to making this conference a success. We would also like to thank our sponsors, the ICCM Laboratory of Lanzhou University, for kindly offering the financial and clerical support that made the conference possible and enjoyable.

August 2007

Franco Preparata
Qizhi Fang

Organization

FAW 2007 was sponsored by the ICCM Laboratory of Lanzhou University, China.

Executive Committee

General Chairs	Mao-cheng Cai (Chinese Academy of Science, China)
	Hao Li (Laboratoire de Recherche en Informatique, France)
Program Co-chairs	Franco P. Preparata (Brown University, USA)
	Qizhi Fang (Ocean University of China, China)
Organization Chair	Lian Li (Lanzhou University, China)
Organization Co-chair	Xiaotie Deng (City University of Hong Kong, China)
Advising Committee Chair	Danny Chen (University of Notre Dame, USA)

Program Committee

Helmut Alt (Freie Universität Berlin)
Cristina Bazgan (Université Paris Dauphine)
Stephan Brandt (Technische Universität Ilmenau)
Hajo Broersma (University of Durham)
Leizhen Cai (The Chinese University of Hong Kong)
Bernard M. Chazelle (Princeton University)
Jianer Chen (Texas A&M University)
Francis Y.L. Chin (The University of Hong Kong)
Genghua Fan (Fuzhou University)
Evelyne Flandrin (Université Paris-Sud)
Anna Gal (University of Texas at Austin)
Ronald L. Graham (University of Californi at San Diego)
Toshihide Ibaraki (Kwansei Gakuin University)
Hiroshi Imai (University of Tokyo)
Kazuo Iwama (Kyoto University)
Kazuo Iwano (IBM Japan, Software Development Laboratory)
Ming-Yang Kao (Northwestern University)
Jean-Claude Latombe (Stanford University)
Der-Tsai Lee (Institute of Information Science Academia Sinica)
Lian Li (Lanzhou University)

Detlef Seese (Universität Karlsruhe (TH))
Yaoyun Shi (University of Michigan)
Lusheng Wang (City University of Hong Kong)
Roger Yu (Thompson Rivers University, also Nankai University)
Louxin Zhang (National University of Singapore)

External Reviewers

Juergen Banke	Cheng-Chung Li	Ingo Schiermeyer
Stephan Chalup	Chung-Shou Liao	Sven Scholz
Tobias Dietrich	Tien-Ching Lin	Xiaoxun Sun
Bin Fu	Ching-Chi Lin	Helen Tu
Gregory Gutin	Bin Ma	Aaron Wong
Horace Ip	Bert Marchal	Mingyu Xiao
Yuval Ishai	Joachim Melcher	Guomin Yang
Ioannis Ivrissimtzis	Gerhard Post	Teng-Kai Yu
Iyad Kanj	Artem Pyatkin	Yong Zhang
Gyula Y. Katona	Amir Safari	Shenggui Zhang
Koji Kojima	Ludmila Scharf	Magnus Bordewich
Arie Koster	Marc Scherfenberg	

Table of Contents

Geometric Algorithms for the Constrained 1-D K-Means Clustering Problems and IMRT Applications*

Danny Z. Chen, Mark A. Healy, Chao Wang, and Bin Xu**

Department of Computer Science and Engineering
University of Notre Dame, Notre Dame, IN 46556, USA
{chen,mhealy4,cwang1,bxu}@cse.nd.edu

Abstract. In this paper, we present efficient geometric algorithms for the discrete constrained 1-D K-means clustering problem and extend our solutions to the continuous version of the problem. One key clustering constraint we consider is that the maximum difference in each cluster cannot be larger than a given threshold. These constrained 1-D K-means clustering problems appear in various applications, especially in intensity-modulated radiation therapy (IMRT). Our algorithms improve the efficiency and accuracy of the heuristic approaches used in clinical IMRT treatment planning.

Keywords: K-means clustering, staircase-Monge property, matrix search algorithm, minimum-weight K-link path algorithm, intensity modulated radiation therapy (IMRT).

1 Introduction

Data clustering is a fundamental problem that arises in many applications (e.g., data mining, information retrieval, pattern recognition, biomedical informatics, and statistics). The main objective of data clustering is to partition a given data set into clusters (i.e., subsets) based on certain optimization criteria and subject to certain clustering constraints.

In this paper, we consider the discrete and continuous constrained 1-D K-means clustering problems and their applications in intensity-modulated radiation therapy (IMRT). The definitions of these two data clustering problems are given as follows.

The **discrete constrained 1-D K-means clustering problem**: We are given a positive bandwidth parameter $\delta \in \mathbb{R}$, integers K and n with $1 < K < n$, n real numbers $x_1, x_2, \ldots, x_n$ with $x_1 < x_2 < \cdots < x_n$, and a positive real-valued probability function $P : \{x_1, x_2, \ldots x_n\} \to (0, 1)$ such that $\sum_{i=1}^{n} P(x_i) = 1$. For any j and l with $0 \le l < j \le n$, we define

$$\mu[l, j] = \frac{\sum_{i=l+1}^{j} P(x_i) * x_i}{\sum_{i=l+1}^{j} P(x_i)}$$

---

* This research was supported in part by NSF Grant CCF-0515203.
** Corresponding author.

F.P. Preparata and Q. Fang (Eds.): FAW 2007, LNCS 4613, pp. 1–13, 2007.

and

$$V[l, j] = \begin{cases} +\infty, & \text{when } x_j - x_{l+1} > \delta \\ \sum_{i=l+1}^{j} P(x_i)(x_i - \mu[l,j])^2, & \text{when } x_j - x_{l+1} \leq \delta \end{cases}$$

We seek a sequence $q = (q_1, q_2, \ldots, q_{K-1})$ of $K-1$ integers with $0 < q_1 < q_2 < \cdots < q_{K-1} < n$ (for convenience, $q_0 \triangleq 0$ and $q_K \triangleq n$) such that the total error $E(q) = \sum_{k=1}^{K} V[q_{k-1}, q_k]$ is minimized. This discrete constrained problem arises in IMRT applications [4].

The **continuous constrained 1-D K-means clustering problem**: We are given a positive bandwidth parameter $\delta \in \mathbb{R}$, integers K and n with $1 < K < n$, and a positive real-valued (density) function $f : [x_b, x_e] \rightarrow [0, 1]$, where x_b and x_e ($x_b < x_e$) are real numbers and $\int_{x_b}^{x_e} f(x)dx = 1$. For any values α and β with $x_b \leq \alpha < \beta \leq x_e$, we define

$$\tilde{\mu}[\alpha, \beta] = \frac{\int_{\alpha}^{\beta} f(x) * x dx}{\int_{\alpha}^{\beta} f(x)dx}$$

and

$$\tilde{V}[\alpha, \beta] = \begin{cases} +\infty, & \text{when } \beta - \alpha > \delta \\ \int_{\alpha}^{\beta} f(x)(x - \tilde{\mu}[\alpha, \beta])^2 dx, & \text{when } \beta - \alpha \leq \delta \end{cases}$$

We seek a sequence $\theta = (\theta_1, \theta_2, \ldots, \theta_{K-1})$ of $K-1$ real numbers with $x_b < \theta_1 < \theta_2 < \cdots < \theta_{K-1} < x_e$ (for convenience, $\theta_0 \triangleq x_b$ and $\theta_K \triangleq x_e$), such that the total error $\tilde{E}(\theta) = \sum_{j=1}^{K} \tilde{V}[\theta_{j-1}, \theta_j]$ is minimized.

Note that in the above definitions, $V[q_{j-1}, q_j] = +\infty$ when $x_{q_j} - x_{q_{j-1}+1} > \delta$, and $\tilde{V}[\theta_{j-1}, \theta_j] = +\infty$ when $\theta_j - \theta_{j-1} > \delta$. Thus both the problems actually have a common constraint, i.e., the maximum difference in any cluster cannot be greater than the bandwidth parameter δ.

Algorithms for these two problems without the above constraint have been widely used in many areas such as signal processing, data compression, and information theory. Various techniques have been applied to solve the unconstrained versions, such as quantization [8,17], K-means clustering [9], and computational geometry [1,2,3,7,16]. However, the constrained versions are also important to some applications such as IMRT, which motivate our study.

Efficient algorithms for the discrete *unconstrained* 1-D K-means clustering problem have been known. Wu [17] modeled the optimal quantization problem (a variation of the unconstrained version) as a K-link shortest path problem and gave an $O(Kn)$ time algorithm based on dynamic programming and Monge matrix search techniques [1,2]. Aggarwal *et al.* [3] showed that the K-link shortest path problem on directed acyclic graphs (DAG) that satisfy the Monge property [15,1,2] can be solved in $O(n\sqrt{K \log n})$ time by using a refined parametric search paradigm, and Schieber [16] improved the time bound to $O(n2^{\sqrt{\log K \log \log n}})$. All these three algorithms exploit the Monge property to find an optimal solution efficiently. A related work due to Hassin and Tamir [10] formulated a class of location problems using a general facility location model; their solution for the

p-median problem is also based on dynamic programming and matrix search techniques.

The continuous *unconstrained* 1-D K-means problem was studied independently by Lloyd [11] and Max [13], who gave algorithms based on iterative numerical methods. The convergence speed of their algorithms was improved in [12]. These algorithms, however, are able to find only a local minimal solution instead of a global minimum. Wu [17] showed that by discretizing the continuous input function f, a computationally feasible global search algorithm could be obtained.

The constrained 1-D K-means clustering problems appear in the radiation dose calculation process of intensity-modulated radiation therapy (IMRT) for cancer treatment [4,14,18]. In IMRT treatment planning, a radiation dose prescription (i.e., a dose function) is first computed for a cancer patient by a treatment planning system. The initially computed dose function is in either a discrete form (i.e., a piecewise linear function) or a continuous form (e.g., a certain continuous smooth function), which is often too complicated to be deliverable. Hence, the initial dose function needs to be processed or simplified into a deliverable form called *intensity profile*. The resulting intensity profile should be as close as possible to the initial dose function both locally and globally to minimize error in the treatment plan. The constrained 1-D K-means clustering problems model the problem of computing an intensity profile from an initial dose function [4,14,18], in which the bandwidth parameter δ specifies the allowed local deviation error and the number K of clusters indicates the (delivery) complexity of the resulting intensity profile (e.g., see Section 4 and Figure 2).

Several algorithms for the constrained 1-D K-means clustering problems have been given in medical literature and used in clinical IMRT planning systems [4,18]. However, these algorithms use only heuristic methods to determine the clusters iteratively [4,14,18], and can be trapped in a local minimal (as shown by our experimental results in Section 5). Also, no theoretical analysis was given for their convergence speed. The clustering constraint defined at the beginning of this section is often used [4]. But, in certain medical settings, another clustering constraint (e.g., in [18]; see Section 2.5) that is quite similar to yet slightly more restricted than the constraint defined in this section is also used. As we will show later, such seemingly slight constraint variations can have quite different computational and algorithmic implications.

Our results in this paper are summarized as follows.

1. We present efficient geometric algorithms for computing optimal solutions to the discrete constrained 1-D K-means problem that is defined in this section. Depending on the relative values of n and K, our algorithms run in $O(Kn)$ or $O(n2^{\sqrt{\log K \log \log n}})$ time, by exploiting the Monge property [15,1,2,16,17] of the problem (see Section 2).

2. We also consider a similar yet slightly more restricted clustering constraint [18] (to be defined and discussed in Section 2.5), and show that the Monge property does not hold for this constraint variation. Our algorithm for this discrete constrained problem version takes $O(K(n + |E'|))$ time, where $|E'|$

is the number of edges in a graph that models the problem ($|E'| \leq n^2$, but in practice $|E'|$ can be much smaller than $O(n^2)$).

3. We extend our solutions to the continuous constrained 1-D K-means problem, by transforming the continuous case to the discrete case (see Section 3).
4. We show that our constrained 1-D K-means algorithms are useful in IMRT applications (see Section 4), and give some experimental results on comparing our solutions with those computed by the heuristic methods in medical literature [4,14,18] (see Section 5).

2 Algorithms for the Discrete Constrained 1-D K-Means Problem

This section presents our algorithms for the discrete constrained 1-D K-means clustering problem.

2.1 Computing the Minimum Number K of Clusters

The number K of clusters can be an input value. However, in some applications such as IMRT, K needs to be as small as possible and thus needs to be computed. Wu *et al.* [18] gave a greedy algorithm for finding the minimum cluster number K, as follows. The input values $x_1, x_2, \ldots, x_n$ are scanned in ascending order and partitioned into K groups:

$$(x_1, \ldots, x_{q_1}), (x_{q_1+1}, \ldots, x_{q_2}), \ldots, (x_{q_{K-1}+1}, \ldots, x_n)$$

such that for any k with $1 \leq k \leq K$, $x_{q_k} - x_{q_{k-1}+1} \leq \delta$ and $x_{q_k+1} - x_{q_{k-1}+1} > \delta$. Clearly, K can be computed in $O(n)$ time.

2.2 Reformulation of the Discrete Constrained 1-D K-Means Problem

In this section, we model the discrete constrained 1-D K-means clustering problem as a K-link shortest path problem on a directed acyclic graph (DAG) $G = (U, E)$, which is defined as follows (see Figure 1). The vertex set $U = \{u_0, u_1, u_2, \ldots, u_n\}$. For any two vertices u_l and u_j ($l < j$), we put an edge in G from u_l to u_j with a weight $V[l, j]$.

Clearly, G is a complete DAG with edge weights. Any K-link path from u_0 to u_n in G, say $p = u_0 \rightarrow u_{q_1} \rightarrow u_{q_2} \rightarrow \cdots \rightarrow u_{q_{K-1}} \rightarrow u_n$, corresponds to a feasible solution $q = (q_1, q_2, \ldots, q_{K-1})$ for the discrete constrained 1-D K-means clustering problem, and *vice versa*. For any path in G, define its *weight* as the sum of the weights of all its edges. It is easy to see that an optimal solution for the discrete constrained 1-D K-means clustering problem corresponds to a shortest K-link path from u_0 to u_n in the DAG G.

The DAG G thus defined has $O(n)$ vertices and $O(n^2)$ edges. In Section 2.3, we will show that the weight of any edge in G can be computed in $O(1)$ time after an $O(n)$ time preprocess. Hence, the graph G can be represented *implicitly*, that is, after the $O(n)$ time preprocess, any vertex and edge of G can be obtained in $O(1)$

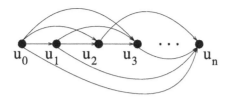

Fig. 1. A weighted, complete directed acyclic graph (DAG)

time. Thus, a K-link shortest path in G can be computed in $O(K(|U|+|E|)) = O(Kn^2)$ time by using the standard dynamic programming approach. In Section 2.4, we will show that by exploiting the underlying geometric properties of the DAG G, a K-link shortest path in G can be computed in $O(n2^{\sqrt{\log K \log \log n}})$ or $O(Kn)$ time.

2.3   Computing the Weights of the Edges in G

In this section, we show that for any l and j ($l < j$), the weight of the edge in G from u_l to u_j can be computed in $O(1)$ time after an $O(n)$ time preprocess. It suffices to consider the case when $x_j - x_{l+1} \leq \delta$. In this case, we have

$$
\begin{aligned}
V[l,j] &= \sum_{i=l+1}^{j} P(x_i)(x_i - \mu[l,j])^2 \\
&= \sum_{i=l+1}^{j} P(x_i)x_i^2 - 2\sum_{i=l+1}^{j} P(x_i) * x_i * \mu[l,j] + \sum_{i=l+1}^{j} P(x_i)(\mu[l,j])^2 \\
&= \sum_{i=l+1}^{j} P(x_i)x_i^2 - \frac{(\sum_{i=l+1}^{j} P(x_i)x_i)^2}{\sum_{i=l+1}^{j} P(x_i)}
\end{aligned}
$$

Therefore, if we precompute all prefix sums of $\sum_{i=1}^{g} P(x_i)$, $\sum_{i=1}^{g} P(x_i)x_i$, and $\sum_{i=1}^{g} P(x_i)x_i^2$, $g = 1, 2, \ldots, n$, which can be easily done in $O(n)$ time, then we can compute any $V[l,j]$ in $O(1)$ time.

2.4   The Staircase-Monge Property of the Problem

This section shows that the discrete constrained 1-D K-means clustering problem satisfies the staircase-Monge property [2].

Lemma 1. *(1a) If $V[l,j] = +\infty$, then $V[l,j'] = +\infty$ for any $j' > j$ and $V[l',j] = +\infty$ for any $l' < l$.*
(1b) For any $0 < l+1 < j < n$, if the four entries $V[l,j], V[l+1,j], V[l,j+1]$, and $V[l+1,j+1]$ are all finite, then $V[l,j]+V[l+1,j+1] \leq V[l+1,j]+V[l,j+1]$.

Proof. (1a) If $V[l,j] = +\infty$, then $x_j - x_{l+1} > \delta$. Since the sequence of $x = (x_1, x_2, \ldots, x_n)$ is in ascending order, for any $j' > j$, we have $x_{j'} - x_{l+1} > x_j - x_{l+1} > \delta$. Hence $V[l,j'] = +\infty$. We can similarly argue that for any $l' < l$, $V[l',j] = +\infty$ holds.
(1b) Fix l and j. We can view $V[l+c, j+d]$'s and $\mu[l+c, j+d]$'s ($c = 0, 1$ and $d = 0, 1$) as multi-variable functions of $x_{l+1}, x_{l+2}, \ldots, x_{j+1}$ and $P(x_{l+1})$, $P(x_{l+2})$, ..., $P(x_{j+1})$. Let $W = V[i+1,j]+V[i,j+1]-V[i,j]-V[i+1,j+1]$. It is sufficient to show $W \geq 0$.

When $P(x_{j+1}) = 0$, we have $\mu[i,j] = \mu[i,j+1]$ and $\mu[i+1,j] = \mu[i+1,j+1]$; further, $V[i,j] = V[i,j+1]$ and $V[i+1,j] = V[i+1,j+1]$. Thus $W = 0$ when $P(x_{j+1}) = 0$.

To show $W \geq 0$, it suffices to show that $\frac{\partial W}{\partial P(x_{j+1})} \geq 0$. Since

$$V[l,j+1] = \sum_{i=l+1}^{j+1} P(x_i)(x_i - \mu[l,j+1])^2$$

we have

$$\begin{aligned}
\frac{\partial V[l,j+1]}{\partial P(x_{j+1})} &= (x_{j+1} - \mu[l,j+1])^2 + \sum_{i=l+1}^{j+1} 2P(x_i)(x_i - \mu[l,j+1])(-\frac{\partial \mu[l,j+1]}{\partial P(x_j)}) \\
&= (x_{j+1} - \mu[l,j+1])^2 + (-2\frac{\partial \mu[l,j+1]}{\partial P(x_j)})\sum_{i=l+1}^{j+1} P(x_i)(x_i - \mu[l,j+1]) \\
&= (x_{j+1} - \mu[l,j+1])^2 + (-2\frac{\partial \mu[l,j+1]}{\partial P(x_j)}) \cdot 0 \\
&= (x_{j+1} - \mu[l,j+1])^2
\end{aligned}$$

Similarly, $\frac{\partial V[l+1,j+1]}{\partial P(x_{j+1})} = (x_{j+1} - \mu[l+1,j+1])^2$. Hence

$$\begin{aligned}
\frac{\partial W}{\partial P(x_{j+1})} &= \frac{\partial V[l,j+1]}{\partial P(x_{j+1})} - \frac{\partial V[l+1,j+1]}{\partial P(x_{j+1})} \\
&= (x_{j+1} - \mu[l,j+1])^2 - (x_{j+1} - \mu[l+1,j+1])^2
\end{aligned}$$

Observing that $\mu[l,j+1] \leq \mu[l+1,j+1] \leq x_{j+1}$, we have $\frac{\partial W}{\partial P(x_{j+1})} \geq 0$.   □

Lemma 1 implies that the $(n+1) \times (n+1)$ matrix $V = (V[l,j])$ is a staircase-Monge matrix [2]. (For convenience, an entry of V is filled with $+\infty$ if the corresponding $V[l,j]$ is undefined.) Using the results of Aggarwal et al. [1], Schieber [16], and Aggarwal and Park [2], it is easy to show that the K-link shortest path problem on G can be solved in $O(n2^{\sqrt{\log K \log \log n}})$ time. Of course, Wu's algorithm [17] based on dynamic programming and Monge matrix search can also be applied to directly solve the discrete constrained 1-D K-means clustering problem in $O(Kn)$ time. Thus, depending on the relative values of n and K, we can choose a more efficient algorithm for the discrete constrained 1-D K-means clustering problem.

2.5   A Slightly Stronger Clustering Constraint

Up to this point, the clustering constraint we have used is that the maximum distance between any two input elements in the same cluster cannot be more than a given threshold value δ [4], which we call the *weak constraint*. A slightly stronger clustering constraint also used in IMRT treatment planning [18] requires that the maximum distance between the "centroid" of each cluster C and any input element in C be no more than $\delta/2$. Since in some IMRT applications, the "centroid" of a cluster C is used to approximate each input element in C (i.e., each element in C is "replaced" by its centroid), this constraint requires that the approximation error between the centroid and any element in C cannot be larger than $\delta/2$. We call this constraint the *strong constraint*. Note that if a cluster satisfies the strong constraint, then it also satisfies the weak constraint; but, the other way around is not true.

To reflect the strong clustering constraint, the definition of the discrete constrained 1-D K-means clustering problem (defined in Section 1) needs a slight change, i.e., we replace $V[l, j]$ with

$$
V'[l, j] = \begin{cases} +\infty, & \text{when } x_j - \mu[l, j] > \delta/2 \text{ or} \\ & \mu[l, j] - x_{l+1} > \delta/2 \\ \sum_{i=l+1}^{j} P(x_i)(x_i - \mu[l, j])^2, & \text{when } x_j - \mu[l, j] \leq \delta/2 \text{ and} \\ & \mu[l, j] - x_{l+1} \leq \delta/2 \end{cases}
$$

It is easy to see that in the same way as in Section 2.2, we can reformulate this constrained clustering problem as computing a K-link shortest path in a DAG G', except that the weights of the edges in this DAG G' are somewhat different. Therefore, the discrete constrained 1-D K-means clustering problem under this slightly stronger constraint can be solved in $O(Kn^2)$ time (using dynamic programming).

It would be interesting to check if the new DAG G' satisfies the staircase-Monge property as the weak constraint case did. Unfortunately, the answer is *no*. Below we give a counterexample. Let a value ϵ be much smaller than $\frac{\delta}{n}$, and consider the following data sequence x of n items such that n is an even number: $\epsilon, 2\epsilon, \ldots, \frac{n}{2}\epsilon, \delta - \frac{n}{2}\epsilon, \delta - (\frac{n}{2} - 1)\epsilon, \ldots, \delta - \epsilon$. For $0 \leq l < \frac{n}{2}$, it is easy to show that the entry $V'[l, j]$ is finite if and only if $l < j \leq \frac{n}{2}$ or $j = n - l$. Thus, the finite entries in a row of V' no longer guarantee to form a consecutive interval (i.e., there can be entries of $+\infty$ between finite entries in a row). We conclude that V' (and further, the DAG G') is neither staircase-Monge nor Monge.

Therefore, a seemingly small change in the constraint causes significant differences in the algorithm and its running time. While the case with the weak constraint can be solved in $O(Kn)$ or $O(n2^{\sqrt{\log K \log \log n}})$ time, the case with the strong constraint takes $O(Kn^2)$ time in the worst case.

It should be pointed out that for some situations (to be discussed in detail below), it is possible to improve the running time of the case with the strong constraint. The observation is that if $V'[l, j]$ is finite, then $V[l, j]$ must also be finite. (Recall that we are using a stronger clustering constraint.) Denote by F_V (resp., $F_{V'}$) the number of finite entries in V (resp., V'). Since V is staircase-Monge, it is easy to determine all its finite entries (using an implicit interval representation) in $O(n)$ time. Note that when using the standard dynamic programming technique to compute a K-link shortest path in G', we need to visit only those edges of G' with finite weights, each of which corresponds one-to-one to a finite entry in V'. It is then clear that we can compute a K-link shortest path in G' in $O(K(n + F_V))$ time (with $O(n)$ space), improving the worst case $O(Kn^2)$ time bound when $F_V = o(n^2)$. Further, if $F_{V'} = o(F_V)$, it will be beneficial to first record all finite entries of V' after a scan of the DAG G'; then the algorithm takes $O(F_V + K(n + F_{V'}))$ time and $O(n + F_{V'})$ space. In practice, $F_{V'}$ can be much smaller than $O(n^2)$, specially in the IMRT clustering settings (e.g., see Section 4.1).

3   Extension to the Continuous Constrained 1-D K-Means Problem

In this section, we briefly discuss how to solve the continuous constrained 1-D K-means clustering problem. Our key idea is to transform the continuous constrained 1-D K-means problem to the discrete problem solved in Section 2. We first partition the domain $[x_b, x_e]$ of $f(x)$ into n intervals of length L each (with $L = (x_e - x_b)/n$), and use the middle point x_i of the i-th interval as the representative point of that interval [17], thus producing a sequence of n values $x_1, x_2, \ldots, x_n$. The *weight* $P(x_i)$ of the value x_i, $i = 1, 2, \ldots, n$, is defined as $P(x_i) = \int_{x_i - \epsilon}^{x_i + \epsilon} f(x) dx$, with $\epsilon = L/2$. We call $P(x_i)$ the "weight" instead of the probability since $P(x_i)$ may not be in $[0, 1]$ and $\sum_{i=1}^{n} P(x_i)$ may not be equal to 1 (e.g., when $f(x)$ is not a density function). We can certainly normalize $P(x_i)$ to a probability, and then apply our algorithm for the discrete constrained 1-D K-means clustering problem to the problem instance on δ, the x_i's, and the $P(x_i)$'s. (In practice, the normalization step can be omitted since it does not affect the final solution.) It is easy to show that in this way, we can approximate the optimal solution for the continuous constrained 1-D K-means clustering problem within any desired precision by choosing a sufficiently large n.

4   IMRT Applications

In this section, we discuss the applications of our clustering algorithms in IMRT. We consider two approaches: the discrete and continuous approaches. In the discrete approach, we apply the algorithms given in Section 2 to the clustering process in IMRT. In the continuous approach, we consider simplifying the continuous dose function curves into (deliverable) intensity profiles, and transform the continuous case to the discrete constrained 1-D K-means problem.

4.1   The Discrete Approach

Figure 2(a) shows a dose function represented by the rectilinear curve of beam intensity vs. position. The curve consists of vertical and horizontal line segments specified by a sequence of intensity values $x_1, x_2, \ldots, x_N$. These intensity values are defined on the corresponding coordinate values of the position axis. Suppose this dose function is not deliverable, and thus these N intensity values need to be grouped into K clusters for the smallest possible number K, such that the maximum difference between any two intensity values in each cluster is no bigger than a given bandwidth parameter δ and the total sum of variances of the K clusters is minimized. The horizontal strips in Figure 2(b) represent the $K = 4$ clusters, each containing its mean intensity value μ_i. The resulting clusters are actually used to specify an intensity profile as shown in Figure 2(c). The intensity profile thus obtained is deliverable and is further converted into a set of delivery operations by a radiation treatment planning system.

One possible way to determine the value of the bandwidth parameter δ is as follows [18]. First, the maximum deviation tolerance e_{max} is defined as the

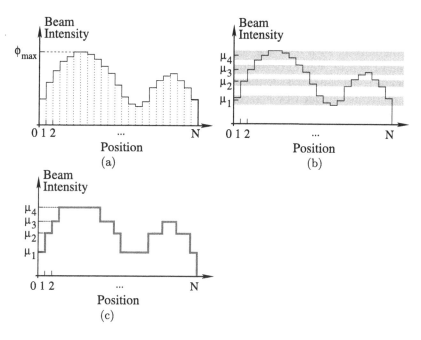

Fig. 2. Illustrating an IMRT application of the discrete constrained 1-D K-means clustering algorithms: (a) A rectilinear input dose function specified by a sequence of discrete intensity values x_1, x_2, ..., x_n; (b) the four clusters of the intensity value sequence (indicated by the four strips); (c) the resulting intensity profile

product of a user-specified percentage error tolerance (ET) and the maximum intensity value $\phi_{max} = \max_{i=1}^{N}\{x_i\}$, i.e., $e_{max} = ET \times \phi_{max}$. Then δ is defined as $2 * e_{max}$. Of course, δ can also be a user-specified input value.

After we perform a sorting plus a "group by value" operation on the sequence $x = (x_1, x_2, \ldots, x_N)$, the sequence x becomes a sequence of n ($n \leq N$) *distinct* intensity values, say $\chi_1, \chi_2, \ldots, \chi_n$. For each i ($1 \leq i \leq n$), $P(\chi_i)$ is defined as the ratio of the number $oc(\chi_i)$ of occurrences of χ_i in x over N, i.e., $P(\chi_i) = oc(\chi_i)/N$. We then directly apply the clustering algorithms given in Section 2 to the χ_i's and $P(\chi_i)$'s for the above clustering process. Once the clustering result is given, the corresponding intensity profile can be easily produced.

Comparing to the known heuristic clustering algorithms in medical literature [4,14,18], our clustering algorithms are very efficient and guarantee to find a globally optimal solution instead of a locally minimal solution.

4.2   The Continuous Approach

In Section 4.1, a sequence of discrete intensity values x_1, x_2, ..., x_n is used to describe approximately the curve of a dose function (i.e., as a piecewise linear function). However, for a better accuracy, the given dose function can also be a continuous and even smooth function $x = g(a)$ defined on an interval $[x_b, x_e]$ on the position axis. Since such a dose function $g(a)$ is in general not deliverable, we must convert it into a deliverable intensity profile subject to a specified deviation error tolerance δ, such that the complexity of the resulting intensity profile is small and the total sum of deviation error is minimized. This can be solved as a continuous constrained 1-D K-means clustering problem.

We illustrate our idea by showing how to handle a unimodal dose function $x = g(a)$ (see Figure 3(a)). The domain of $g(a)$ on the position axis can be partitioned into two intervals $[a_1, a_3]$ and $[a_3, a_5]$, on each of which the curve of $g(a)$ is monotone. Suppose in the clustering process, the range of $g(a)$ on the x-axis (for the beam intensity) is partitioned into two intervals $[x_1, x_2]$ and $[x_2, x_3]$, and their corresponding mean values are μ_1 and μ_2, respectively (see Figure 3(b)). Then clearly, the total mean-square error E incurred by clustering the intensities of the dose function $g(a)$ is:

$$E = \int_{a_1}^{a_2} (g(a) - \mu_1)^2 da + \int_{a_2}^{a_4} (g(a) - \mu_2)^2 da + \int_{a_4}^{a_5} (g(a) - \mu_1)^2 da.$$

The clustering process aims to minimize E for a specified bandwidth parameter δ and a given cluster number (2 in this case). Below we show that this is actually a continuous constrained 1-D K-means clustering problem.

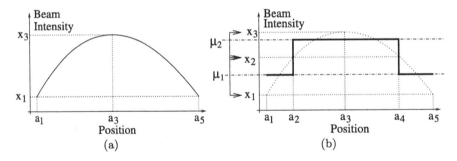

Fig. 3. Illustrating an IMRT application of the continuous constrained 1-D K-means clustering algorithms: (a) The curve of an input continuous and smooth dose function; (b) the clustering result (the mean values of the two clusters are indicated by the dash-dot lines) and the final intensity profile (the heavy solid rectilinear curve)

We write $g(a)$ as

$$g(a) = \begin{cases} g_1(a), & \text{when } a_1 \leq a \leq a_3 \\ g_2(a), & \text{when } a_3 \leq a \leq a_5 \end{cases}$$

Since g_1 and g_2 are both monotone, $g_1^{-1}(x)$ and $g_2^{-1}(x)$ exist. Define $f(x) = (g_1^{-1}(x))' - (g_2^{-1}(x))'$. (For the moment, we assume that both $g_1^{-1}(x)$ and $g_2^{-1}(x)$ are differentiable.) Then we have

$$E = \int_{a_1}^{a_2} (g(a) - \mu_1)^2 da + \int_{a_2}^{a_4} (g(a) - \mu_2)^2 da + \int_{a_4}^{a_5} (g(a) - \mu_1)^2 da$$
$$= \int_{a_1}^{a_2} (g_1(a) - \mu_1)^2 da + \int_{a_2}^{a_3} (g_1(a) - \mu_2)^2 da + \int_{a_3}^{a_4} (g_2(a) - \mu_2)^2 da$$
$$+ \int_{a_4}^{a_5} (g_2(a) - \mu_1)^2 da$$
$$= \int_{x_1}^{x_2} (x - \mu_1)^2 (g_1^{-1}(x))' dx + \int_{x_2}^{x_3} (x - \mu_2)^2 (g_1^{-1}(x))' dx$$
$$+ \int_{x_3}^{x_2} (x - \mu_2)^2 (g_2^{-1}(x))' dx + \int_{x_2}^{x_1} (x - \mu_1)^2 (g_2^{-1}(x))' dx$$
$$= \int_{x_1}^{x_2} (x - \mu_1)^2 (g_1^{-1}(x) - g_2^{-1}(x))' dx + \int_{x_2}^{x_3} (x - \mu_2)^2 (g_1^{-1}(x) - g_2^{-1}(x))' dx$$
$$= \int_{x_1}^{x_2} (x - \mu_1)^2 f(x) dx + \int_{x_2}^{x_3} (x - \mu_2)^2 f(x) dx$$

Further, it is not difficult to verify that to minimize E, μ_1 and μ_2 need to satisfy

$$\mu_1 = \frac{\int_{x_1}^{x_2} f(x) * x dx}{\int_{x_1}^{x_2} f(x) dx} \text{ and } \mu_2 = \frac{\int_{x_2}^{x_3} f(x) * x dx}{\int_{x_2}^{x_3} f(x) dx}$$

The above expressions of the mean-square error E and μ show the close relation between the continuous constrained clustering problem in IMRT dose calculation and the continuous constrained 1-D K-means clustering problem. The clustering result and the final intensity profile for this example are given in Figure 3(b).

If the given dose function $g(a)$ has more than one peak, then the corresponding function $f(x)$ will be defined as $f(x) = -(L(x))'$, where $L(x)$ is Lebesque measure [6] of the set $\{a \mid g(a) \geq x\}$. Geometrically, for any t, $L(t)$ is the total "length" of the intersection of the horizontal line $x = t$ with the region under the dose function curve $x = g(a)$. Clearly, $f(x) \geq 0$, and we can always normalize $f(x)$ to be a dense function such that $\int_{x_b}^{x_e} f(x) dx = 1$ holds, where x_b and x_e are the minimum and maximum intensities on the curve g, respectively. In practice, the normalization step is usually omitted since it does not affect the final clustering result.

5   Implementation and Experiments

To study the quality and performance of our new constrained 1-D K-means clustering algorithms with respect to clinical IMRT applications, we implemented our discrete algorithms using the C programming language on Linux and UNIX systems. For the purpose of comparison, we also implemented the discrete constrained 1-D K-means clustering algorithm by Wu *et al.* [18] which is based on heuristic methods. We used clinical data from the Department of Radiation Oncology, the University of Maryland School of Medicine in order to generate simulations for clustering. We used the clinical intensity fields thus obtained, and added random values to the positive entries in these intensity fields to create simulated unclustered intensity fields for our experiments.

We conducted an extensive comparison study with the algorithm in [18]. The mean-square errors for generating 63 clustered intensity fields using our algorithms and the algorithm in [18] were calculated for a percentage range of the e_{max} value (note that the bandwidth parameter $\delta = 2 * e_{max}$). Table 1 shows some of the comparison results. For each of our test cases, our algorithm always provides a lower mean-square error than the algorithm in [18]. For the range

Table 1. Comparisons of the mean-square errors for clustering 63 intensity fields with the e_{max} values ranging from 1% to 5%

e_{max}	Wu et $al.$'s algorithm [18]	Our algorithm	Improvement
1%	99.834	91.102	8.7%
2%	464.088	422.743	8.9%
3%	1084.847	1009.205	7.0%
4%	1882.953	1747.737	7.2%
5%	2833.705	2621.713	7.5%
Total:	6365.427	5892.500	7.4%

of 1% – 5% of the e_{max} values, our algorithm shows an average improvement of about 7.4% over the algorithm in [18]. While our algorithm always produces an optimal clustering solution, it is important to note that the improvement over the algorithm in [18] varies in terms of the given data. For individual instances in our testing, while the minimum improvement was 0.0%, the maximum improvement was 37.7%.

Regarding the execution time of our discrete algorithms, our experiments actually showed that our algorithm for the strong constraint case did not execute as fast as the heuristic algorithm in [18]. However, our algorithm still executes very fast, running less than one second in all tested cases.

References

1. Aggarwal, A., Klawe, M., Moran, S., Shor, P., Wilbur, R.: Geometric applications of a matrix-searching algorithm. Algorithmica 2, 195–208 (1987)
2. Aggarwal, A., Park, J.: Notes on searching in multidimensional monotone arrays. In: Proc. 29th IEEE Annual Symp. Foundations of Computer Science, pp. 497–512. IEEE Computer Society Press, Los Alamitos (1988)
3. Aggarwal, A., Schieber, B., Tokuyama, T.: Finding a minimum weight k-link path in graphs with Monge property and applications. In: Proc. 9th Annual ACM Symp. on Computational Geometry, pp. 189–197. ACM Press, New York (1993)
4. Bär, W., Alber, M., Nüsslin, F.: A variable fluence step clustering and segmentation algorithm for step and shoot IMRT. Physics in Medicine and Biology 46, 1997–2007 (2001)
5. Bruce, J.D.: Optimum quantization, Sc.D. Thesis, MIT (1964)
6. Craven, B.D.: Lebesque Measure & Integral, Pitman (1982)
7. Du, Q., Faber, V., Gunzburger, M.: Centroidal Voronoi tessellations: Applications and algorithms. SIAM Review 41, 637–676 (1999)
8. Gray, R., Neuhoff, D.: Quantization. IEEE Transactions on Information Theory 44, 2325–2383 (1988)
9. Har-Peled, S., Sadri, B.: How fast is the k-means method? Algorithmica 41, 185–202 (2005)
10. Hassin, R., Tamir, A.: Improved complexity bounds for location problems on the real line. Operations Research Letters 10, 395–402 (1991)
11. Lloyd, S.P.: Least squares quantization in PCM, unpublished Bell Laboratories technical note, 1957. IEEE Transactions on Information Theory 28, 129–137 (1982)

12. Lu, F.S., Wise, G.L.: A further investigation of Max's algorithm for optimum quantization. IEEE Transactions on Communications 33, 746–750 (1985)
13. Max, J.: Quantizing for minimum distortion. IEEE Transactions on Information Theory 6, 7–12 (1960)
14. Mohan, R., Liu, H., Turesson, I., Mackie, T., Parliament, M.: Intensity-modulated radiotherapy: Current status and issues of interest. International Journal of Radiation Oncology, Biology, Physics 53, 1088–1089 (2002)
15. Monge, G.: Déblai et remblai. Mémories de I'Académie des Sciences, Paris (1781)
16. Schieber, B.: Computing a minimum-weight k-link path in graphs with the concave Monge property. In: Proc. 6th Annual ACM-SIAM Symp. on Discrete Algorithms, pp. 405–411. ACM Press, New York (1995)
17. Wu, X.: Optimal quantization by matrix searching. Journal of Algorithms 12, 663–673 (1991)
18. Wu, Y., Yan, D., Sharpe, M.B., Miller, B., Wong, J.W.: Implementing multiple static field delivery for intensity modulated beams. Medical Physics 28, 2188–2197 (2001)

A Fast Preprocessing Algorithm to Select Gene-Specific Probes of DNA Microarrays*

Seung-Ho Kang , In-Seon Jeong, Mun-Ho Choi, and Hyeong-Seok Lim

Dept. of Computer Science, Chonnam National University, Yongbong-dong 300,
Buk-gu, Gwangju 500-757, Korea
kinston@natural.chonnam.ac.kr, isjung0@hotmail.com, howork@paran.com,
hslim@chonnam.ac.kr

Abstract. The performance of a DNA microarray is dependent on the quality of the probes it uses. A good probe is uniquely associated with a particular sequence that distinguishes it from other sequences. Most existing algorithms to solve the probe selection problem use the common approach that directly filters out "bad" probes or selects "good" probes of each gene. However, this approach requires a very long running time for large genomes. We propose a novel approach that screens out a "bad" gene(not probe) set for each gene before filtering out bad probes. We also provide a $O(1/4^q N^2)$ time preprocessing algorithm for this purpose using q-gram for a length-N genome, guaranteeing more than 95% sensitivity. The screened bad gene sets can be used as inputs to other probe selection algorithms in order to select the specific probes of each gene.

Keywords: DNA microarray, probe selection algorithm, preprocessing.

1 Introduction

The DNA microarray is a widely used tool to perform experiments rapidly on a large scale. It is able to monitor the whole genome on a single chip, so that researchers can obtain a better picture of the interactions of various genes simultaneously. The range of application of microarrays extends from gene discovery and mapping to gene regulation studies, diagnosis, drug discovery, and toxicology[1].

The DNA microarray consists of a solid surface, usually a microscope slide, onto which DNA molecules(called probes) have been chemically bonded. And a probe is a short piece of single-stranded DNA complementary to the target gene whose expression is measured on the microarray by that probe[13]. The performance of a microarray is fairly dependent on the quality of the selected probes. Good probes should have similar reaction temperature (*homogeneity*), should be totally specific to their respective targets to avoid any cross-hybridization (*specificity*), and should not form stable secondary structures that may interfere with the probes by forming heteroduplexes during hybridization (*sensitivity*) [2, 6, 13].

---

* This work was supported by the Korea Research Foundation Grant funded by the Korean Government(MOEHRD)(KRF-2005-041-D00747).

F.P. Preparata and Q. Fang (Eds.): FAW 2007, LNCS 4613, pp. 14–25, 2007.

Among these properties, the homogeneity and sensitivity of a probe can be determined in linear time[13]. However, the specificity step is computationally expensive and takes up the most time in the probe selection process. The specificity identifies probes that are unique to each gene in the genome. This condition minimizes the cross-hybridization of the probes with other gene sequences. The brute force approach for specificity checking scans through the whole length-N genome for every length-m probe and determines if the distances between them exceed some pre-specified limit under some predefined distance measurement. Such a process requires a time of $O\left(mN^2\right)$ if the Hamming distance[3] is used as the specificity measurement. The Hamming distance is the number of positions at which the characters at corresponding positions of the two strings differ. If a probe has a Hamming distance greater than some constant, the probe is said to be sufficiently specific.

In the mean time, there have been many attempts to find efficient algorithms which can select specific probe sets for each gene in a large genome. Existing algorithms usually select probes using the criteria of homogeneity, sensitivity and specificity proposed by Lockhart et al[6]. Li and Stormo[5] proposed a heuristic algorithm to solve the probe selection problem. To improve its time efficiency, their algorithm uses advanced data structures such as suffix array and landscape. However, this algorithm is still not fast enough for the computation of large genome sets. By considering the thermodynamic property, Kaderali and Schliep[4] attempted to design a probe set by heuristic dynamic programming. Although this solution based on the Nearest Neighbor Model[11] has higher accuracy, since it considers the thermodynamic property, their algorithm is very slow and is unsuitable for large genomes. Rahmann[9, 10] presented a fast algorithm that is practical for designing short probes of up to 30 nucleotides. His algorithm approximates the unspecificity of a probe by computing its longest common contiguous substring. This algorithm allows the selection of probes for large genomes like Neurospora crassa in 4 hours. However, his approach can only design short probes. Furthermore, the approximation used is not very accurate and some good probes may be missed out[13]. Sung and Lee[13] presented a filtering algorithm based on the Pigeon Hole Principle to select probes, in which several hashing techniques were used to reduce the search space for the probes[2]. The Hamming distance is used as the specificity measure of the probes. By using the Pigeon Hole Principle, they avoided redundant comparisons for probes and greatly improved the time efficiency. Recently, Gasieniec et al. [2] presented another algorithm. Their algorithm selects just a small probe set instead of all possible probes using randomization.

These algorithms share a common approach that selects "good" probes or eliminates "bad" probes for each gene by directly comparing the probe candidates with other gene sequences. The probe candidates are every pre-defined length substring of a gene. If a candidate does not satisfy the condition of the specificity measurement, it is considered to be a "bad" probe and so is eliminated. This process causes the probe selection algorithm to be inefficient and makes it unsuitable for large genomes. However, we observed that there are a very small number of other genes(far less than 1%) which actually cause the probe candidates of a gene to be "bad" probes. Bad probes are probe candidates that we want to eliminate. So, first, if we can screen out the bad genes for each gene rapidly using a preprocessing procedure before filtering out the bad probes, the running time of the probe selection process will be greatly reduced. We only need to compare a gene with the screened bad genes to select the good probes using any of the previously proposed algorithms.

In this paper, we propose a new approach to solve this problem efficiently and provide a $O\left(1/4^q N^2\right)$ time preprocessing algorithm for screening out gene sets which other probe selection algorithms can use as an input to select specific probes for each gene. The average size of a screened gene set for each gene is less than 1% of other genes which need to be compared to filter out bad probes in previously common approach. And our algorithm has a sensitivity of more than 95% to the bad genes. The basic idea behind the proposed algorithm is that it first finds a small number of high matching regions between two genes by q-gram counting and verifies these regions. If one of these regions satisfies our pre-defined condition, the two genes are considered to have a "bad" relation, and two genes having a bad relation must be subjected to further processing to eliminate the bad probes. Our algorithm can be used as a preprocessing procedure by other probe selection algorithms.

The remainder of this paper is organized as follows. In the next section, we review the existing methods of dealing with the probe selection problem and discuss some of their properties. In Section 3, we give the necessary definitions and notations for our discussion. We also prove some properties of these methods. In Sections 4 and 5, we present our new approach, test it on several real datasets and compare its performance to that of the existing algorithms. We conclude in Section 6 with a discussion and future research directions.

2   Definitions and Lemmas

In this section, we define some problems and describe our observations. Our algorithm uses the Hamming distance as the similarity measurement. The probe selection problem is defined as follows.

Definition 1 (Probe Selection Problem). *Given a set of genes $G = \{g_1, g_2, g_3, ..., g_n\}$ and a parameter m which specifies the length of the probes, the probe selection problem finds, for every gene g_i, all length–m probes s which satisfy $HD(s, t) \geq k$ for all t and some constant k.*

k is a pre-specified threshold under the assumption of the Hamming distance. t is a length-m substring of gene $g_j(1 \leq j \leq n, j \neq i)$. The Hamming distance $HD(s, t)$ is the number of positions at which the characters at corresponding positions of the two strings differ. If the Hamming distances between s and all ts are greater than the threshold k, the probe p is said to be a "good" probe, or otherwise a "bad" probe.

Definition 2 (Good Relations and Bad Relations). *If two genes g_i and g_j have no s and t which satisfy $HD(s,t) < k$, we say that they have a "good relation". Otherwise, we say they have a "bad relation".*

Lemma 1. *Let $mHD(g_i, g_j)$ be the minimum $HD(s,t)$ of two genes g_i and g_j. If mHD is no less than the threshold k, the two genes g_i and g_j have a good relation.*

Proof. If mHD is not less than the threshold k, then every other $HD(s,t)$ of the two genes is at least k, because mHD is the smallest HD. There are no bad genes between them. Therefore, they have a good relation by Definition 2. $\square$

Corollary 1. *If two genes have a bad relation, the number of match characters of the maximally matching length-m substring pairs of the two genes is greater than $m - k$.*

Proof. If two genes have a bad relation, there must be a pair of probes having a match count greater than $m - k$. Therefore, if the HD of a pair of probes s and t is mHD, the match count of s and t must be greater than $m - k$. $\square$

Lemma 1 and Corollary 1 provide the basis of an approach which can be used to solve the probe selection problem efficiently. We concisely describe the overall scheme as follows.

1) Screen out the bad relation gene set for each gene using a preprocessing procedure.

2) Eliminate the bad probes of each gene against the corresponding bad gene set obtained from 1) using any probe selection algorithm.

This approach requires a few conditions to be met for it to succeed. In effect, the size of the bad gene set for each gene must be small and the preprocessing procedure should be efficient and accurate. If the number of bad genes for each gene is very large, there is no need to use the preprocessing procedure, because the second phase is almost as same as the original problem. Also, if the preprocessing procedure is not efficient or accurate, it will itself constitute an overhead and deteriorate the performance of the DNA microarray. However, we observed that there is a very small number of bad genes for each gene. The following lemma confirms this observation.

Lemma 2. *If 4 nucleotides are evenly distributed in two genes g_i and g_j, then the probability of two genes having a bad relation is given by*

$$\frac{1}{4^m} \sum_{a=0}^{k-1} \binom{m}{a} 3^a (|g_i| - m + 1)(|g_j| - m + 1)$$

where m is the probe size and k is a threshold.

Proof. By Definition 2, two genes having a bad relation must have at least a pair of length-m subsequences which satisfy $HD < k$. Let s be a length-m

subsequence in g_i. For consecutive m-postions in g_j, there can be 4^m DNA subsequences. Among them, we count the number of cases to which the HD of s is less than k. Because there are 3 different nucleotides for each mismatch position, the number of cases is $\sum_{a=0}^{k-1} \binom{m}{a} 3^a$. And the total number of subsequence pairs of two genes is $(|g_i|-m+1)(|g_j|-m+1)$. Therefore, the above formula holds.   □

From Lemma 2, we can ascertain that the probability of two genes having a bad relation is very low. This means there is a very small number of bad genes for each gene. So if we can find an efficient and accurate preprocessing algorithm to screen the bad genes for each gene, the proposed approach will be successful in solving the probe selection problem for a large genome such as the human genome efficiently.

From Lemma 1 and Corollary 1 we can define a useful problem to derive a preprocessing algorithm.

Definition 3 (Maximum matching substring finding problem). *Given two genes g_i and g_j and a parameter m which specifies the length of the probes, the MMSF problem finds length-m substrings s and t for each gene which satisfy*

$$\arg\min_{s \in g_i, t \in g_j} HD(s,t).$$

If the number of match characters of the two probe candidates s and t is greater than $m - k$, then the two genes g_i and g_j have a bad relation or otherwise they have a good relation. We propose a simple and efficient preprocessing algorithm which searches the regions assumed to have such pairs of s and t and determines if the two genes have a bad relation by q-gram counting.

3   A New Preprocessing Approach

This section presents a preprocessing algorithm used to screen out bad relation gene sets for each gene. The main idea behind our algorithm is to make use of q-gram counting to locate those regions which are expected to contain probe candidate pairs having a Hamming distance of mHD between two genes. If a length-m substring pair of two genes has many matching q-grams, it has a high probability of having a Hamming distance of mHD between the two genes. Therefore, for the purpose of finding the maximally matching length-m substring pairs, we focus only on those regions with more than a certain threshold number of q-grams between the two genes. Figure 1 presents the data structures needed to find such regions. Initially our algorithm converts every gene sequence to corresponding bit vectors and builds a list of locations of q-grams. It also builds $|g_q| + |g_t| - 2q + 1$ circular queues of size m to count the matching q-grams.

3.1   Algorithm

After building the data structures, our algorithm starts by finding regions which have more than a certain threshold number of q-grams between the query gene

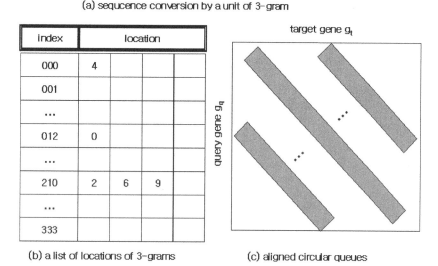

AGTGAAATGATGATC...   ⟶   012100021021023...

(a) sequcence conversion by a unit of 3-gram

index	location			
000	4			
001				
...				
012	0			
...				
210	2	6	9	
...				
333				

(b) a list of locations of 3-grams   (c) aligned circular queues

Fig. 1. Data structures

g_q and target gene g_t. Our algorithm inserts the position of the q-gram of the target gene obtained from a location list into the corresponding queue at the position following the consecutive q-grams($=h_i$) of the query gene and increases the q-gram counter of the corresponding queue by one. Subsequently, it needs to adjust the queue such that only those q-grams whose distance from the newly inserted q-gram is not more than m remain in it. The q-gram counter also needs to be updated in accordance with the queue updating procedure. Also, when the counter exceeds the pre-specified threshold w, the verification procedure is performed. The pseudo code of our algorithm is described below. It runs on each gene against every other gene.

3.2   Verification

The verification procedure determines whether a pair of genes has a bad relation. After counting the number of match characters of the region with more than w q-grams, we compute the match ratio by dividing the number of matches by the length of the region. If the match ratio of the region is greater than the predetermined threshold MT, these two genes are classified as having a bad relation and the algorithm stops. When our algorithm terminates its iterations without successful verification, the two genes are classified as having a good relation. The region of verification is shown in Figure 2. We performed many experiments with different values of the parameters(w: 6, 5, 4, 3, 2 and MT: 0.4, 0.5, 0.6). The q-gram threshold w has a trade-off between the number of regions which need verification and the sensitivity of the algorithm. However, we observed that although w is small, the number of regions is very low and, therefore, we can

Algorithm BadGeneJudgementbyQ-gramCounting
Input: hash sequence H_q of query g_q
   location list T_t of target g_t
   verification condition w
   match ratio threshold MT
Output: Bad or Good relation between g_q and g_t

Begin
 For each q-gram $h_i \in H_q$
  For all location j (whose index equals to h_i)of T_t
   Insert($Queue_{i-j}, j$)
   $Qgram\_Counter_{i-j} \leftarrow Qgram\_Counter_{i-j} + 1$
   For $j - front_{i-j} > m$ //adjust queue
    Delete($Queue_{i-j}$)
    $Qgram\_Counter_{i-j} \leftarrow Qgram\_Counter_{i-j}$ - 1
   if $Qgram\_Count_{i-j} > w$
    if Verify($i, front_{i-j}, rear_{i-j}$) = 1
     return bad // bad relation
 return good // good relation
End

Function Verify($i, front_{i-j}, rear_{i-j}$)
Begin
 determine the region which needed verity using arguments
 $mr \leftarrow$ compute the ratio of match characters of this region
 if $mr > MT$
  return 1
 return 0
End

use a small value of w to increase the sensitivity. We show the average number of these regions between two genes of S. cerevisiae in Table 1. To speed up and additionally increase the sensitivity of our algorithm, we use the matching ratio of the verification region rather than checking the exact Hamming Distance of every probe pair in the region.

3.3   Complexity

The time required to build the data structures is linearly proportional to their size. With the assumption that the genome sequence is random, the expected number of matching q-grams of the target gene to a q-gram of the query gene is $1/4^q(|g| - q + 1)$ and each q-gram is inserted into the corresponding queue only once. If the total length of all of the genes is N, the complexity of the algorithm is $O(1/4^q N^2)$. However, this does not include the running time spent verifying the regions. The running time of verification is computed as follows. The length of a region cannot be more than $3m$ and we also use the hash table technique proposed by [13] to speed up the match counting of the region. The additional running time is $O(av(q, w)n^2 3m/\alpha)$, where $av(q, w)$ is the average value listed

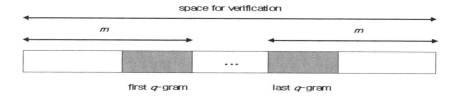

Fig. 2. Region to be verified

Table 1. The number of regions needing to be verified (S. cerevisiae with MT 0.6)

	8-gram Total Average	7-gram Total Average	6-gram Total Average	5-gram Total Average
$w = 4$	375,950 0.01	1,172,147 0.05	4,474,925 0.22	24,408,177 1.21
$w = 3$	3,629,061 0.18	12,894,386 0.64	50,664,722 2.51	254,610,930 12.65
$w = 2$	43,667,018 2.17	158,522,470 7.88	607,207,038 30.18	2,661,324,557 132.31

in Table 1, n is the number of genes and α is the hash size of the hash table(we use $\alpha = 10$). As shown in Table 1, the number of regions requiring verification is very small. Therefore, their effect on the complexity is low enough to be ignored.

4   Analysis of Experimental Results

The proposed algorithm is implemented in C language and tested on a single 32-bit system (Xeon 2.45Ghz with 1GB RAM). The genomes involved in the experiments are listed in table 2. We exclude the result of the human genome because it takes an extremely long time to obtain the exact bad relation data of human genes. In the case of N. crassa, it takes a few weeks. The experiments were performed using a probe length $m = 50$ and mismatch threshold $k = 12$ following the findings of Li and Stormo. They used the fact that, in general, the hybridization free energy of a near-match is sufficiently large when the near-match contains more than 4 errors for 25mer oligonucleotides and 10 errors for 50mer oligonucleotides[13].

Table 3 shows the number of exact bad gene relations of each genome obtained by the brute force approach. This number constitutes the actual number of pairs which need to be compared to filter out bad probes. From this, we know that there is a very small number of gene pairs which need to be compared. In the case of S. cerevisiae, the value 10037 is only about 0.5% of the total number of comparisons(6343*6342/2) between the genes. The experiments were done under 8, 7, 6 and 5-gram and q-gram thresholds of 4, 3, and 2. To evaluate

Table 2. Information of datasets used in experiments

	E. coli	S. pombe	S. cerevisiae	N. crassa	Homo sapiens
Length (bps)	3,469,168	7,278,949	8,865,725	16,534,812	32,297,711
# of genes	3466	5,487	6,343	10,620	25,560

Table 3. The number of total bad relations of each genome

	E. coli	S. pombe	S. cerevisiae	N. crassa
Total bad relations	522	2444	10037	31489

Table 4. Experimental Results for the genome of S. cerevisiae with MT 0.5

	8-gram	7-gram	6-gram	5-gram
	Sensitivity	Sensitivity	Sensitivity	Sensitivity
	Efficiency	Efficiency	Efficiency	Efficiency
	Run time(min)	Run time(min)	Run time(min)	Run time(min)
	0.6862	0.7705	0.8611	0.9679
$w = 6$	0.9996254	0.9995518	0.9993949	0.998764
	6	8	19	57
	0.756	0.8349	0.9166	0.9887
$w = 5$	0.9995673	0.9994504	0.9991398	0.997775
	6	8	19	56
	0.8245	0.8975	0.9594	0.9961
$w = 4$	0.9994752	0.9992599	0.998621	0.996045
	5	8	19	58
	0.8848	0.9446	0.9771	0.9981
$w = 3$	0.99933	0.998932	0.997762	0.993843
	6	8	21	64
	0.9266	0.9686	0.9851	0.9985
$w = 2$	0.9991163	0.998447	0.996777	0.99223
	7	12	32	106

the capability of our algorithm, we use measurements of the sensitivity and efficiency. The sensitivity represents not only the ratio of true positive to bad gene relations, but also the accuracy of our algorithm. The efficiency represents the ratio: $1 - (true\ positive + false\ positive)/total\ relations(= n(n-1)/2)$. This ratio provides an indication of how much our pre-processing procedure can contribute to any probe selection algorithm. As Table 4 shows, its contribution is great and the accuracy is very high.

Table 5 shows the running time of the previous probe selection algorithms and our preprocessing algorithm. We only include the running time for the case where the sensitivity is greater than 0.95. We tested the result of S. cerevisiae obtained by using our preprocessing algorithm as the input of the brute force approach. It took about 7 minutes. Therefore, the total running time is about $19(12 + 7)$

Table 5. Comparison between our algorithm and other algorithms

	Li and Stormo	Rouillard, Herbvert and Zuker	Rahmann	Sung and Lee	Our algorithm (>0.95)
E. coli	1.5 days			3.1 mins	1.2 mins
S. cerevisiae	4 days	1 day		49 mins	12 mins
N. crassa			4 hours	3.5 hours	28 min
Homo sapiens			20 hours		1.8 hours

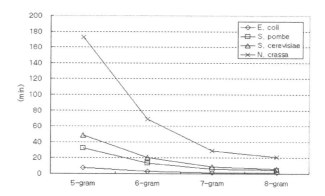

Fig. 3. Running times of our preprocessing algorithm for various datasets with MT 0.5

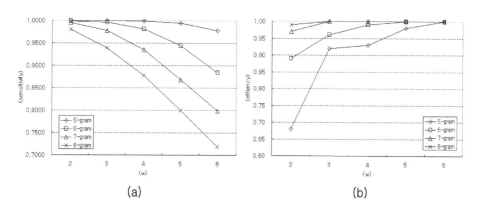

(a)   (b)

Fig. 4. (a) Sensitivity and (b) Efficiency of our algorithm for S. cerevisiae with MT 0.5

minutes. We also measured how long it would take to preprocess the human genome with 25560 genes. It took about 110 minutes and screened out 300408 pairs of genes which need to be compared to select the probes. This number, 300408, is less than the number of gene comparisons of E. coli with 3466 genes which needed by other algorithms. If our algorithm is used as a preprocessing procedure of other algorithms to find specific probes, then the running time of other algorithms can be greatly improved.

Below, we show the results of our experiments for all data sets except for the human genome under various q-grams and w values with MT 0.5. From the result, we can observe that the performance of our algorithm is very well under parameters of 7-gram and w 4 or 3 with MT 0.5. However, the parameter values can be selected as needed.

5   Conclusion

The crux of the microarray design for large genomes lies in how to select unique probes efficiently that distinguish a given genomic sequence from other sequences.

We proposed a new approach to the probe selection problem. Based on the observation that there is very small number of genes which cause the probe candidates of a gene to be bad probes, our approach involves a novel preprocessing procedure that screens out bad relation genes (not probes) for each gene in the commonly used scheme. It screens out bad genes by verifying those regions with more than a certain threshold number of q-grams. Our preprocessing algorithm helps other probe selection algorithms to perform very efficient probe selection by providing them with relation sets obtained by the preprocessing procedure as inputs. We also demonstrated that our new approach greatly reduces the running time and guarantees high accuracy for large genomes.

In further research, we will extend our experiment to very large genomes such as tree genomes. In addition, we will generalize our approach so that the edit distance can be also used as a specificity measure. Since the probe selection problem using the edit distance as a specificity measure is a more time consuming problem, it is important to have such a new approach which can tackle the problem more efficiently.

References

1. Debouck, C., Goodfellow, P.N.: DNA microarrays in drug discovery and development. Nature Genetics 21, 48–50 (1999)
2. Gasieniec, L., Li, C.Y., Sant, P., Wong, P.W.H.: Efficient Probe Selection in Microarray Design. In: IEEE Symposium on Computational Intelligence in Bioinformatics and Computational Biology, IEEE Computer Society Press, Los Alamitos (2006)
3. Hamming, R.W.: Error-detecting and error-correcting codes. Bell System Technical Journal 29(2), 147–160 (1950)
4. Kaderali, L., Schliep, A.: Selecting signature oligonucleotides to identify organisms using DNA arrays. Bioinformatics 18(10), 1340–1349 (2002)
5. Li, L., Stormo, G.: Selection of optimal DNA oligos for gene expression analysis. Bioinformatics 17(11), 1067–1076 (2001)
6. Lockhart, D.J., Dong, H., Byrne, M.C., Follettie, M.T., Gallo, M.V., Chee, M.S., Mittmann, M., Wang, C., Kobayashi, M., Horton, H., Brown, E.L.: Expression monitoring by hybridization to high-density oligonucleotide arrays. Nature Biotechnology 14, 1675–1680 (1996)

7. Manber, U., Myers, G.: Suffix arrays: a new method for online string serarches. SIAM Journal of Computing 22(5), 935–948 (1993)
8. Mount, D.W.: Bioinformatics Sequence and Genome Analysis, 2nd edn. CSHLPress (2004)
9. Rahmann, S.: Rapid large-scale oligonucleotide selection for microarrays. In: Proc. IEEE Computer Society Bioinformatics Conference, vol. 1, pp. 54–63. IEEE Computer Society Press, Los Alamitos (2002)
10. Rahmann, S.: Fast and sensitive probe selection for DNA chips using jumps in matching statistics. In: Proc. IEEE Computational Systems Bioinformatics, vol. 2, pp. 57–64. IEEE Computer Society Press, Los Alamitos (2003)
11. SantaLucia, J.J., Allawi, H.T., Seneviratne, P.A.: Improved Nearest-Neighbor Parameters for Predicting DNA Duplex Stability. Biochemistry 35, 3555–3562 (1996)
12. Stekel, D.: Microarray Bioinformatics. Cambridge (2003)
13. Sung, W., Lee, W.: Fast and accurate probe selection algorithm for large genomes. Proc. IEEE Computational Systems Bioinformatics 2, 65–74 (2003)

Approximation Algorithms for a Point-to-Surface Registration Problem in Medical Navigation

Darko Dimitrov, Christian Knauer*, Klaus Kriegel*, and Fabian Stehn*

Institut für Informatik, Freie Universität Berlin
{darko,knauer,kriegel,stehn}@inf.fu-berlin.de

Abstract. We present two absolute error approximation algorithms for a point-to-surface registration problem in $3D$ with applications in medical navigation systems. For a given triangulated or otherwise dense sampled surface $\mathcal{S}$, a small point set $P \subset \mathbb{R}^3$ and an error bound μ we present two algorithms for computing the set $\mathcal{T}$ of rigid motions, so that the directed Hausdorff distance of P transformed by any of these rigid motions to $\mathcal{S}$ is at most the Hausdorff distance of the best semioptimal[1] matching plus the user chosen error bound μ.

Both algorithms take advantage of so called characteristic points $S_c \subset \mathcal{S}$ and $P_c \subset P$ which are used to reduce the search space significantly. We restrict our attention to scenarios with $|P_c| = 2$. The algorithms are implemented and compared with respect to their efficiency in different settings.

1   Introduction

Most neurosurgical operations nowadays are supported by *medical navigation systems*. The purpose of these systems is to provide the surgeon with additional information during the surgery, like a projection of the instrument into a $3D$ model (of the part) of the patient (where the actual operation is taking place). These models are computed from computer tomography (CT) or magnetic resonance tomography (MRT) scans of the patient. Once the rigid transformation that correctly maps the operation field into the model is known, this problem can be solved easily. Thus, the central task is the computation of that transformation, the so–called *registration process*. A common approach to registration in current medical navigation systems makes use of so called *landmarks*. These are special markers fixed on the patient (from the model acquisition until the beginning of the surgery), such that their positions can be automatically recognized in the model. After gaging the marker positions with the tracking system at the beginning of the surgery (or at least the positions of a subset of four or five markers), the correct rigid transformation mapping the two point configurations

---

* Supported by the German Research Foundation (DFG), grand KN 591/2-1.

[1] We call a matching semioptimal, if the directed Hausdorff distance of P to $\mathcal{S}$ is minimized under the restriction that P_c is aligned centrally with any two points of S_c.

F.P. Preparata and Q. Fang (Eds.): FAW 2007, LNCS 4613, pp. 26–37, 2007.
© Springer-Verlag Berlin Heidelberg 2007

to each other can be computed. Since the total number of markers is small, one can find the correct correspondence between the points of the two configurations even by brute force techniques. A more advanced approach making use of geometric hashing techniques is presented in [3].

Frequently surgeons would like to avoid the use of markers for medical reasons, and because in many cases (e.g. spinal surgery) it is very hard or even impossible to fix markers before the surgery. Thus, the design of algorithms for registration procedures without markers is an important and challenging task. In that case the surgeon would gage only a small number of arbitrary points from the relevant anatomic region and the algorithm has to find the best matching of such a point pattern onto the surface of the corresponding region in the model.

A lot of research has been done in the last years in the domain of registration algorithms with medical applications. E.g. geodesics and local geometry are used in [4] to determine point correspondence between point-pairs of 3D-surfaces, another approach uses thin splines to solve the point registration problem [5], applications in transcranial magnetic stimulation by point-to-surface registration using ICP are presented in [6].

In this paper we present a new approach to the registration problem without markers. To understand the background, we have to discuss the drawbacks of two alternative approaches. Firstly, there are several heuristic methods that could be used directly, for example ICP (Iteratve Closest Point) or simulated annealing combined with randomly generated starting configurations. Both methods perform very well, but they might get stuck in a local optimum far from the global optimium and, thus, they do not guarantee that the computed transformation is close to the optimal one. Secondly, one could try to reconstruct a surface from the tracked points and to apply a surface matching algorithm. However, this would require a rather dense pattern of tracked points and, moreover, most of these surface matching algorithms contain also some heuristic routines.

Minimising the distance between a point set and a surface under rigid motions is usually a hard algorithmic problem, mostly because rigid motions in $\mathbb{R}^3$ have six degrees of freedom. We try to reduce that huge search space making use of few anatomical landmarks in addition to the arbitrarily gaged points. We call such points *characteristic points*. They are known in the model (e.g the root of the nose) and play the role of natural markers. In a first step the surgeon has to gage as many characteristic points as possible followed by a few arbitrary, additional points. If the surgeon can manage to gage at least three (non-coalligned) characteristic points, one can apply the established landmark approach using the other points as additional control points. In this paper we study the nontrivial, but very realistic case, that the point set contains only two characteristic points.

The distance function considered in this paper is the so-called *directed Hausdorff distance* defined by

$$\boldsymbol{H}(A, B) = \sup_{a \in A} \operatorname{dist}(a, B) = \sup_{a \in A} \inf_{b \in B} \|a - b\|.$$

where $\|a - b\| = \operatorname{dist}(a, b)$ is the Euclidean distance between a and b. Some basic ideas of our general approach can be found in [1] where we mainly discussed the problem of approximating the minimal directed Hausdorff distance of the measured points on the patient to the model with a constant factor $\lambda > 1$. Such approximations can be computed in polynomial time, but the algorithm is not very useful in practice because of large constants and a $(\frac{1}{(\lambda-1)})^5$ factor in the run time.

Here, we will present two solutions for a version of the problem that meets more the practical requirements: the approximation with a constant, additive error. The first solution is basicly an adaption of a sweep line algorithm in angle space that can be implemented easily. The second solution uses an augmented *segment tree* as an auxiliary data structure. Although in the theoretical analysis and in our experiments both algorithm require nearly the same run time there is also a strong argument in favor of the second approach: If the input data does not suffice to guarantee a unique solution, the update with one or more additionally gaged points is much faster with the segment tree approach.

Formal problem description. We assume that the model is given as a triangulated surface $\mathcal{S}$ consisting of n triangles together with a set of n_c characteristic points $S_c \subset \mathcal{S}$, where n_c is a small number. To avoid scaling factors in the analysis, we assume that the surface is a subset of the unit cube $[0,1]^3$. Furthermore, the second part of the input is a set of points $P \subset \mathbb{R}^3$ with a distinguished subset of characteristic points $P_c \subset P$. The quality of a rigid transformation $t : \mathbb{R}^3 \to \mathbb{R}^3$ is defined by

$$\epsilon(t) = \max\{\, \boldsymbol{H}(t(P), \mathcal{S}), \boldsymbol{H}(t(P_c), S_c)\,\}.$$

A rigid transformation t that minimizes $\epsilon(t)$ is called an *optimal transformation* or an *optimal matching* and the quality of an optimal transformation is denoted by ϵ_{opt}. Given an absolute error bound μ we ask for the set $\mathcal{T}$ of rigid motions, so that

$$\forall t \in \mathcal{T} : \ \boldsymbol{H}(t(P), \mathcal{S}) \leq \epsilon_{opt} + \mu \ \land \ \boldsymbol{H}(t(P_c), S_c) \leq \epsilon_{opt} + \mu.$$

Our results. Let t_{init} be the rigid transformation that aligns the points of $P_c = \{p, p'\}$ with two points $s, s' \in S_c$ such that $\|p - s\| = \|p' - s'\|$. Let t_α be the rigid motion that rotates around the line through s and s' by an angle of α, then

$$\epsilon_{sem} = \min_{\alpha \in [0,2\pi)} \boldsymbol{H}(t_\alpha \circ t_{init}(P), \mathcal{S})$$

is called value for the best *semioptimal* matching.

As described in [1] the value for the best semioptimal matching ϵ_{sem} and the ratio of the largest distance of any point in P to the rotation axis to the distance of the points in P_c can be used to find a lower bound to ϵ_{opt}. This bound together with an approximation value $\lambda > 1$ can be used to introduce a grid structure around the axis, s.th. the best semioptimal matching for all perturbations of the rotation axis through the grid points is guaranteed to be at most $(\lambda - 1)\epsilon_{opt}$ away from the optimal matching position.

In this paper we present two algorithms for computing the set of rigid motions t which fulfill the criteria that $\boldsymbol{H}(t(P), \mathcal{S}) \leq \epsilon_{sem} + \mu$. Both implementations require a pre-processing time of $\mathcal{O}(n/\mu^3)$. The first implementation is a modified sweep line variant having a run time of $\mathcal{O}\left((k/\mu)\log k\right)$ for $k = |P \setminus P_c|$. The second implementation uses an augmented segment tree and has a run time of $\mathcal{O}\left((k/\mu)\log\left(k/\mu\right)\right)$. Measuring and adding a point to P to increase the quality of a computed solution costs $\mathcal{O}\left((k+1)/\mu\right)$ time in case of the sweep line version. The same operation costs $\mathcal{O}((k+1)/\mu\log\left(k/\mu\right))$ for the segment tree variant, which in practice is faster than the sweep line method, as our evaluation results show.

Notes on assumptions. This paper concentrates on algorithms computing a set of rigid motions, which applied to the measured point set minimize the directed Hausdorff to the model. Due to the flexibility of the cranial bone and the way how a patient, more precisely his head is fixed during the operation, the skull may undergo slight non-rigid deformations. We consider these effects on hard tissues like bones to be marginal so that only rigid motions instead of affine transformations are taken into consideration. A lot of research has been done for non-rigid registration for soft tissue surgery, like in [10,11].

Measuring characteristic points[2] instead of auxiliary landmarks whose position in the operation field is exactly defined, causes an additional error. This error is due to the distance of the characteristic points as defined in the model and the actual measured landmarks on the patient. In this paper this error is not analysed separately, but influences the position of the rotation axis relative to $P \setminus P_c$ and therefore the quality of the semioptimal matching.

Organisation of this paper. This paper is structured as follows: Section 2 describes the two approximation methods and their analysis. It is furthermore described how these algorithms have to be modified in order to handle points added later on. Section 3 compares the implementation of the two introduced registration methods and evaluates their performance on several test settings. Section 4 concludes with summarizing the presented results and gives an overview of the questions and tasks that remain.

2   The Registration Process

In this chapter we first present a general strategy for solving instances of the considered registration problem and then describe two different implementations for this strategy.

2.1   A Strategy for Finding Semioptimal Registrations

Let p and p' be the two characteristic points in P_c. For every transformation t the distance $\boldsymbol{H}(t(P_c), \mathcal{S}_c)$ is a lower bound for $\epsilon(t)$ and there is always a pair (s, s') of two characteristic points $s, s' \in S_c$ that realizes this distance. Thus, we fix such a pair (s, s') and construct a rigid transformation t_{init} that minimizes $\max\{\text{dist}(t(p), s), \text{dist}(t(p'), s')\}$. Note that such an initial transformation can

---

[2] Landmarks which correspond to distinctive features in the area of interest on the patient.

be obtained by translating first the center of the line segment $[p, p']$ onto the center of the line segment $[s, s']$ and rotating the translated segment $[tr(p), tr(p')]$ around the common center such that all four points are on the same line and p gets on the same side of the center as s, see figure 1.

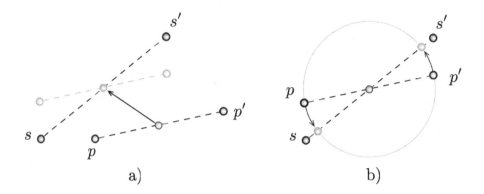

Fig. 1. The initial transformation t_{init} which centrally aligns $s, s' \in S_c$ and $p, p' \in P_c$

Let $\mathcal{T}_{(s,s')}$ denote the family of all rigid transformations with the same property as the inintial one, i.e.,

$$\mathcal{T}_{(s,s')} = \{ t \mid t(p), t(p'), s \text{ and } s' \text{ are co–aligned}$$
$$\text{and } \mathrm{dist}(t(p), s) = \mathrm{dist}(t(p'), s') < \mathrm{dist}(t(p), s') \}$$

It is clear, that any $t \in \mathcal{T}_{(s,s')}$ can be obtained by starting with the initial transformation t_{init} and then rotating around the axis spanned by the four points. A transformation $t \in \mathcal{T}_{(s,s')}$ is called *semioptimal* (with respect to the pair (s, s')) if it minimizes the distance $\boldsymbol{H}(t(P \setminus P_c), \mathcal{S})$. In [1] it is shown how to approximate an optimal transformation starting from semioptimal ones. The idea is to define a pertubation scheme for mapping p and p' and computing semioptimal matchings with respect to the rotation axis spanned by the images of p and p'. Since these ideas also apply for absolute error approximations, we concentrate here on the problem how to approximate a semioptimal matching, provided that the images of p and p' are fixed.

Recall that $\mathcal{S}$ is a subset of the unit cube and that an error bound μ is given. We subdivide the unit cube into a set $\mathcal{G}$ of subcubes of side length $\mu/\sqrt{3}$. This way μ is an upper bound for the distane between two points in the same subcube and the total number of subcubes is $O(1/\mu^3)$. Consider a pair (s, s') and an initial transformation $t_{init} \in \mathcal{T}_{(s,s')}$, the line l containig $s, s', t_{init}(p), t_{init}(p')$ and the point set $P' = \{ t_{init}(p) \mid p \in P \setminus P_c \}$. Let t_α be the rigid motion which corresponds to an rotation around l by an angle of α. For each $p \in P'$ the sets C_p of all subcubes are determined which are intersected by the trajectory of p around l:

$$C_p = \{ c \in \mathcal{G} \mid \exists \alpha \in [0, 2\pi) : t_\alpha(p) \in c \} .$$

Let $\left[\alpha_{c,p}, \alpha'_{c,p}\right)$ be the angle interval for a cell $c \in C_p$, such that $\alpha_{c,p}$ denotes the angle for which $t_{\alpha_{c,p}}(p)$ enters the subcube c and $\alpha'_{c,p}$ denotes the angle for which p leaves c. After a whole rotation of p around l the intervals of the collected set C_p form a partition of the angle space:

$$\forall p \in P' : \bigcup_{c \in C_p} \left[\alpha_{c,p}, \alpha'_{c,p}\right) = [0, 2\pi).$$

The distance of a subcube c to the surface is defined as

$$\text{dist}(c, \mathcal{S}) := \min_{q \in c} \min_{s \in \mathcal{S}} \|q - s\|.$$

Let $\mathcal{A}(\bigcup_{p \in P'} C_p)$ be the refined subdivision induced by the intervals of the collected subcubes. To find the best rotation angles we determine the set of cells A_{min} of this arrangement, for which the largest distance value of a covering subcube is minimal. Let $\hat{A} = \bigcup_{a \in A_{min}} a$ be all angles contained in A_{min}, then the following inequality holds $\boldsymbol{H}(t_\alpha(P'), \mathcal{S}) \leq \epsilon_{sem} + \mu$ for all $\alpha \in \hat{A}$, see figure 2.

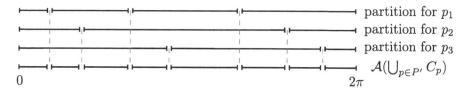

Fig. 2. The refined subdivision $\mathcal{A}(\bigcup_{p \in P'} C_p)$ of the interval sets for $P' = \{p_1, p_2, p_3\}$

For any $\alpha \in \hat{A}$ let B_α be the set of all subcubes containing at least one point of P' rotated by α around l:

$$B_\alpha = \{c \in \mathcal{G} \,|\, \exists p \in P' : t_\alpha(p) \in c\}$$

and let $c_\alpha = \arg\max_{c \in B_\alpha} \text{dist}(c, \mathcal{S})$ be one of the subcubes with the largest distance to $\mathcal{S}$. As the largest distance of any two points within one subcube is at most μ, the directed Hausdorff distance of $t_\alpha(P')$ to $\mathcal{S}$ is at most $\text{dist}(c_\alpha, \mathcal{S}) + \mu$. And as ϵ_{sem} is an upper bound to $\text{dist}(c_\alpha, \mathcal{S})$, the following condition holds:

$$\forall \alpha \in \hat{A} : \boldsymbol{H}\left(t_\alpha(P'), \mathcal{S}\right) \leq \text{dist}(c_\alpha, \mathcal{S}) + \mu \leq \epsilon_{sem} + \mu$$

2.2   The Implementation

We present two methods for evaluating the cells $A_{min} \subset \mathcal{A}(\bigcup_{p \in P'} C_p)$ that satisfy this condition, the first is a standard sweep line approach and the second uses a data structure called *counting segment tree*. Even though the second is slightly more complex than the first we will see in the next chapters that the segment tree method outperforms its competitor in practice when it comes to adding points to a solution in order to increase its quality.

In both cases the sets of subcubes C_p for all $p \in P'$ are collected by tracing the trajectory of p through $\mathcal{G}$. If $\mathcal{G}$ is organised as a three dimensional array it takes $\mathcal{O}(1)$ time to locate the cell $c_p \in \mathcal{G}$ containing p. By intersecting the six sides of c_p with the trajectory of p around l one can find the succeeding cell to the current cell also in $\mathcal{O}(1)$ time. By continuing that way until c_p is reached again, one can gather C_p in $\mathcal{O}(1/\mu)$ time, as each C_p contains at most this many subcubes. While walking along the trajectory the intervals for each point and inspected subcube are also evaluated and organised as a linked list, so that each interval stores a pointer to the succeeding interval of the next subcube on the trajectory. Moreover we compute in a pre-processing step the distances of each subcube of $\mathcal{G}$ to the surface $\mathcal{S}$ and store a copy of this value in each angle interval that is later evaluated. This can be done naively by evaluating for all $\mathcal{O}(1/\mu^3)$ subcubes their shortest distance to any of the n triangles of the model, taking $\mathcal{O}(n/\mu^3)$ time.

Looking at the application in which these registration process are used this time consuming pre-processing can be done directly after the model is constructed by the MRT- or CT scan and therefore it is not crucial for the matching time during the surgery.

2.3   The Sweep Line Variant

In this variant we sweep over the $|P'|$ interval sets from 0 to 2π and keep track of the largest distance value of the corresponding subcubes under the sweep line. These distances are stored in a max-heap which is initialised with the distance values of the subcubes whose corresponding intervals contain the rotation angle $\alpha = 0$. The event queue for the sweep line process consists of all intervals ordered increasing by their largest contained angle.

Suppose that $i = [\alpha_{c,.}, \alpha'_{c,.})$ is the current element in the event queue. We then assign the current top value of the heap to the cell of the arrangement which right limit is $\alpha'_{c,.}$, remove the value dist$(c, \mathcal{S})$ from the heap[3] and add the distance value of the subcube corresponding to successor of i(see figure 3). While processing the event queue the algorithm keeps track of the cells A_{min} of the arrangement which have the smallest distance value.

2.4   The Counting Segment Tree Variant

An alternative approach for finding the best cells of the refinement is to use a data structure called counting segment tree (cST). A cST is a slightly modified variant of the well known segment tree data structure as e.g. described in [2]. This structure stores and organizes an arrangement of intervals and keeps track of of the number of intervals that cover a cell of the arrangement. Moreover a cST is able to report in $\mathcal{O}(1)$ time the cell t_{max} which is covered by the largest number of intervals.

---

[3] Note that this is not necessarily the current maximal value of the heap.

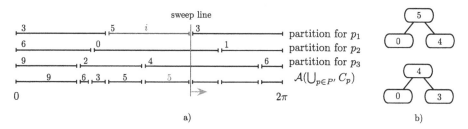

Fig. 3. a) the next event in the sweep line is the interval with distance 5 **b)** top: heap before handling i, bottom: after handling i

We first build a counting segment tree upon the refined subdivision $\mathcal{A}(\bigcup_{p \in P'} C_p)$ of all collected intervals and then add intervals to this tree in a certain order. Therefore all intervals are merged and sorted in an ascending manner with respect to the distance values of their corresponding subcubes. Finally the intervals are inserted into the counting segment tree in the introduced order starting with the interval which corresponding subcube has the smallest distance value.

Let i be the first interval added to the segment tree that causes t_{max} to be covered by $|P'|$ intervals and $\mathrm{dist}(i, \mathcal{S})$ be the distance value of its corresponding subcube. We then keep inserting intervals to the tree in the described order as long as their corresponding distance values are equal to $\mathrm{dist}(i, \mathcal{S})$. As soon as the first interval has a corresponding distance value larger than $\mathrm{dist}(i, \mathcal{S})$ the algorithm terminates and all cells of the refinement are reported that are covered by $|P'|$ intervals.

As each C_p forms a partition of the angle space, the intersection of two intervals gathered by the rotation of the same point p is always empty. In other words if a cell is covered by $|P'|$ intervals, all points in P' are moved into subcubes that have an distance value of at most $\mathrm{dist}(i, \mathcal{S})$ when rotated by any angle taken from this intervals. As the intervals are inserted in ascending order of their corresponding distance values, the reported cells are equal to the A_{min} cells described in the previous section.

2.5   Increasing the Quality by Adding Points to a Solution

Motivated by the medical application of these algorithms it is desired to be able to increase the quality of a matching by measuring additional points during the operation. Therefore one has modify the previous methods so that they dynamically support adding points to P, the set of measured points.

In this section we compare the two methods stated above regarding to this demand. Let q be the additional measured point and C_q the set of all subcubes which are intersected by the trajectory of q around the rotation axis l.

Adding points with the sweep line method. Let $\mathcal{A}(\bigcup_{p \in P'} C_p)$ be the refined subdivision of all intervals collected for the point set $P' = \{t_{init}(p) | p \in P \setminus P_c\}$, where each cell stores its assigned distance value, computed during the sweep line process.

The new event queue is emulated by taking the next interval either from C_q, or $\mathcal{A}(\bigcup_{p \in P'} C_p)$, depending on which has the smaller right interval limit. This means all intervals of the old arrangement together with the intervals of the additionally collected intervals corresponding to cells in C_q have to be inspected to find the new set of best rotation angles. That, apart from the $\log |P'|$ factor for the max heap is as expensive as restarting the whole sweep line process with $P \cup \{q\}$.

Adding points with the segment tree variant. To add a point to a solution which has been computed with the segment tree method one has to sort the collected intervals corresponding to C_q with respect to the distance values of their subcubes.

As before we now add intervals into the tree as long as the number of intervals covering t_{max} is smaller then $|P'| + 1$. The interval added to the tree is either the head of the old event queue or the head of the sorted list of intervals for q, depending on the smaller distance value of their corresponding subcubes. If an interval i_q of C_q is added to the tree it might happen, that a leaf of the segment tree containing an interval limit of i_q has to be splitted. This corresponds to refining the underlying arrangement. Note that the segment tree is now not guaranteed to be balanced anymore, but even if all intervals of q would be added the height of the segment tree can only increase by at most two.

This method of adding points to a solution has the advantage, that all previously gathered information can directly be reused to find the next optimum without inspecting all cells of the arrangement.

2.6   Analysis

The pre-processing step computes for all subcubes their distance to the surface. For $n = |\mathcal{S}|$, this can be done naively in $\mathcal{O}(n/\mu^3)$ time.

In a regular grid with mesh size $\mu/\sqrt{3}$ embedded in the unit cube any trajectory of a point rotating around an axis can intersect $\mathcal{O}(1/\mu)$ subcubes. For $k = |P \setminus P_c|$ it takes $\mathcal{O}(k/\mu)$ time to determine all inspected subcubes and their corresponding angle intervals.

The sweep line variant merges all k interval sets in $\mathcal{O}(k/\mu \log k)$ and sweeps through all $\mathcal{O}(k/\mu)$ events and updates the heap in $\mathcal{O}(\log k)$ time, this results in a matching time without pre-processing of

$$\mathcal{O}\left(\frac{k}{\mu} \log k\right)$$

for this method.

The run time of the segment tree method is dominated by the time needed to order all intervals with respect to the distance value of their corresponding subcubes, which is also the time needed to build the segment tree. Adding an interval to the tree takes $\mathcal{O}(\log k/\mu)$ time, therefore the run time of this variant is

$$\mathcal{O}\left(\frac{k}{\mu} \log \frac{k}{\mu}\right)$$

Adding points to a solution. In case of the sweep line method all $\mathcal{O}(k/\mu)$ intervals of the refined subdivision and all $\mathcal{O}(1/\mu)$ intervals of the added point are inspected. As both sets are given as a linked list it takes $\mathcal{O}((k+1)/\mu)$ time to merge them and to update the refined subdivision.

To add a point p with the counting segment tree variant one has to sort all intervals collected while rotating p around l in $\mathcal{O}((1/\mu)\log(1/\mu))$ time. Adding an interval to the segment tree costs $\mathcal{O}(\log k/\mu)$ time. In the worst case, all intervals that are not added to the tree during the computation of the previous solution and all new intervals have to be added to find a cell of the refinement that is covered by $k+1$ intervals. Therefore the worst case run time of adding a point with this method is $\mathcal{O}\left((k+1)/\mu\log(k/\mu)\right)$. In practice actually only a small number of intervals are added before a new solution is found.

The counting segment tree variant has the advantage that additional points can be *weaved* into current solution without having to touch all cells of the refinement of the previous solution.

3   Evaluation

We implemented both, the segment tree and the sweep line, variants of the presented algorithms, and compared their performances. The evaluation was performed on ten different test configurations, each consisting of: a triangulated surface of a skull model $\mathcal{S}$ (comprised of nearly 3000 triangles), a set of four characteristic points $\mathcal{S}_c \subset \mathcal{S}$, and a set of eight points P manually chosen from the patient, with a subset of two characteristic points $P_c \subset P$. In all configurations the grid density μ was set to 0.01. The evaluation of the implementations was performed on a 2.33 GHz Intel Core 2 Duo processor computer equipped with 2GB of main memory.

We measured the *matching time* (*MT*) (the time needed to compute the best rotation angle for each possible assignment of the two points of P_c to any two points of $\mathcal{S}_c$), and the *matching time for the best assignment* (*MT-BA*) (the time needed to compute the best angle for the right assignment of P_c to $\mathcal{S}_c$). The results, see Table 1, show that the differences of the matching times of the both variants are negligible.

However, the segment tree variant has an obvious advantage in enhancing the quality of the registration. Usually, after the best registration between P and $\mathcal{S}$ is determined, one or more additional chosen points are added to the point set P and the registration is reevaluated. In our tests, we added three such additional points. In the case of the sweep line variant, each time we add a new additional point, a sweep over the interval sets of all points of P must be restarted. In contrast, in the case of the segment tree variant, the already built segment tree, with all existing information about added intervals, can be reused, which makes the operation of adding a new point quite fast (for experimental results see the values for 1.AP-3.AP in Table 1).

The screenshot of figure 4 shows a triangulated model of a skull, a set P of ten points and three characteristic points $\mathcal{S}_c$. The two characteristic points $P_c \subset P$

Table 1. Comparison of of the segment tree and the sweep line method

case	counting segment tree method					sweep line method				
	MT	MT-BA	1.AP	2.AP	3.AP	MT	MT-BA	1.AP	2.AP	3.AP
1	1.53	0.16	0.07	0.07	0.07	1.48	0.16	0.3	0.38	0.48
2	0.47	0.26	0.09	0.07	0.08	0.46	0.25	0.33	0.44	0.55
3	0.13	0.05	0.04	0.04	0.04	0.12	0.05	0.1	0.16	0.21
4	1.76	0.47	0.07	0.03	0.11	1.75	0.46	0.53	0.61	0.68
5	1.1	0.19	0.04	0.07	0.03	1.08	0.2	0.32	0.39	0.45
6	3.95	0.47	0.09	0.09	0.07	3.83	0.48	0.6	0.7	0.81
7	0.97	0.19	0.07	0.08	0.08	0.94	0.19	0.33	0.43	0.55
8	1.19	0.48	0.07	0.09	0.08	1.17	0.48	0.44	0.85	0.99
9	0.86	0.35	0.07	0.11	0.09	0.86	0.36	0.56	0.69	0.79
10	2.15	0.59	0.08	0.09	0.08	2.11	0.58	0.71	0.85	0.98
AVG	1.41	0.32	0.07	0.07	0.07	1.38	0.32	0.42	0.55	0.65

MT=matching time, MT-BA=matching time for the best assignment,
i.AP=time for adding the i-th point   (measured in seconds)

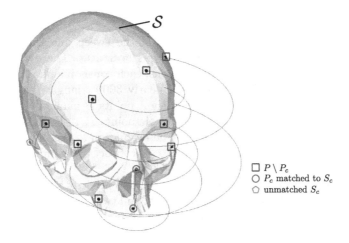

$\square\ P \setminus P_c$
$\bigcirc\ P_c$ matched to S_c
$\varheartsuit$ unmatched S_c

Fig. 4. The model S and the pointset P in the approximated semioptimal position ($\mu = 0.01$), along with the trajectories of the points in $P \setminus P_c$

are located at the top of the nasal bone and on the center of the maxilla (upper jaw) respectively.

4   Conclusion and Future Work

We have presented two algorithms for a point-to-surface matching problem that is motivated by medical applications. The presented approach does not have the drawbacks of the methods currently used in medical navigation systems, like the necessity of landmarks or the chance to get stuck in a local minimum and it achieves a fast matching time that is competitive with existing approaches.

An interesting problem, especially from theoretical point of view, is to inspect the matching scenario with only one characteristic point, a scenario that increases the search space and therefore the matching time. Applying the algorithms directly on the voxel data, instead of applying them on a triangulated approximation reconstructed from this voxel data, could further increase the precision of the matching and significantly reduce the preprocessing time. On the practical side, it is planned to embed this algorithm in a real medical navigation system and test its performance in real world applications.

References

1. Dimitrov, D., Knauer, C., Kriegel, K.: Registration of 3D - Patterns and Shapes with Characteristic Points. In: Proceedings of International Conference on Computer Vision Theory and Applications - VISAPP 2006, pp. 393–400 (2006)
2. de Berg, M., van Kreveld, M., Overmars, M., Schwarzkopf, O.: Computational Geometry: Algorithms and Applications. Springer, Heidelberg (1997)
3. Hoffmann, F., Kriegel, K., Schönherr, S., Wenk, C.: A simple and robust algorithm for landmark registration in computer assisted neurosurgery. Technical report B 99-21, Freie Universität Berlin (1999)
4. Wang, Y., Peterson, B.S., Staib, L.H.: Shape-Based 3D Surface Correspondence using geodesics and local geometry. In: IEEE Computer Society Conference on computer vision and pattern recognition, vol. II, IEEE Computer Society Press, Los Alamitos (2000)
5. Chui, H., Rangarajan, A.: A new algorithm for non-rigid point matching. In: Computer Vision and Pattern Recognition Proceedings, IEEE Computer Society Press, Los Alamitos (2000)
6. Matthäus, L., Giese, A., Trillenberg, P., Wertheimer, D., Schweikard, A.: Solving the Positioning Problem in TMS GMS CURAC, vol. 1, Doc06 (2006)
7. van Laarhovn, P.J.M., Aarts, E.H.L.: Simulated Annealing: Theory and Applications. D. Reidel Publishing Company, Kluwer (1987)
8. Besl, P.J., McKay, N.D.: A method for registration of 3-d shapes. IEEE Trans. on Pattern Recognition and Machine Intelligence 14(2), 239–256 (1992)
9. Rusinkiewicz, S., Levoy, M.: Efficient variants of the ICP algorithm. In: Third Intern. Conference on 3D Digital Imaging and Modeling, pp. 145–152 (2001)
10. Bodensteiner, C., Darolti, C., Schweikard, A.: 3D-reconstruction of bone surfaces from sparse projection data with a robotized c-arm by deformable 2D/3D registration. In: Computer Assisted Radiology and Surgery, Germany (June 27-30, 2007)
11. Clatz, O., Delingette, H., Talos, I., Golby, A.J., Kikinis, R., Jolesz, F.A., Ayache, N., Warfield, S.K.: Robust nonrigid registration to capture brain shift from intraoperative MRI. In: Computer Assisted Radiology and Surgery, Berlin, Germany (June 27-30, 2007)

Biometric Digital Signature Key Generation and Cryptography Communication Based on Fingerprint

Je-Gyeong Jo, Jong-Won Seo, and Hyung-Woo Lee

Div. Computer Information of Software, Hanshin University,
411, Yangsan-dong, Osan, Gyunggi, 447-791, Korea
aiking@hs.ac.kr, seo0207@hs.ac.kr, hwlee@hs.ac.kr

Abstract. A digital signature is a term used to describe a data string which associates a digital message with an assigned person only. Therefore, the main goal (or contribution) of this work is to study a simple method for generating digital signatures and cryptography communication using biometrics. The digital signature should be generated in the way that it can be verified by the existing cryptographic algorithm such as RSA/ElGamal without changing its own security requirement and infrastructure. It is expected that the proposed mechanism will ensure security on the binding of biometric information in the signature scheme on telecommunication environments.[1]

Keywords: Biometrics Digital Key, Fuzzy Vault, Secure Communication, Digital Signature, Digital Key Generation.

1 Introduction

A digital signature has many applications in information security such as authentication, data integrity, and non-repudiation. One of the most significant advances in digital signature technologies is the development of the first practical cryptographic scheme called RSA[1], while it still remains as one of the most practical and versatile digital signature techniques available today[2].

It is often desirable to generate a digital signature by deriving a signature key (or a semantically equivalent value) from human source(biometrics) in today's communications environment rather than keeping the key in an external hardware device. Therefore, biometrics is the science of using digital technologies to identify a human being based on the individual's unique measurable biological characteristics[4,5].

This work proposes *a general framework of biometric digital signature key generation from both technical and practical perspectives* in order to establish safe environment using telebiometric systems and protect individual privacy.

---

[1] This work was supported by the Korea Research Foundation Grant funded by the Korean Government (MOEHRD, Basic Research Promotion Fund) (KRF-2006-311-D00857).

F.P. Preparata and Q. Fang (Eds.): FAW 2007, LNCS 4613, pp. 38–49, 2007.

From the technical point of view, this work proposes several biometric digital key and signature generation frameworks to ensure data integrity, mutual authentication, and confidentiality. From the practical perspective, this work describes overall framework that allow protection of biometric data as related to their enrolment, signature generation and verification. This work also outlines measures for protection of the biometric information as related to its generation, storage, and disposal.

Actually digital key generation from biometric has many applications such as automatic identification, user authentication with message encryption, etc. Therefore, this work analysis the related schemes and proposes a simplified model where a general signature scheme (including an RSA scheme that requires a large signature key) can be applied without losing its security.

This work also can be applicable into authentication frameworks for protection of biometric systems as related to their operational procedures, roles and responsibilities of the personnel involved in system design. It is expected that the proposed countermeasures will ensure security and reliability on the flow of biometric information in the telecommunication environment.

2   Overview on Existing Scheme

2.1   Biometric Digital Key

Digital Signature Key Generation for Telebiometrics. Fig. 1 depicts the common component or modules on Telebiometrics system with proposed key generation module, which commonly includes a step to extract features through signal processing after acquiring biometric data from a biometric device such as a sensor. The features are then compared or matched against the biometric data, which were already obtained through the same processes and saved in a database, and the result is decided on decision step.

Based on existing Telebiometrics model, this work proposes digital signature keys (both private key and public key pair) generation framework from biometric information. Therefore, it is possible to combine existing public key infrastructure such as RSA[1] or ElGamal[3] to generate digital signature key on biometric data.

Therefore, the main goal (or contribution) of this work is to study a simple method for generating digital signatures using biometrics by exploiting the existing Fuzzy Vault[4] scheme.

Biometric Template. Biometric templates are processed measurement feature vectors. Biometrics of different individuals is independent realizations of a random process that is equal for all individuals. We assume that the processing of biometrics results in templates that can be described as a sequence of n independent identically distributed random variables. Noisy measurements of biometrics are modeled as observations through a memoryless noisy channel. It is assumed that the enrollment measurements of the biometric templates are noise free.

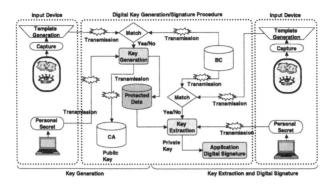

Fig. 1. Biometric Digital Key Generation on Telebiometrics

We examine the existing biometrics-based digital signature scheme and analysis them on the fly. First this document can classify those schemes into *key derivation(generation)* and *signature generation and verification* framework. The key derivation schemes imply that the signature key is derived directly from biometrics while the key authentication schemes mean that the signature key is accessed by biometric authentication.

Common Digital Signature Scheme. A cryptographic primitive that is fundamental in authentication, authorization, and non-repudiation is the digital signature. The process of signing entails transforming the message and some secret information held by the entity into a tag called a signature[5].

- $\mathcal{M}$ is the set of messages which can be signed.
- $\mathcal{S}$ is a set of elements called *signature*, possibly binary strings of a fixed length.
- $\mathcal{S}_{\mathcal{A}}$ is a transformation from the message set $\mathcal{M}$ to the signature set $\mathcal{S}$, and is called a *signing transformation* for communication entity $\mathcal{A}$. The transformation $\mathcal{S}_{\mathcal{A}}$ is kept secret by sender $\mathcal{A}$, and will be used to create signatures for messages from $\mathcal{M}$.
- $\mathcal{V}_{\mathcal{A}}$ is a transformation from the set $\mathcal{S} \times \mathcal{M}$ to the set {true; false}. $\mathcal{V}_{\mathcal{A}}$ is called a verification transformation for $\mathcal{A}$'s signatures, is publicly known, and is used by other entities to verify signatures created by $\mathcal{A}$.

Therefore, the transformations $\mathcal{S}_{\mathcal{A}}$ and $\mathcal{V}_{\mathcal{A}}$ provide a *digital signature scheme* for $\mathcal{A}$. Occasionally the term *digital signature mechanism* is used. The size of the key space is the number of encryption/decryption key pairs that are available in the cipher system. A key is typically a compact way to specify the encryption transformation (from the set of all encryption transformations) to be used. Each can be simply described by a permutation procedure which is called the key. It is a great temptation to relate the security of the encryption scheme to the size of the key space.

RSA is the first practical cryptographic scheme for digital signature and still remains as one of the most practical and versatile techniques available today

[1]. This document supposes to use a simple hash-and-sign RSA primitive in a probabilistic manner (with k-bit random numbers).

The public-private keys are respectively $\langle e, N \rangle$ and $\langle d, N \rangle$ where N is the product of two distinct large primes p and q, and $ed \equiv 1 \bmod \phi(N)$ for the Euler totient function $\phi(N) = (p-1)(q-1)$ [1]. The public key is postulated to be certified by the $\mathcal{CA}$. We assume signer $\mathcal{S}$ returns signature on a message m; $\langle s, r \rangle$ where $s \leftarrow H(m, r)^d \bmod N$ and $r \leftarrow R\{0, 1\}^k$.

The ElGamal public-key encryption scheme can be viewed as Diffie-Hellman key agreement in key transfer mode[3]. Its security is based on the intractability of the discrete logarithm problem and the Diffie-Hellman problem. The ElGamal signature scheme is a randomized signature mechanism. It generates digital signatures with appendix on binary messages of arbitrary length, and requires a hash function $h: \{0, 1\}^* \rightarrow Z_p$ where p is a large prime number. Each entity creates a public key and corresponding private key.

Public Key Infrastructure and Biometric Certificate. For an authorized assertion about a public key, we commonly use digital certificates issued by a trusted entity called the certificate authority (CA) in the existing public key infrastructure(PKI).

Biometric identification process is combined with digital certificates for electronic authentication as biometric certificates. The *biometric certificates* are managed through the use of a biometric certificate management system. Biometric certificates may be used in any electronic transaction for requiring authentication of the participants.

Biometric data is pre-stored in a biometric database of the biometric certificate management system by receiving data corresponding to physical characteristics of registered users through a biometric input device. Subsequent transactions to be conducted over a network have digital signatures generated from the physical characteristics of a current user and from the electronic transaction. The electronic transaction is authenticated by comparison of hash values in the digital signature with re-created hash values. The user is authenticated by comparison against the pre-stored biometric certificates of the physical characteristics of users in the biometric database.

2.2   Existing Biometric Key Mechanisms

Recently several methods have been proposed to use biometrics for generating a digital signature. In 2001, P. Janbandhu and M. Siyal studied a method for generating biometric digital signatures for Internet based applications [12]. Their scheme was actually focused on using a 512-byte iris code invented by J. Daugman [9,10], and deriving a singature key from the iris code.

In 2002, R. Nagpal and S. Nagpal proposed a similar method except that they used a multi modal technique combining iris pattern, retina, and fingerprint in order to derive RSA parameters [7]. In 2002, P. Orvos proposed a method for deriving a signature key from a biometric sample and a master secret kept securely in a smart card [14]. In the commercial fields, several products that generate a digital signature only by accessing the server or smart card through biometric authentication, are being announced [8].

We could observe that the first two schemes are far from practice due to their inadequate assumption on acquiring deterministic biometrics [7,12], while the remaining results eventually use biometrics as only a means to access the signature key stored in some hardware devices [8].

Recently in 2004, Y. Dodis, L. Reyzin, and A. Smith showed a method of using biometric data to securely derive cryptographic keys which could be used for authentication by introducing a secure sketch which allows recovery of a shared secret and a fuzzy extractor which extracts a uniformly distributed string from the shared secret in an error-tolerant way [8].

In 2005, X. Boyen, Y. Dodis, J. Katz, R. Ostrovsky, and A. Smith improved this result in a way that resists an active adversary and provides mutual authentication and authenticated key exchange [9]. There is an approach of template-protecting biometric authentication proposed by P. Tuyls and J. Goseling in 2004 [6], but it does not provide a method for deriving a cryptographic key.

2.3   Secret Hiding Function : Fuzzy Vault Scheme

Fuzzy vault is a simple and novel cryptographic construction. A player Alice may place a secret value k in a fuzzy vault and 'lock' it using a set $\mathcal{A}$ of elements from some public universe $\mathcal{U}$. If Bob tries to 'unlock' the vault using a set $\mathcal{B}$ of similar length, he obtains k only if $\mathcal{B}$ is close to $\mathcal{A}$, i.e., only if $\mathcal{A}$ and $\mathcal{B}$ overlap substantially[4].

Thus, a fuzzy vault may be thought of as a form of error tolerant encryption operation where keys consist of sets. Fuzzy vault like error-tolerant cryptographic algorithms are useful in many circumstances such as privacy protected matching and enhancement, authentication with biometrics and in which security depends on human factors like fingerprint, etc[4].

Follow diagram(Fig. 2) shows the general *Fuzzy Vault* model for providing biometric digital key generation and protection functions.

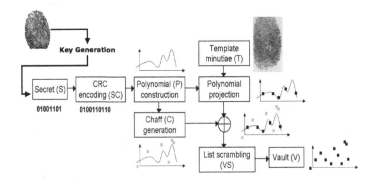

Fig. 2. Fuzzy Vault Scheme for Biometric Digital Key Protection

3 Proposed Telebiometrics Digital Key Generation

Biometric data can be input by using diverse application such as fingerprint reader, etc. Therefore, this work can use transformation module on input biometric data. In this section, the proposed key generation framework is described in detail. For example, follow diagram(Fig. 3) shows the abstracted flow on biometric digital key pairs from the fingerprint data. On the Alice's fingerprint data, key pairs can be generated by using existing common public key cryptosystem such as RSA or ElGamal.

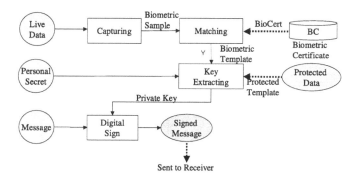

Fig. 3. Biometric Digital Key Generation Framework

In the key generation framework, the identity of the user is verified by comparing the captured biometric data with *biometric certificate*(BC) that is stored in BCA. Therefore, no one can act for the original user in key generation mechanism. Only the user who has registered hisher own biometric template on BCA can make public and private key pairs.

3.1 Biometric Digital Key Generation

Capturing Function. This function is a sensing and input module with feature enhancement function. In fingerprint, ridge regions in the image are identified and normalized. In detail ridge orientations are determined, local ridge frequencies calculated, and then contextual filters with the appropriate orientation and frequency are applied.

This function makes the biometric template from the biometric raw data. The noise on the captured image is reduced through image processing. Then the set of minutiae are extracted from the enhanced image. Finally, the biometric template is made from location and angle values of minutiae set.

Matching Function. The role of adjustment function is finding helper data that is used for revising location and angle values of minutiae set. A user's biometric data is always different whenever the biometric data is captured from the device. Thus the adjustment function is needed to revise newly captured biometric data.

Biometric Digital Key Generation Function. The private key is generated by hashing a user's personal secret and a biometric template. If we use only the biometric template for private key generation, we always get to generate same key since the biometric data is unique. Moreover, if the private key is disclosed, the user's biometric data can not be used any more. Therefore, in order to cancel and regenerate the private key, the user's personal secret is also required in generation of the private key.

The private key is generated by hash function such as MD5 or SHA-1 on the biometric template with the personal secret. If we use only the biometric template for private key generation, we always get to generate same key since the biometric data is unique. Moreover, if the private key is disclosed, the user's biometric data can't be used any more. Therefore, in order to cancel and regenerate the private key, the user's personal secret is also required in generation of the private key.

3.2   Biometric Digital Key Protection

Biometric template stores the subject's biometric feature data, which is vital to the overall system security and individual security. Once the biometric data is leaked out, the individual authentication is confronted with threat, and individual biometric authentication in other applications may be confronted with security vulnerability, too. So, it is most important to protect the biometric template. Private secret data is stored as a protected form on protected storage.

The confidentiality of biometric private key can be assured by this mechanism. For biometric private key as well as biometric template is a kind of individual private data, the certificate user have right and must delete their biometric private key from certificate database when the biometric certificate is revoked.

For example, we can implement 'Shuffling' module by using fuzzy-vault scheme as follows. Firstly, we generate fake minutiae set and insert them to the user's biometric template. Secondly, for hiding private key, polynomial for real minutiae set(original biometric template) and polynomial for fake minutiae set are constructed. Then, the private key is projected to each polynomial. Finally, the protected data(template) is made by combining these results. It consists of minutiae's (location, angle, result) value set.

Key Shuffling Function. A key shuffling(secret locking) function is used to hide the private key to enforce security of it. Therefore, it is a simplified secret shuffling module on biometric data for hiding and securing the private key.

- **Step 1:** Generate fake minutiae set and insert them to the user's biometric template for protection of the template.
- **Step 2:** For hiding private key, polynomial for real minutiae set and polynomial for fake minutiae set are constructed. Then, we project private key into each polynomial and get results.
- **Step 3:** The protected template is made by combining results from step1 and step 2. It consists of minutiae's (location, angle, result) value set.

Follow diagram(Fig. 4) shows the abstracted model on the secret locking and private key hiding on the fingerprint data set based on Fuzzy Vault scheme.

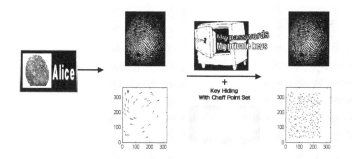

Fig. 4. Biometric Digital Key Protection with Fuzzy Vault

3.3   Public Key Generation and Management

This work uses ElGamal signature scheme to simulate the digital signature generation in our framework. The private key is generated and the public key y is computed as follows. Entity $\mathcal{A}$ generates a large random prime p and a generator g of the multiplicative group Z_p. And then $\mathcal{A}$ selects a random secret a, $a \leq a \leq p - 2$. $\mathcal{A}$ also computes $y \equiv g^a \bmod p$. $\mathcal{A}$'s public key is and private key is $(p; g; y)$. It is computationally infeasible to solve discrete logarithms over GF(p). Generated public key $(p; g; y)$ is stored on DB and certified by CA.

In order to combine biometric authentication with cryptographic techniques, the proposed framework uses adjustment function during the digital signature phase. The adjustment function guarantees that a unique template can be derived from various biometrics of same person. In key generation framework, the private key is concealed in the protected template in order to prevent disclosure of the private key. In signature generation framework, the private key is derived from the user's biometric data and the protected template.

3.4   Biometric Digital Signature Generation and Verification

At first, user authentication is performed by comparing the signer's captured biometric image with his(her) own biometric template in BCA. Due to the property of the proposed key extraction mechanism, the signer can not make other signers do signing a message. The signer gets the private key which is extracted from the key extraction mechanism. Then, the signer generates his(her) own digital signature on the message with the private key and sends it to the verifier.

The verifier gets the signer's public key from CA(Certificate Authority) and verifies the signature on the message with the public key. Signature verification mechanism is as same as that of ordinary digital signature verification scheme. Entity at the receiver(server) can verify the digital signature by using the signer's public key and biometric certificate.

We can generate the digital signature on the message using the private key obtained from the input device. Entity $\mathcal{A}$ signs a binary message m of arbitrary length. Any entity $\mathcal{B}$ can verify this signature by using $\mathcal{A}$'s public key.

Biometric Digital Key Extraction. Follow diagram(Fig. 5) shows the digital key extraction model. User authentication is performed at first as same as in key generation mechanism. The user cannot disguise himself(herself) as a other for extracting the private key stored by a protected data. This requirement should be considered in key extraction mechanism based on biometric data. The private key is extracted from protected data by using 'Key Extracting' function with biometric template and personal secret value. Cryptographic function such as fuzzy vault can be applicable into this mechanism.

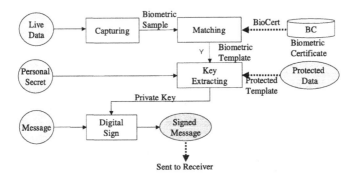

Fig. 5. Biometric Digital Key Extraction Framework with Fuzzy Vault

Digital Signature Generation and Verification. Digital signatures are fast emerging as a viable information security solution, satisfying the objectives of data integrity, entity authentication, privacy, non-repudiation and certification. In this section, digital signature generation mechanism is described using the proposed key generation/extraction mechanisms in telebiometric environment.

The private key extracted from the previous module is used to sign the message. The message and the signature on it are sent to the verifier. For example, we can generate digital signature (r, s) on the message m from the input private key.

Entity $\mathcal{A}$ generates a large random prime p and a generator g of the multiplicative group Z_p. And then $\mathcal{A}$ selects a random integer α, $1 \leq \alpha \leq p - 2$. $\mathcal{A}$ also computes $y \equiv g^\alpha \bmod p$. $\mathcal{A}$'s public key is $(p; q; y)$ and private key is α. And then entity $\mathcal{A}$ signs a binary message m of arbitrary length. Any entity $\mathcal{B}$ can verify this signature by using $\mathcal{A}$'s public key. In detail entity $\mathcal{A}$ selects a random secret integer k, with $gcd(k, p - 1) = 1$. And then $\mathcal{A}$ computes $r \equiv g^k \bmod p$ and $K^{-1} \bmod (p-1)$. Finally entity $\mathcal{A}$ can generates and computes $s \equiv k^{-1}\{h(m) - \alpha r\} \bmod (p - 1)$ and then $\mathcal{A}$'s signature for m is the pair $(r; s)$.

To verify $\mathcal{A}$'s signature $(r; s)$ on m, $\mathcal{B}$ should do the following: $\mathcal{B}$ obtains $\mathcal{A}$'s authentic public key $(p : \alpha; y)$ and verifies that $1 \leq r \leq p - 1$; if not, then reject

the signature. If satisfied, $\mathcal{B}$ computes $v1 \equiv y^r r^s \ mod \ p$ and $v2 \equiv g^{h(m)} \ mod \ p$. $\mathcal{B}$ accepts the signature if and only if $v1 = v2$.

4 Cryptography Communication with Fingerprint

4.1 Message Encrypt with Digital Key

The main objective of biometric encryption is to provide privacy and confidentiality using biometric digital key(a private and a public key pairs). In biometric encryption systems each client receives public key from DB. Any entity wishing to securely send a message to the receiver obtains an authentic copy of public key and then uses the encryption transformation. To decrypt, the receiver applies decryption transformation to obtain the original message after biometric authentication process. Common biometric encryption mechanism with digital key is also possible as follow Fig. 6.

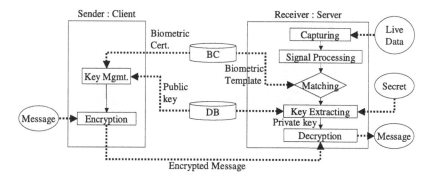

Fig. 6. Biometric Encryption for Cryptography Communication with Fingerprint

For message encryption, private key is extracted from the protected data stored by fuzzy vault scheme. And then we can generate encrypted ciphertext by using public key.

4.2 Implementation Results

We implemented a biometric encryption system with fingerprint in MATLAB. First step is a fingerprint enhancement proceedure for extracting a feature set of fingerprint minutiae. After this fingerprint enhancement function, a digitized feature set of minutiae is generated and it is used for generating private key with personal secret from template registration and key generation step. Follow Fig. 7 show the fuzzy vault set after locking someone's secret(private key) within his/her own biometric template. Developed module provides and generates protected template from input fingerprint template.

Fig. 8 shows the implementation and experimental results on the encryption/decryption mechanism based on ElGamal type of biometric digital key. Using biometric key pairs, we can encrypt and decrypt plaintext message.

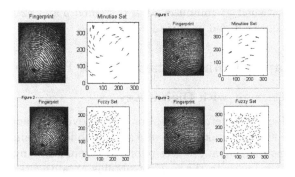

Fig. 7. Fuzzy Vault Result After Locking Private Key within Fingerprint Feature Set

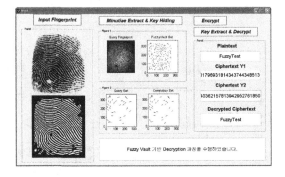

Fig. 8. Fuzzy Vault based Biometric Encryption and Decryption in MATLAB

5   Conclusions

We propose biometric digital key generation/extraction mechanisms for crypto-graphic secure communication, which are essential both for authentication and digital signature protocols on open network environments. The proposed system uses biometric template in Biometric Certificate for user authentication in key generation/extraction mechanisms. This work shows how to generation key from biometrics and message encryption. We can use user authentication and cryptography communication. So we can protect biometrics and communications that using Fuzzy Vault and ElGamal encryption scheme. Someday this work will be used Internet banking system or society that use biometrics mainly.

Acknowledgments. This work was also supported by the University IT Research Center(ITRC) Project(IITA-2006-(C1090-0603-0016)).

References

1. Rivest, R., Shamir, A., Adleman, L.: A method for obtaining digital signatures and public-key cryptosystems. Communications of the ACM 21, 120–126 (1978)
2. Boneh, D.: Twenty years of attacks on the RSA cryptosystem. Notices of the American Mathematical Society (AMS) 46(2), 203–213 (1999)
3. ElGamal, T., Public, A.: Key Cryptosystem and a Signature Scheme based on Discrete Logarithms. IEEE Transactions on Information Theory IT-30(4), 469–472 (1985)
4. Juels, A., Sudan, M.: A Fuzzy Vault Scheme, also available at `http://www.rsasecurity.com/rsalabs/staff/bios/ajuels/publications/fuzzy-vault/fuzzy_vault.pdf`
5. Menezes, A., van Oorschot, P., Vanstone, S.: Handbook of Applied Cryptography, pp.287-291, pp.312-315. CRC Press, Boca Raton, USA (1997)
6. Tuyls, P., Goseling, J.: Capacity and examples of template-protecting biometric authentication systems. In: Maltoni, D., Jain, A.K. (eds.) BioAW 2004. LNCS, vol. 3087, pp. 158–170. Springer, Heidelberg (2004)
7. Nagpal, R., Nagpal, S.: Biometric based digital signature scheme. Internet-Draft, draft-nagpal-biometric-digital-signature-00.txt (May 2002)
8. Dodis, Y., Reyzin, L., Smith, A.: Fuzzy Extractors: How to generate strong keys from biometric identification. In: Proceedings of the Security and Privacy, IEEE Computer Society Press, Los Alamitos (1998)
9. Boyen, X., Dodis, Y., Katz, J., Ostrovsky, R., Smith, A.: Secure remote authentication using biometrics. In: Cramer, R.J.F. (ed.) EUROCRYPT 2005. LNCS, vol. 3494, pp. 147–163. Springer, Heidelberg (2005)
10. Soutar, C.: Biometric system performance and security (2002), Manuscript available at `http://www.bioscrypt.com/assets/biopaper.pdf`
11. Jain, A., Hong, L., Pankanti, S.: Biometric identification. Communications of the ACM (February 2000)
12. Janbandhu, P., Siyal, M.: Novel biometric digital signatures for Internet-based applications. Information Man-agement & Computer Security 9(5), 205–212 (2001)
13. Kwon, T., Lee, J.: Practical digital signature generation using biometrics. In: Laganà, A., Gavrilova, M., Kumar, V., Mun, Y., Tan, C.J.K., Gervasi, O. (eds.) ICCSA 2004. LNCS, vol. 3043, Springer, Heidelberg (2004)
14. Orvos, P.: Towards biometric digital signatures. Networkshop, pp. 26–28. Eszterhazy College, Eger (March 2002)
15. Tuyls, P., Goseling, J.: Capacity and examples of template-protecting biometric authentication systems. In: Maltoni, D., Jain, A.K. (eds.) BioAW 2004. LNCS, vol. 3087, pp. 158–170. Springer, Heidelberg (2004)

New Algorithms for the Spaced Seeds

Xin Gao[1], Shuai Cheng Li[1], and Yinan Lu[1,2,*]

[1] David R. Cheriton School of Computer Science
University of Waterloo
Waterloo, Ontario, Canada N2L 6P7
[2] College of Computer Science and Tecnology of Jilin University
10 Qianwei Road, Changchun, Jilin Province, China 130012
{x4gao,scli}@cs.uwaterloo.ca, luyinan@email.jlu.edu.cn

Abstract. The best known algorithm computes the sensitivity of a given spaced seed on a random region with running time $O((M+L)|B|)$, where M is the length of the seed, L is the length of the random region, and $|B|$ is the size of seed-compatible-suffix set, which is exponential to the number of 0's in the seed. We developed two algorithms to improve this running time: the first one improves the running time to $O(|B'|^2ML)$, where B' is a subset of B; the second one improves the running time to $O((M|B|)^{2.236}log(L/M))$, which will be much smaller than the original running time when L is large. We also developed a Monte Carlo algorithm which can guarantee to quickly find a near optimal seed with high probability.

Keywords: homology search, spaced seed, bioinformatics.

1 Introduction

The goal of homology search is to find similar segments or local alignments between biological molecular sequences. Under the framework of match, mismatch, and gap scores, the Smith-Waterman algorithm guarantees to find the global optimal solution. However, the running time of the Smith-Waterman algorithm is too large to be used on real genome data.

Many programs have been developed to speed up the homology search, such as FASTA [11], BLAST [1,2,16,14], MUMmer [8], QUASAR [5], and PatternHunter [12,9]. BLAST (Basic Local Alignment Search Tool) is the most widely used program to do homology search. The basic idea of BLAST is that by using a length 11 seed, which requires that two sequences have locally 11 consecutive matches, local matches can be found, and a reasonably good alignment can be then generated by extending those local matches. However, Li *et al* [12] found that the homology search sensitivity can be largely improved if long "gapped" seeds are used instead of short "exact" seeds. PatternHunter is developed based

---

* Corresponding author.

F.P. Preparata and Q. Fang (Eds.): FAW 2007, LNCS 4613, pp. 50–61, 2007.

on long "gapped" seeds. PatternHunter applies a dynamic programming (DP) based algorithm to compute the hit probability of a given seed. We refer the algorithm to compute the sensitivity of a seed in PatternHunter as the PH algorithm.

The running time of the PH algorithm is dominated by the product of the length of random region and the size of seed-compatible-suffix set. In [9], Li *et al* proved that computing the hit probability of multiple seeds is NP-hard. In [10], Li *et al* further proved that computing hit probability of a single seed in a uniform homologous region is NP-hard.

The problems of computing the sensitivity of a given spaced seed and finding the most sensitive pattern have been studied for a long time. Choi and Zhang and coworkers [7,13] studied the problem of calculating sensitivity for spaced seeds from computational complexity point of view, and proposed an efficient heuristic algorithm for identifying optimal spaced seeds. Choi *et al.* [6] found that an optimal seed on one sequence similarity level may not be optimal on another similarity level. Yang *et al.* [15] proposed algorithms for finding optimal single and multiple spaced seeds. Brejova *et al.* [3] studied the problem of finding optimal seeds for sequences generated by a Hidden Markov model. Brown [4] formulated choosing multiple seeds as an integer programming problem, and gave a heuristic algorithm. So far, the PH algorithm is still the best running time algorithm for calculating sensitivity of a given spaced seed. And most algorithms for finding the optimal seed can not give any guarantee on the performance.

Here, we develop two algorithms to improve the PH algorithm for some cases. The first algorithm improves the PH algorithm when the size of seed-compatible-suffix set is large, while the second algorithm improves PH algorithm when the region length is large. We further develop a Monte Carlo algorithm which can guarantee to quickly find the optimal seed with high probability.

2 Preliminaries

2.1 Notations and Definitions

The notations are largely followed from those in [9].

Denote the i-th letter of a string s as $s[i-1]$. The length of s is denoted as $|s|$. A spaced seed a is a string over alphabet $\{1, 0\}$. Denote $M = |a|$. For a spaced seed a, we require $a[0] = 1$ and $a[M-1] = 1$. The number of 1's in a is called the weight of a, here denoted as W. A 1 position in a means "required match", while a 0 in a means "do not care".

A homologous region R with length L is defined as a binary string, in which a 1 means a match and a 0 means a mismatch. In this paper, we only focus on random homologous regions with uniform distribution. That is, $Pr(R[i] = 1) = p$, $0 \leq i \leq L-1$, where p is referred to as *similarity level* of R. For a spaced seed a and a homologous region R, if there $\exists j$, $0 \leq j \leq L - M$, such that whenever $a[i] = 1$, we have $R[j+i] = 1$, then we say that a *hits* region R.

This paper studies the following two problems:

1. **Seed Sensitivity:** Given a spaced seed a, and a homologous region R , what is the probability of a hitting R? This probability is referred to as the sensitivity of this seed on the region R. We just call it *sensitivity* of R if the context is clear.
2. **Optimal Seed:** Given a seed length M and weight W, and a homologous region R with similarity level p, what is the seed with the highest sensitivity? A seed with the highest sensitivity is called an *optimal seed*.

2.2   Reviews of the PatternHunter Algorithm for Seed Sensitivity

Li *et al* [9] developed a dynamic programming based algorithm to compute the sensitivity of a given seed.

For a seed a of length M and weight W, we call a string b *compatible* with a if $b[|b| - j] = 1$ whenever $a[|a| - j] = 1$ for $0 < j \leq \min\{|a|, |b|\}$. Suppose the random region R has length L, and similarity level p. For a binary string b, let $f(i, b)$ be the probability that seed a hits region $R[0 : i - 1]$ which has b as the suffix of $R[0 : i - 1]$. Generally, we are only interested in the case $0 \leq |b| \leq M$. There are two cases: 1) the position before b has value 0; and 2) the position before b has value 1. Thus, $f(i, b)$ can be recursively expressed as:

$$f(i, b) = (1 - p)f(i, 0b) + pf(i, 1b) \tag{1}$$

This will generate a $O(L2^M)$ dynamic programming algorithm because length of b is at most M. However, since the only case seed a can hit the "tail" of a region R is that the suffix of the region R is compatible with seed a, instead of considering all possible suffixes, they only consider those suffixes which are compatible with a.

Define B to be the set of binary strings that are not hit by a but compatible with a. Let $B(x)$ denote the longest proper prefix of x that is in B. The PH algorithm thus uses the following recursion function:

$$f(i, b) = (1 - p)f(i - |b| + |b'|, 0b') + pf(i, 1b) \quad 0b' = B(0b). \tag{2}$$

It is clear that any entry in the dynamic programming table depends on two previously computed entries. Therefore, PH algorithm has to consider all the possible $b \in B$ for each i, $0 \leq i \leq L - 1$. The size of B is bounded by $M2^{M-W}$. The running time is thus $O((M + L)M2^{M-W})$. However, the size of B can be reduced.

In this paper, we improve the PH algorithm from two perspectives:

– Instead of considering all possible suffixes $b \in B$, we consider only a small subset of B, which will result in an algorithm with a better running time when $|B|$ is large. That is, following the similar idea applied by PH algorithm, we further reduce the number of suffixes needed.
– We reduce the factor L in the running time of PH algorithm to $\log L$, which will generate an algorithm with much better running time when L is large.

3   Suffix-Recursion-Tree Algorithm

3.1   Algorithmic Details

The basic idea of our suffix-recursion-tree (SRT) algorithm is that if we pre-compute more steps of the recursion function Eq. 2 instead of only one step, we may have a small suffix set B', s.t. any suffix $b \in B'$ is a recursive function of only these suffixes from B', which has the size much smaller than B. Recall that the sensitivity is stored in entry (L, ϵ). Thus, we require ϵ in B'. The SRT of a spaced seed is a tree with root node being labeled as ϵ, and each node of the tree is labeled with $b \in B$. The label of the left child of any node b' is $B(0b')$, where $B(0b')$ is the longest prefix of $0b'$ that belongs to B; and the label of the right child of b' is $1b'$. If $1b'$ is comparable with a and have the same length as seed a, then $1b'$ is "hit" by the seed a, and the corresponding node is labeled as "hit". The SRT is built by a depth first search. A node is a leaf only if 1) b is labeled as a "hit", or 2) the label b has been occurred before. Fig.1 shows an example of the SRT of the spaced seed 1101011.

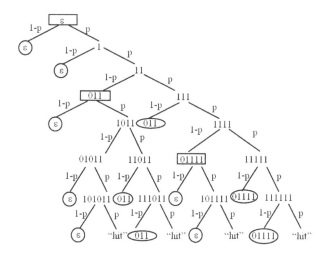

Fig. 1. An Illustration of Suffix-recursion-tree of seed 1101011

As shown in Fig.1, there are only three suffixes that recur more than once, $\epsilon, 011, 01111$. Thus, $B' = \{\epsilon, 011, 01111\}$. Note that the depth of the suffix-recursion-tree is M. For any node in the tree, it depends on its left child with probability $(1 - p)$, while p on its right child, if the children exist. Each path from a node v to its ancestor u represents the probability that u depends on v. Thus, $f(i, \epsilon)$ can be expressed only by those circled entries. Similarly, we can get a recursive relation for each internal node v on all the leaves of the subtree rooted at v. The dependency of b on all $b' \in B'$ is thus presented in the sub-tree rooted at b, and the depth of the sub-tree rooted at b is at most $M - 1$. Thus, by building this SRT, a set of new recursion functions can be found that only depends on a small suffix set B'.

By applying dynamic programming on this new recursion function, we have a new algorithm for computing the sensitivity of a single spaced seed on a random region. First, we construct the SRT for a given seed. Second, we deduce the recurrence relations. Third, we do a dynamic programming based on this set of new recurrence relations. Lastly we output the sensitivity.

Algorithm SRT-DP

Input Seed a, similarity level p, and region length L.
Output Sensitivilty of seed a on a random region of length L at similarity level p.
  1.   Construct the SRT of the seed a, and build set B'
  2.   Deduce all the recursive equations for every $b' \in B'$.
  3.   For $i \leftarrow 0 \dots L$
  4.   For $b' \in B'$ with decreasing lengths
  5.   Compute $f(i, b')$ by the recursive equations from Step 2.
  6.   Output $f(L, \epsilon)$

Fig. 2. Algorithm SRT-DP

Theorem 1. *Let a be a spaced seed and R be a random region. Algorithm SRT-DP computes $Pr(a$ hits $R)$ in $O(|B'|^2 ML + |B|)$ time, where M is the length of seed a, and B' is the suffix subset determined by the suffix-recursion-tree of seed a.*

Proof. The correctness of the algorithm comes from the discussion before the theorem. Line 1 can be done by depth first search. Since each node is determined by its two direct children from Eq. 2, and once the node is traversed, the depth first search will stop if the node recurs somewhere else, the running time is thus $O(|B|)$. Line 2 can be done in $O(|B|)$, which is the size of the tree. After precomputing, line 3 to line 5 takes $O(|B'|^2 ML)$ running time because for each entry in the dynamic programming table, it depends on at most $M|B'|$ entries. Therefore, the total running time for the algorithm SRT-DP is $O(|B'|^2 ML + |B|)$ □

3.2   A Concrete Example

In this section, we will give a concrete example to show our algorithm has a much better running time than the PH algorithm on some cases. First, we prove the following results.

Lemma 1. *Any suffix $b \in B'$ of any spaced seed can be either ϵ or a binary string starting with 0. That is, any $b \in B'$ can not start with 1.*

Proof. By contradiction. Suppose there is a $b \in B'$ that starts with 1, i.e. $b = 1b'$. That means 1) b has occurred more than once, and 2) b' is a suffix in B, and b'

is the parent of b. Considering any two places where b occurs, at each place, b' is the parent of b. Thus, b' has also occurred at least twice in the suffix-recursion tree. For the definition of the suffix-recursion tree, the tree should stop at one b', which contradicts with b is a child of this b'.

Therefore, any suffix $b \in B'$ can be either ϵ or a binary string starting with 0.   □

Recall that the running time of PH algorithm is $O((M + L)|B|)$, in which $|B|$ can be as large as $O(M2^{M-W})$. From lemma 1, we know B' is a subset of B, the size of which is much smaller than B, because all suffixes starting with 1 in B will not be in B'. Furthermore, we construct a simple example illustrate that the algorithm SRT-DP is much better than PH algorithm. Better examples are available, but it is out of the scope.

$$
10 \underbrace{11\cdots11}_{m^3+(m-1)\ 1's} \quad 0\cdots\underbrace{11\cdots11}_{m^3+1\ 1's}0\underbrace{11\cdots11}_{m^3\ 1's} \tag{3}
$$

For a seed as shown in Eq. 3, there are m 0's in a. The size of B for this case is:

$$
\begin{aligned}
|B| &= m^3 + 2 + (m^3 + 1) \times 2 + 4 + \cdots + (m^3 + m - 1) \times 2^{m-1} + 2^m + 2^m \\
&= \sum_{i=1}^{m} 2^i + \sum_{i=0}^{m-1} m^3 2^i + \sum_{i=0}^{m-1} i2^i + 2^m \\
&= (2^{m+1} - 2) + m^3(2^m - 1) + (m2^m - 2^{m+1} + 2) + 2^m \\
&= (m^3 + m + 1)2^m - m^3
\end{aligned}
$$

Lemma 2. *Any suffix $b \in B'$ of our seed a can be either ϵ or a binary string starting with 0 and followed by 1's and at most one 0.*

Proof. From lemma 1, we know that the only possible suffix $b \in B'$ of seed a can be either ε or a binary string starting with 0. By contradiction, suppose there is a suffix $b \in B'$ which starts with 0 and followed by at least two 0's.

From the definition of B', b should be compatible with the seed a. Thus, the 0's in b which follows the first 0 have to be matched to some 0's in a.

$$
\begin{array}{ll}
a: & 10\cdots11\cdots11\,011\cdots11\,011\cdots11011\cdots11\,011\cdots11 \\
   & \qquad\qquad\qquad\quad 1 \qquad\qquad\qquad 2 \\
b: & \qquad\qquad\quad 011\cdots11\,011\cdots11111\cdots11\,011\cdots11
\end{array}
$$

Recall that any suffix $b \in B'$ is generated by taking the longest compatible prefix of some $0b'$. Thus, b in the above figure is the result of cutting the tail of some binary string. Thus, there are two different pairs of 0's in a, which have the same distance between each other. Suppose the first pair of 0's (position 1 and position 2 in seed a) contains region $\underbrace{11\cdots11}_{l_1+(n_1-1)\ 1's}\ 0\ \underbrace{11\cdots11}_{l_1+(n_1-2)\ 1's}\ 0\cdots\underbrace{11\cdots11}_{l_1+1\ 1's}0\underbrace{11\cdots11}_{l_1\ 1's}$, and the second pair of 0's (position 1 and position 2 in suffix b) contains region which corresponds to the region $\underbrace{11\cdots11}_{l_2+(n_2-1)\ 1's}\ 0\ \underbrace{11\cdots11}_{l_2+(n_2-2)\ 1's}\ 0\cdots\underbrace{11\cdots11}_{l_2+1\ 1's}0\underbrace{11\cdots11}_{l_2\ 1's}$

in seed a. Note here $n_1 \le m$ and $n_2 \le m$. Since the distances between these two pairs are the same. We have

$$l_1+(l_1+1)+\cdots+(l_1+n_1-1)+(n_1-1) = l_2+(l_2+1)+\cdots+(l_2+n_2-1)+(n_2-1)$$

From this equation, we have

$$(2l_1 + n_1 + 1)n_1 = (2l_2 + n_2 + 1)n_2$$
$$2(l_1n_1 - l_2n_2) = (n_2 + 1)n_2 - (n_1 + 1)n_1$$

From the construction of the seed a, for any i, we have $l_i = m^3 + j$, where $0 \le j \le m - 1$. If n_1 and n_2 are different, without loss of generality, assume $n_1 > n_2$. Let $n_1 = n_2 + h$, where $h \ge 1$. The left part of the above equation is then:

$$2(l_1n_1 - l_2n_2)$$
$$= 2[(m^3 + j_1)(n_2 + h) - (m^3 + j_2)n_2]$$
$$= 2[m^3h + j_1(n_2 + h) - j_2n_2]$$
$$\ge 2(m^3h - m^2)$$
$$\ge m^3 \qquad (when \quad m \ge 2)$$

Thus, the absolute value of the left part of the above equation is at least m³, while the right part is at most m². Thus, $n_1 = n_2$, and $l_1 = l_2$. This contradicts to the assumption that these two regions are different.

Therefore, any suffix $b \in B'$ of seed a can be either ϵ or a binary region starting with 0 and followed by 1's and at most one 0.   □

Combining lemma 1 and lemma 2, the number of suffixes $b \in B'$ for seed a is at most:

$$m + \binom{m}{2}$$
$$= O(m^2)$$

In this seed a, the total length M is $m^4 + \frac{m^2}{2} + \frac{m}{2} + 1$.

Therefore, the total running time for the algorithm SRT-DP is $O(|B'|^2ML + |B|) = O(m^8L + m^32^m)$, while the running time for PH algorithm is $O((M + L)|B|) = O(m^72^m + Lm^32^m)$. The dominant term here is L or 2^m. Thus, the SRT-DP algorithm is much faster than PH algorithm because it is the sum of the two dominant terms instead of the product of them.

4   Block-Matrix Algorithm

Now, we develop another algorithm to solve the problem of calculating the sensitivity of a given seed, which mainly handle the case when the length of the homology region is long.

Recall Eq. 2, in which $0b'$ is the longest prefix of $0b$ in B. Any entry in the dynamic programming table depends on only two previously computed entries.

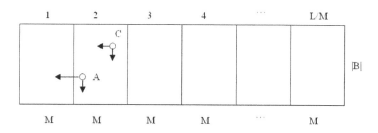

Fig. 3. An Illustration

For any $b \in B$, $|b| - |b'|$ is bounded by M. Thus, if we divide random region R into blocks, each of which has length M, all entries in one block will only depend on entries from itself or entries from another block. Fig. 3 shows an illustration of dividing region R into blocks.

For any entry in block 1, it depends on two previously computed entries in block 1. Thus, all entries in block 1 can be pre-computed by only using entries from itself. For an entry in block 2, it depends on only two entries, one of which must be in the same column as this entry in the same block (corresponding to $f(i, 1b)$). For the other dependent entry, there are two cases: 1) the entry is also in block 2 (as point C in Fig.3); and 2) the entry is in block 1 (as point A in Fig.3). In either case, the entry has been computed already. Let $F(i, j)$ denote the dependency relationship of block i on block j and i. From previous discussion, we can easily compute $F(1, 1)$ and $F(2, 1)$. Thus, for any entry in block 3, we can consider block 3 as block 2, and block 2 as block 1, then use $F(2, 1)$ on the entry. Clearly, $F(3, 2)$ is the same as $F(2, 1)$. If we further substitute any entry in block 2 used in $F(3, 2)$ by its $F(2, 1)$ relationship, we can have a dependency relationship between any entry in block 3 and entries in block 1 and block 3, i.e. $F(3, 1)$. For any entry in block 5, we can apply $F(3, 1)$ twice, which will result in $F(5, 1)$. Generally speaking, assume $L = (1 + 2^r)M$, we can apply this idea to reduce the region length dimension of the dynamic programming, L, to $\log L$ by using the algorithm shown in Fig.4.

We now analyze the running time of this algorithm. Line 1 is just for illustration purpose, in practice, we don't need to divide R into blocks, instead we just need to compute $[i/M] + 1$ for a given index i; line 2, 3, and 4 can be done in $O(M|B|)$; line 5 can be done in $O(M|B|)$ because each entry in block 2 depends on only 2 other entries from block 1 or block 2.

Theorem 2. *The Block-matrix Algorithm has a running time* $O((M|B|)^{2.236} \log(L/M))$.

Proof. If an entry in block $(1 + 2^i)$ depends only on previously computed entries in the same block, the algorithm takes $O(M|B|)$ running time. Thus, we can assume an entry in block $(1 + 2^i)$ depends on entries in block $(1 + 2^{i-1})$ by using previously computed $F(1 + 2^{i-1}, 1)$. Let $a_i, i = 1 \ldots M|B|$ denote all entries in block 1 with the condition that an entry with smaller line index is indexed

---

Block-matrix Algorithm

Input Seed a, similarity level p, and region length L.
Output Sensitivilty of seed a on a random region R of length L at similarity level p.
  1.   Divide length L into blocks, each of which has length M. Index these blocks
         as block 1, block 2, $\ldots$, block $1 + 2^r$.
  2.   For $i \leftarrow 0 \ldots M - 1$
  3.   For $b \in B$ with decreasing lengths
  4.   Compute $f(i, b)$ by the recursive function Eq. 2.
  5.   Compute $F(2, 1)$.
  6.   For $i \leftarrow 1 \ldots r$
  7.   Compute $F(1 + 2^i, 1)$ by using $F(1 + 2^{i-1}, 1)$.
  8.   Output $f(L, \epsilon)$

Fig. 4. Block-matrix Algorithm

before any entry with larger line index, and an entry with smaller column index is indexed before any entry with larger column index and the same line index. Let $b_i, i = 1 \ldots M|B|$ and $c_i, i = 1 \ldots M|B|$ denote all entries in block $(1 + 2^{i-1})$ and $(1 + 2^i)$, respectively, with the same indexing rule as a_i. Thus, any b_j is a linear combination of all a_i, and any c_k is a linear combination of all b_j. That means:

$$b_j = \sum_{i=1}^{M|B|} w_{ji} a_i$$

$$c_k = \sum_{j=1}^{M|B|} w'_{kj} b_j$$

Note that $F(1 + 2^i, 1)$ is the same as $F(1 + 2^{i-1}, 1)$, because $(1 + 2^i) - (1 + 2^{i-1}) = (1 + 2^{i-1}) - 1 = 2^{i-1}$. Thus, $w_{ij} = w'_{ij}$.
Thus,

$$c_k = \sum_{j=1}^{M|B|} w_{kj} b_j = \sum_{j=1}^{M|B|} w_{kj} \sum_{i=1}^{M|B|} w_{ji} a_i = \sum_{j=1}^{M|B|} \sum_{i=1}^{M|B|} w_{kj} w_{ji} a_i.$$

This means that the dependency relationship between any c_k and all a_i can be calculated by matrix multiplication. Thus, the running time for line 6 and 7 is bounded by $O((M|B|)^{2.236})$.

Therefore, the total running time of Block-matrix algorithm is $O((M|B|)^{2.236} \log(L/M))$. □

It's not difficult to extend our algorithm when L is not in $(1 + 2^r)M$ format. Recall the running time of PH algorithm is $O((M + L)|B|)$, our Block-matrix Algorithm can provide a much better running time if L is large.

5 Monte Carlo Algorithm for Finding Optimal Seed

In this section, we propose a Monte Carlo algorithm to solve the optimal seed problem. Given seed length M, weight W, a random region R of length L with distribution $Pr(R[i] = 1) = p$, $0 \leq i \leq L - 1$, we want to find a seed a of good enough sensitivity with high probability. The deterministic algorithm for finding the optimal seed has a running time of $O(\binom{M}{W}(M + L)|B|)$. Here, we provide a $O((k\binom{M}{W}M + t|B|)L)$ Monte Carlo algorithm that can find a near-optimal seed with high probability, where k and t are parameters to adjust the errors of the sensitivity and to control the probability.

The intuition behind our algorithm is that if we randomly generate k binary regions of length L with similarity level p, the number of hits for a seed a on these regions will be relevant to the sensitivity of a. The higher the sensitivity is, the more expected hits are. The outline of the algorithm is displayed in Fig.5. First, we randomly generate k regions. Then we count the occurrence of each pattern. If a seed occurs multiple times in a random region, it counts only once. After that, we select top t patterns with the largest numbers of occurrences. Lastly, we employ an deterministic algorithm to compute the sensitivity of the t seeds, and output the seed with the highest sensitivity.

	Monte Carlo Algorithm for Finding Optimal Seed
Input	Integer M, W, similarity level p, and random region R of length L.
Output	A seed a of length M, weight W with the highest sensitivity on R.
1.	Randomly generate k binary regions with similarity level p.
2.	For each possible binary pattern a' of length M and weight W.
	2.1 Let $cnt[a']$ be the number of random regions that contain a'.
3.	Let C be the set of t patterns with largest cnt values, ties break randomly.
4.	Compute the sensitivity of the patterns in C by a deterministic algorithm.
5.	Output the pattern with the highest sensitivity in C.

Fig. 5. Monte Carlo Algorithm for Finding Optimal Seed

Theorem 3. *The running time for the algorithm in Fig. 5 takes time $O((k\binom{M}{W}M + t|B|)L)$*

Proof. Step 1 takes $O(kL)$. The running time of step 2 is $(kML\binom{M}{W})$. The running time for step 3 is dominated by step 2. Step 4 takes $O(t(M + L)|B|)$ if HP algorithm is used to compute the sensitivity of a single seed. Thus, the overall running time is: $O((k\binom{M}{W}M + t|B|)L)$. □

We further estimate the probability that the Monte Carlo algorithm can find out the optimal seed. Let s_o be the sensitivity of an optimal seed on the random region R, s be the sensitivity of the seed on R found by our Monte Carlo algorithm. We have the following results:

Theorem 4. *The Monte Carlo algorithm ensures $s_o - s \leq \epsilon$ with high probability $1 - e^{-\frac{\epsilon^2}{3(1-\epsilon)}kt}$. If $kt \geq \frac{3(1-\epsilon)}{\epsilon^2} \log \frac{1}{\sigma}$, the Monte Carlo algorithm can guarantee to find out an optimal seed with probability $1 - \sigma$.*

Proof. Define three random variables: C, C_o, and X. C is the sum of C_i which is defined to be the number of hits of a seed found by our algorithm on random region i, (values can be 0 or 1 with probability $1 - s$ and s, respectively)); C_o is the sum of C_{oi} which is defined to be the number of hits of an optimal seed on random region i, (values can be 0 or 1 with probability $1 - s_o$ and s_o, respectively); X is the sum of X_i which is defined to be $C_i - C_{oi} + 1$, (values can be 0, 1, 2 with probability $(1 - s)s_o$, $ss_o + (1 - s)(1 - s_o)$, and $s(1 - s_o)$, respectively). Since the number of random regions is k, $X > k$ means the real number of hits of the optimal seed is smaller than the number of hits of an arbitrary seed. Let T denote the set of the indices of the top t seeds chosen by our Monte Carlo algorithm.

Since both s_o and s lay in $[0, 1]$, we can assume that when k and t are large enough, our algorithm can find out seeds with $s_o - s < 0.5$. Let C^i be random variable C for seed i. Thus,

$\Pr(an\ optimal\ seed\ is\ in\ top\ t\ seeds) = 1 - \Pr(no\ optimal\ seed\ is\ in\ top\ t\ seeds)$
$= 1 - \Pr(C_o\ is\ smaller\ than\ C^i,\ i \in T)$
$= 1 - \prod_{i \in T} \Pr(C_o\ is\ smaller\ than\ C^i)\ (because\ of\ independency)$

We now compute $\Pr(C_o\ is\ smaller\ than\ C^i)$ by Chernoff bounds. For $0 < \delta \leq 1$, $\Pr(X \geq (1 + \delta)\mu) \leq e^{-\mu\delta^2/3}$, where X is the sum of independent Poisson trials, and $\mu = E[X]$. It is obvious that $X_i = C_i - C_{oi} + 1$ is independent Poisson trials. Thus,

$$\mu = E[X] = k\{0 \times (1 - s)s_0 + 1 \times [ss_0 + (1 - s)(1 - s_o)] + 2 \times (1 - s_0)s\}$$
$$= (1 + s - s_o)k$$

Let $(1 + \delta)\mu = k$, we get $\delta = \frac{k}{\mu} - 1 = \frac{s_o - s}{1 + s - s_o}$.
By Chernoff bounds, we have

$$\Pr(C_o\ is\ smaller\ than\ C^i) = \Pr(X \geq k)$$
$$\leq e^{-(1+s-s_o)k\frac{(s_o-s)^2}{3(1+s-s_o)^2}} = e^{-\frac{(s_o-s)^2}{3(1+s-s_o)}k}$$

Let $\epsilon = s_o - s$, we have $\Pr(C_o\ is\ smaller\ than\ C^i) \leq e^{-\frac{\epsilon^2}{3(1-\epsilon)}k}$.
Thus, the probability that our Monte Carlo algorithm guarantees to find out an optimal seed is:

$$\Pr(an\ optimal\ seed\ is\ in\ top\ t\ seeds) = 1 - \prod_{i \in T} \Pr(C_o\ is\ smaller\ than\ C^i)$$
$$\geq 1 - (e^{-\frac{\epsilon^2}{3(1-\epsilon)}k})^t = 1 - e^{-\frac{\epsilon^2}{3(1-\epsilon)}kt}$$

Thus, when k and t increases, the probability increases quickly. If we require the probability that our Monte Carlo algorithm fails to find out an optimal

seed is smaller than σ, $0 < \sigma < 1$, we can have the requirement on k and t:
$kt \geq \frac{3(1-\varepsilon)}{\varepsilon^2} \log \frac{1}{\sigma}$.   □

Acknowledgements. We are grateful to Dongbo Bu for his thought provoking discussion and comments. This work was supported by the Application Foundation Project of Technology Development of Jilin Province, Grant 20040531.

References

1. Altschul, S., Gish, W., Miller, W., Myers, E., Lipman, D.: Basic local alignment search tool. J.Mol.Biol. 215, 403–410 (1990)
2. Altschul, S., Madden, T., Schäffer, A., Zhang, J., Zhang, Z., Miller, W., Lipman, D.: Gapped blast and psi-blast: a new generation of protein database search programs. Nucleic Acids Res. 25, 3389–3402 (1997)
3. Brejova, B., Brown, D., Vinar, T.: Optimal spaced seeds for hidden markov models, with application to homologous coding regions. In: CPM2003: The 14th Annual Symposium on Combinatorial Pattern Matching, Washington, DC, USA, pp. 42–54. IEEE Computer Society Press, Los Alamitos (2003)
4. Brown, D.: Optimizing multiple seeds for protein homology search. IEEE/ACM Trans. Comput. Biol. Bioinformatics 2(1), 29–38 (2005)
5. Burkhardt, S., Crauser, A., Lenhof, H., Rivals, E., Ferragina, P., Vingron, M.: q-gram based databse searching using a suffix array. In: Third Annual International Conference on Computational Molecular Biology, pp. 11–14 (1999)
6. Choi, K., Zeng, F., Zhang, L.: Good spaced seeds for homology search. Bioinformatics 20(7), 1053–1059 (2004)
7. Choi, K., Zhang, L.: Sensitivity analysis and efficient method for identifying optimal spaced seeds. Journal of Computer and System Sciences 68, 22–40 (2004)
8. Delcher, A., Kasif, S., Fleischmann, R., Peterson, J., White, O., Salzberg, S.: Alignment of whole genomes. Nucleic Acids Res. 27, 2369–2376 (1999)
9. Li, M., Ma, B., Kisman, D., Tromp, J.: Patternhunter ii: highly sensitive and fast homology search. JBCB 2(3), 417–439 (2004)
10. Li, M., Ma, B., Zhang, L.: Superiority and complexity of the spaced seeds. In: Proceedings of the seventeenth annual ACM-SIAM symposium on Discrete algorithms (SODA 2006), pp. 444–453. ACM Press, New York (2006)
11. Lipman, D., Pearson, W.: Rapid and sensitive protein similarity searches. Science 227, 1435–1441 (1985)
12. Ma, B., Tromp, J., Li, M.: Patternhunter: faster and more sensitive homology search. Bioinformatics 18(3), 440–445 (2002)
13. Preparata, F., Zhang, L., Choi, K.: Quick, practical selection of effective seeds for homology search. JCB 12(9), 1137–1152 (2005)
14. Tatusova, T., Madden, T.: Blast 2 sequences - a new tool for comparing protein and nucleotide sequences. FEMS Microbiol. Lett. 174, 247–250 (1999)
15. Yang, I., Wang, S., Chen, Y., Huang, P.: Efficient methods for generating optimal single and multiple spaced seeds. In: BIBE 2004: Proceedings of the 4th IEEE Symposium on Bioinformatics and Bioengineering, Washington, DC, USA, p. 411. IEEE Computer Society Press, Los Alamitos (2004)
16. Zhang, Z., Schwartz, S., Wagner, L., Miller, W.: A greedy algorithm for aligning dna sequences. J.Comput.Biol. 7, 203–214 (2000)

Region-Based Selective Encryption for Medical Imaging*

Yang Ou[1], Chul Sur[2], and Kyung Hyune Rhee[3]

[1] Department of Information Security, Pukyong National University,
599-1, Daeyeon3-Dong, Nam-Gu, Busan 608-737, Republic of Korea
ouyang@pknu.ac.kr
[2] Department of Computer Science, Pukyong National University
kahlil@pknu.ac.kr
[3] Division of Electronic, Computer and Telecommunication Engineering,
Pukyong National University
khrhee@pknu.ac.kr

Abstract. The confidential access to medical images becomes significant in recent years. In this paper, we propose two types of region-based selective encryption schemes to achieve secure access for medical images. The first scheme randomly flips a subset of the bits belonging to the coefficients in a Region of Interest inside of several wavelet sub-bands, which is performed in compression domain but only incurs little loss on compression efficiency. The second scheme employs AES to encrypt a certain region's data in the code-stream. The size of encrypted bit-stream is not changed and there is no compression overhead generated in the second scheme. Moreover, both of two schemes support backward compatibility so that an encryption-unaware format-compliant player can play the encrypted bit-stream directly without any crash.

Keywords: Medical Imaging, Selective Encryption, Region of Interest, Bit Flipping, Quality Layer Organization.

1 Introduction

The organization of today's health systems often suffers from the fact that different doctors do not have access to each others patient data. The enormous waste of resources for multiple examinations, analysis, and medical check-ups is an immediate consequence. In particular, multiple acquisition and related private rights of medical image data should be protected. An obvious solution to these problems is to create a distributed database system where each doctor has his/her own digital access to the existing medical data related to a patient. Hence, there is an urgent need to provide the confidentiality of medical image

* This work was partially supported by the Korea Science and Engineering Foundation(KOSEF) grant funded by the Korea government(MOST) (No. R01-2006-00-10260-0), and the MIC of Korea, under the ITRC support program supervised by the IITA(IITA-2006-C1090-0603-0026).

F.P. Preparata and Q. Fang (Eds.): FAW 2007, LNCS 4613, pp. 62–73, 2007.

data related to the patient when medical image data is stored in databases and transmitted over networks.

The medical image data is different from other visual data for multimedia applications. Since the lossy data may cause some negative misdiagnosis, it is constrained by the fact that a diagnosis should be based on a lossless compressed image which holds a much larger amount of data than the lossy compressed image. A possible solution to this problem is to use selective compression where parts of the image that contain crucial information are compressed in a lossless way whereas regions containing unimportant information are compressed in a lossy manner. Fortunately, the JPEG2000 still image compression standard [17] provides a significant feature called Region of Interest (ROI) coding, which allows both lossless and lossy compression in one image. Also, the lossless encoded part can be decoded or reconstructed first.

The JPEG2000 is an emerging research field for medical imaging. Many techniques are exploited to create efficient encryption schemes based on JPEG2000. Scrambling of the codewords on the compressed bit-stream proposed in [1,4] is often computed with low complexity, but cannot provide high security. The permutations on wavelet coefficients [9,13] can be simply implemented. However, these techniques might cause some efficiency losses in the subsequent entropy coding. Wavelet domain based encryption schemes proposed by Uhl et al. [3,5,10] only protect the wavelet filters or wavelet packet structures. The encrypted data of these wavelet-based algorithms is lightweight but the security cannot be totally confirmed. Since some parameters or structures are encrypted, the actual coefficients values are still plaintext. A selective encryption scheme for JPEG2000 bit-stream is introduced by Norcen et al. [8], which encrypts 20% of the compressed bit-stream. Whereas Lian et al. [6] proposed a scheme to encrypt some sensitive data in frequency sub-bands, bit-planes or encoding-passes selectively, the total encrypted data is no more than 20%.

These techniques described so far achieve that the whole image is distorted. However, for the selective compressed medical images, especially in radiography and biomedical imaging, encryption in the spatial domain of the whole image is unnecessary to some extent.

Our Contributions. In this paper, we propose two types of schemes for region-based selective encryption of medical images, which concentrate on the security of crucial parts in medical images. The first scheme randomly invert the most significant bits of ROI coefficients in several high frequency sub-bands in the transform domain. This scheme is simply implemented and low complex-computed, and further, has little compression overhead. The second scheme employs AES cipher as the basic cryptographic building block to selectively encrypt the ROI data in the final code-stream, which provides sufficient confidentiality and can be used for medical image storage. Furthermore, our proposed schemes are backward compatible so as to ensure a standard bitstream compliant decoder can reconstruct the encrypted images without any crash. Note that, throughout the paper, we assume that an existing key management protocol is used to distribute the key, and we focus on how the key is used to achieve secrecy.

The remainder of the paper is organized as follows. In Section 2, we review the coding mechanism of JPEG2000 and ROI coding. Section 3 presents the proposed schemes for region-based selective encryption of medical images. The experimental results are shown and analyzed in Section 4. We discuss the security and efficiency of our schemes in Section 5. Finally, we conclude the paper in Section 6.

2   Preliminaries

In general, wavelet-based image processing methods have gained much attention in the biomedical imaging community. Applications range from pure biomedical image processing techniques such as noise reduction, image enhancement to computed tomography (CT), magnetic resonance imaging (MRI), and functional image analysis [14]. Image compression methods that use wavelet transforms have been successful in providing high compression ratios while maintaining good image quality. The JPEG2000 is not only one of the most popular standard, but might be also widely applied in medical imaging in the near future.

2.1   The Coding Mechanism of JPEG2000

The JPEG2000 image coding standard is based on a scheme originally proposed by Taubman, which is known as Embedded Block Coding with Optimized Truncation (EBCOT) [11].

In JPEG2000, the Discrete Wavelet Transform (DWT) is firstly applied to the original image to decompose into different resolution levels. The sub-bands in each resolution level are partitioned into smaller non-overlapping rectangular blocks called code-blocks. The wavelet coefficients inside a code-block are processed from the most significant bit-plane (MSB) towards to the least significant bit-plane (LSB). Furthermore, in each bit-plane the bits are scanned in a maximum of three passes called coding passes. Finally, during each coding pass, the scanned bits with their context value are sent to a context-based adaptive arithmetic encoder that generates the code-block's bitstream.

The rate-distortion optimal merging of these bit-streams into the final one is based on a sophisticated optimization strategy. This last procedure carries out the creation of the so-called layers which roughly stand for successive qualities at which a compressed image can be optimally reconstructed. These layers are built in a rate-allocation process that collects, in each code-block, a certain number of coding-passes codewords. Hence, the final JPEG2000 code-stream is organized from the embedded contributions of each code-block with the formal syntax information to depict the bit-stream structure. If an ROI is defined in JPEG2000, its information may appear in the front of the code-stream or the earlier quality layers.

2.2 Region of Interest Coding

The ROI coding scheme in part I of JPEG2000 standard [2,17] is based on coefficients scaling. An ROI mask is generated in each wavelet sub-band to identify the ROI coefficients as shown in Fig.1. The principle of ROI coding is to scale (shift) coefficients so that the bits associated with the ROI are placed in higher bit-planes than the bits associated with the background. During the embedded

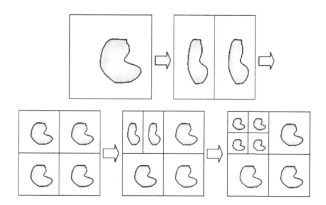

Fig. 1. ROI Mask Generation

coding process, the most significant ROI bit-planes are placed in the bit-stream before any background bit-planes of the image (Note that some bits of ROI coefficients might be encoded together with non-ROI coefficients). Thus, the ROI will be decoded, or refined, before the rest of the image. If the bit-stream is truncated, or the encoding process is terminated before the whole image is fully encoded, the ROI will be of higher fidelity than the rest of the image [12].

3 Region-Based Selective Encryption

In this section, we present two types of region-based selective encryption schemes to provide the security of crucial part in medical images. Generally, the multimedia encryption can be performed before, in the middle of, or after the compression. Regarding the first way, encryption will destroy the strong correlation and statistical information of multimedia data. Therefore, most encryption schemes are applied within or after compression [16]. We also follow this principle to construct our schemes.

3.1 Scheme-I: Encryption During Compression

The first region-based selective encryption scheme is addressed during compression. In particular, the encryption is performed in the transform-domain. The main idea is to mask a pseudo-random noise which can be implemented by randomly inverting some most significant bits of ROI coefficients belonging to certain wavelet sub-bands. As mentioned in Section 2, ROI masks are generated in each wavelet sub-band to identify the ROI coefficients after quantization. And then, the bit-planes belonging to ROI are upshifted to a maximum value which has the largest magnitude of the background coefficients. Finally, the bit inverting is operated after bit-plane upshifting.

The encryption should have a minimal impact on compression efficiency. As the wavelet coefficients are strongly correlated, encrypting all bits of the coefficients would reduce coding efficiency. However, the sign bits of wavelet coefficients are typically weakly correlated, and appropriate for inverting. Also, since the most significant bits of each coefficient holds the most important information, encryption of them can achieve the fuzziest visual quality. Furthermore, the lowest frequency sub-band (LL-band) reserves larger energy and statistical information of the image than other sub-bands, hence we only consider protecting the sub-bands in high frequency level.

In Scheme-I, the quantized wavelet coefficients belonging to high frequency sub-bands and corresponding to the ROI are encrypted by randomly inverting their sign bits and some most significant bits. Note that this method modifies few bits of each ROI coefficients and can be performed on-the-fly during entropy coding. A Pseudo Random Number Generator (PRNG) is used to drive the flipping process, characterized by its seed value. Thus, for each coefficient, a new pseudo-random value (one-bit value) is generated. If this value is 1, the related bit is inverted. Otherwise, the bit is left unchanged. The amount of inverted bits can be adjusted by restricting the inversion to fewer wavelet sub-bands. It should be noticed that the seed value represents the encryption key, which is the only information we need to protect.

With above scheme, encrypted regions can have arbitrary shapes. The shape of the ROI has to be available at both the encoder for flipping and decoder for inversely flipping. The shape information can be transmitted in a separate channel or implicitly embedded using the ROI mechanism of JPEG2000.

3.2   Scheme-II: Encryption After Compression

The second region-based selective encryption scheme is performed after compression, i.e., encryption on the compressed data. We exploit the flexible quality layer mechanism in JPEG2000 standard to selectively encrypt ROI. Each quality layer has a collection of contributions from all the code-blocks. The first layer is formed from the optimally truncated code-block bit-streams such that the target bit-rate achieves the highest possible quality in terms of minimizing distortion. Accordingly, the compressed ROI data in the first quality layer is encrypted in Scheme-II.

Since the medical images always have a large amounts of data, public key techniques are not suitable for them. Therefore, we employ a symmetric block cipher as the basic cryptographic building block to construct our second scheme. Especially, the Advanced Encryption Standard (AES) is a recent symmetric block cipher which is used in most applications for providing confidentiality.

We make use of AES with Cipher Feed Back (CFB) mode which satisfies the requirement for encryption of arbitrary sized data.

The basic encryption cell is the contribution of each ROI code-block to the first quality layer. In order to achieve format compliance, we need to only encrypt the purely compressed data and leave the syntax information of the code-stream untouched. Parsing and extracting for the necessary information from the code-stream must have high computational complexity, hence they cause much handling time. To avoid these overhead, we encrypt ROI data with synchronizing the rate allocator, which can be applied directly after the optimal rate-distortion calculation of each code-block, but before writing the code-block's codewords to each quality layer. It is worthily mentioned that in case there is only one ROI coefficient in a code-block, this code-block should be identified as a ROI code-block. That means the ROI boundary is aligned with code-blocks, but not exact to each coefficient. Also, the identification of ROI code-blocks should be informed to decryption side to correctly decrypt the ciphertext.

4   Experimental Results

4.1   Experimental Environments

The proposed schemes have been implemented based on the publicly available JPEG2000 implementation JJ2000 (Version 5.1) [18]. The encryption is simulated on following monochrome medical images: an ultrasound image (US), a computer radiology chest image (CR), a computerized tomography image (CT) and a magnetic resonance image (MRI). All of them are in gray level with 8 bit per pixel (bpp) (see Table 1. for detailed information of these images). In our experiments, we use the JPEG2000 default parameter settings: five-level DWT decomposition, 64×64 code-blocks, 20-layers, layer progressive bitstream etc.

The peak signal to noise ratio (PSNR) was used for objective measure of the image distortion. PSNR for an image with bit depth of n bits/pixel is defined as

$$PSNR = 10 log_{10} \frac{(2^n - 1)^2}{MSE}$$

and MSE is given by

$$MSE = \frac{\sum (\hat{x} - x)^2}{A},$$

where x denotes the original pixel values, $\hat{x}$ is the reconstructed pixel values, and A is the area of ROI. In the following, ROI PSNR is used to measure the rate distortion (RD) performance of ROI, while whole PSNR is used to assess the background quality. Note that higher the PSNR values reveal the better quality of images. Consequently, the lower PSNR values imply the higher security.

4.2   Experimental Results and Analysis

Our first experiment is implemented on Scheme-I: randomly flipping the bits belonging to ROI coefficients. We independently encrypt the bits in each ROI

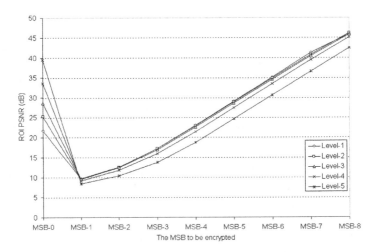

Fig. 2. Comparison of ROI PSNR performance with encrypting different MSBs in each resolution level

bit-plane related to each resolution level. Fig.2 illustrates the PSNR performance of ROI in the reconstructed encrypted image with the first nine MSBs from the ultrasound image, where the MSB-0 is the sign bit-plane. Only the high frequency sub-bands are tested, i.e., the level-1 only includes HL1, LH1 and HH1 sub-bands. We look for the low PSNR quality to make selective encryption secure in Fig.2. It can be clearly noticed that image quality is low if we randomly flip the bits in the first MSBs (except MSB-0), especially in the higher resolution levels. However, inverting the sign bits results an opposite relationship to resolution levels and has less degradation on image quality. The benefit of encrypting sign bits is shown in the discussion of Fig.3.

We use the lossless encoding mode in order to test the file size changes which explicitly indicate the compression efficiency overhead shown in Fig.3, with the same parameters as Fig.2. The compression overhead is calculated by file size increasing ratio. Obviously, sign bits encryption generates the least compression overhead, particularly in lower resolution levels (Level 1-4) where the overheads are almost close to zero. In addition, the overheads generated from encrypting the lowest three resolution levels (Level 1-3) are no more than 0.5%, whereas encryption of the highest level (Level 5) results the largest compression overhead.

There should be a good trade-off between the degradation of image visual quality and compression efficiency overhead. Hence, depending on the previous experimental results, randomly flipping of the bits belonging to the MSB-0 and MSB-1 in resolution level 1-3 can achieve the compatible purpose of selective encryption. The visual performance of the encrypted image is shown in Fig.4.

In the following, we implement our second scheme of ROI encryption. The AES cipher is applied to encrypt ROI data in the first quality layer. Table 1. shows the encrypted data ratio and ROI PSNR for different size medical images. The ROI's area is chosen as one-fifth of the whole image's. We calculate the

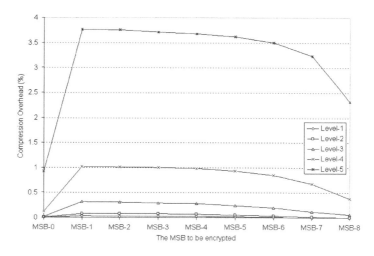

Fig. 3. Comparison of compression overhead with encrypting different MSBs in each resolution level

PSNR value of both ROI and whole images, as well the encrypted data ratio. Note that the ratio is calculated from the data we selectively encrypted to the total ROI part, but not to the whole image.

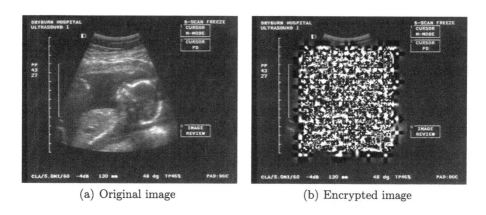

(a) Original image   (b) Encrypted image

Fig. 4. Visual quality of the reconstructed encrypted image US

In Table 1, We find that even though the encrypted data ratio is very low (less than 2%), the degradation of PSNR performance is fairly large - almost less than 10dB. These two bounds can be sufficiently accepted in most security systems. Also, the values in the second and fourth columns reveal that naive increasing of the amount of encrypted data may not achieve a large degradation. For example, comparing MRI and US images, the data ratio of MRI (1.258%) is lower but it has a higher PSNR performance (10.002dB) than US's (0.96%, 9.812dB).

Table 1. Encrypted Data Ratio with ROI as 20% of the Whole Image

Images and Size	ROI PSNR (dB)	Whole PSNR (dB)	Encrypted data ratio
MRI(1024×1024)	10.002	12.793	1.258%
US(1215×948)	9.812	12.396	0.960%
CT(1134×1133)	7.780	12.814	0.727%
CR(1275×1067)	6.347	11.353	1.393%

This case illustrates that there is no direct proportion between the amount of encrypted data and the PSNR performance. Moreover, proper selection of the crucial data to be encrypted is important for selective encryption. In our second scheme, the employment of the quality layer organization in JPEG2000 helps us to immediately and efficiently select the significant information to be encrypted. The visual performance results of reconstructed MRI are shown in Fig.5.

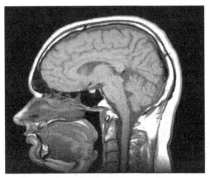
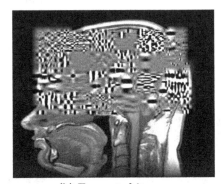

(a) Original image   (b) Encrypted image

Fig. 5. Visual quality of the reconstructed encrypted MRI

5   Discussions on Security and Efficiency

Security of multimedia encryption involves two different aspects. One aspect is the visual effect, i.e., how much visual information leaks after encryption. Another aspect is the system security which means the robustness of the system against cryptanalysis. Tolerance to visual content leakage and judgement of the success of an attack depends on applications. For medical imaging system, content leakage is somewhat acceptable since an image with lower quality cannot provide enough technical information to medical analysis and diagnosis. In the following we will discuss the security of our schemes by conducting the simple ciphertext-only attack.

   The randomly bit-flipping in our first scheme is not sufficiently secure to error-concealment attack. One way to conceal bit error (encrypted bits) is to replace the error bits with a constant value, such as 0 or 1. Here we simulate that using 0

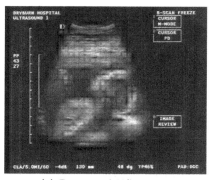

(a) Recovered US image

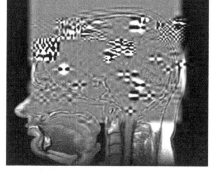

(b) Recovered MRI image

Fig. 6. Visual quality of the attacked recovered images

to replace all ROI MSB-0 and MSB-1 in resolution level 1-3 in order to attack our encrypted image. It is simulated that the recovered images' PSNR performance is up to 20.22dB (Fig.6(a)), and the visual quality is correspondingly higher than Fig.4(b). To avoid this attack, the bit-flipping can be combined with a permutation algorithm to improve robustness against error-concealment attack [15]. On the other hand, the larger space of the encryption key, i.e. the seed value for PRNG in this scheme, can achieve a more robustness towards the brute-force attack. As we mentioned at the beginning, we focus on how the seeds are used, but the definition of a specific random number generation is outside the scope of this paper. There are a lot of secure PRNGs are introduced until so far, and we can use one of them properly depending on application purposes. In our simulation, we present the seed value with 32 bits which allows to employ 2^{32} different keys.

Our second scheme uses AES as the underlying cipher to selectively encrypt the data in the first quality layer which holds the furthest optimized rate-distortion. Obviously, the security of AES is indubitable. Since the bit-stream values which are arithmetically decoded depend on earlier results, the constant number-based replacement attack does not have the desired effect as discussed above. However, the error resilience feature in JPEG2000 makes some sense to attack this scheme. We insert an error resilience segmentation symbol in the codewords at the end of each bit-plane. Decoders can use this information to detect and conceal errors [7]. Fig.6(b) shows the visual performance of MRI after using error resilience symbol (compared to Fig.5(b)). To enhance the security of this scheme, an efficient way is to increase the bit-rate of the first layer, i.e., implicitly increase the amount of encrypted data which has been optimized allocated into the first layer.

Regarding on the efficiency of the proposed schemes, as mentioned in Section 4, even though Scheme-I destroys some statistical information of the original image, the compression overhead is less than 0.5%. Furthermore, the bit inverting can be simply implemented and it has very low computational complexity. Nevertheless, using AES cipher incurs a little compression time overhead in Scheme-II, but the output file size is not changed, also it is resistable against several kinds of attacks. Consequently, Our schemes have their own advantages and are suitable for different medical imaging systems.

6   Conclusion

In this paper, we have presented two novel region-based selective encryption schemes for medical imaging. The first scheme is to randomly invert the first two MSBs of ROI coefficients in wavelet transform domain. It can be efficiently implemented and only incurs little compression efficiency overhead, also it can be extended to other motion formats. The second scheme, selective encryption of the compressed ROI data, provides a high level security and has no file size changes. Both of them are format backward compatible and have their own advantages so that they are applicable to different medical security imaging systems to satisfy different requirements.

References

1. Ando, K., Watanabe, O., Kiya, H.: Partial-scambling of images encoded by JPEG2000. The Institute of Electronics Information and communication engineers transactions 11(2), 282–290 (2002)
2. Bradley, P.A., Stentiford, F.W.M.: JPEG2000 and region of interest coding. Digital Image Computing Techniques and Applications(DICTA) (2002)
3. Engel, D., Uhl, A.: Lightweight JPEG2000 encryption with anisotropic wavelet packets. In: International Conference on Multimedia & Expo, pp. 2177–2180 (2006)
4. Grosbois, R., Gerbolot, P., Ebrahimi, T.: Authentication and access control in the JPEG2000 compressed domain. In: Processing SPIE 46th annual Meeting, Applications of Digital Image Processing XXIV, pp. 95–104 (2001)
5. Kockerbauer, T., Kumar, M., Uhl, A.: Lightweight JPEG 2000 confidentiality for mobile environments. In: Proceedings of the IEEE International Conference on Multimedia and Expo(ICME), pp. 1495–1498. IEEE Computer Society Press, Los Alamitos (2004)
6. Lian, S., Sun, J., Zhang, D., Wang, Z.: A selective image encryption scheme based on JPEG2000 codec. In: Aizawa, K., Nakamura, Y., Satoh, S. (eds.) PCM 2004. LNCS, vol. 3332, pp. 65–72. Springer, Heidelberg (2004)
7. Moccagatta, L., Soudagar, S., Liang, J., Chen, H.: Error-resilient coding in JPEG2000 and MPEG-4. IEEE Journal on Selected Areas in Communications 18(6), 899–914 (2000)
8. Norcen, R., Uhl, A.: Selective encryption of the JPEG2000 bitstream. In: Lioy, A., Mazzocchi, D. (eds.) CMS 2003. LNCS, vol. 2828, pp. 194–204. Springer, Heidelberg (2003)
9. Norcen, R., Uhl, A.: Encryption of wavelet-coded imagery using random permutations. In: International Conference on Image Processing(ICIP), pp. 3431–3434 (2004)
10. Pommer, A., Uhl, A.: Selective encryption of wavelet-packet encoded image data - efficiency and security. ACM Multimedia Systems (Special issue on Multimedia Security), pp. 279–287 (2003)

11. Taubman, D.: High performance scalable image compression with EBCOT. IEEE Transactions On Image Processing 9(7), 1158–1170 (2000)
12. Taubman, D., Marcellin, M.: JPEG2000 image compression fundamentals, standards and practice. Kluwer Academic Publishers, Boston (2002)
13. Uehara, T., Safavi-Naini, R., Ogunbona, P.: Securing wavelet compression with random permutations. In: IEEE Pacific Rim Conference on Multimedia, pp. 332–335. IEEE Computer Society Press, Los Alamitos (2000)
14. Unser, M., Aldroubi, A.: A review of wavelets in biomedical applications. In: Proceeding IEEE, vol. 84, pp. 626–638. IEEE Computer Society Press, Los Alamitos (1996)
15. Wen, J., Severa, M., Zeng, W.: A format-compliant configurable encryption framework for access control of video. IEEE Trans. 12(6), 545–557 (2002)
16. Zeng, W., Yu, H., Lin, C.: Multimedia security technologies for digital rights management. Elsevier, Amsterdam (2006)
17. ISO/IEC 15444-1: Information technology-JPEG2000 image coding system-Part 1: Core Coding System (2000)
18. JJ2000 - Java implementation of JPEG2000, http://jpeg2000.epfl.ch/

Extracting Information of Anti-AIDS Inhibitor from the Biological Literature Based on Ontology*

Chunyan Zhang[1], Jin Du[1], Ruisheng Zhang[1,2,**], Xiaoliang Fan[1],
Yongna Yuan[2], and Ting Ning[1]

[1] Department of Computer Science, Lanzhou University,Gansu, China
[2] Department of Chemistry and Chemical Engineering, Lanzhou University,
Gansu, China
{zhangchy05,dujin04}@lzu.cn, zhangrs@lzu.edu.cn

Abstract. Nowadays, it is still the primary problem to find the inhibitors of retrovirus, protease and integrase in anti-AIDS drug design. However, the research and experimental results about anti-AIDS inhibitors mainly exist in large numbers of scientific literature, not in readable format for computer. In this paper, we introduce an Ontology-based Information Extraction (OIE) approach to extract anti-AIDS inhibitors from literature. Key to the approach is the construction of anti-AIDS inhibitors ontology, which provides a semantic framework for information extraction, and annotation of corpus. Consequently, this paper primarily focuses on the architecture of OIE, on which we construct the anti-AIDS ontology using Protégé tool and annotate corpus. Finally, we employ a demonstrated application scenario to show how to annotate the PubMed articles based on the ontology we have constructed.

Keywords: Information Extraction, Ontology, Annotation, anti-AIDS, Inhibitor.

1   Introduction

Acquired Immune Deficiency Syndrome(AIDS) is caused by Human Immunodeficiency Virus (HIV) infection. Due to the lack of an effective vaccine to prevent HIV infection, the development of efficacious anti-AIDS drugs is urgent in the field of AIDS research. Currently, seeking and developing the new anti-AIDS drugs with new structure types or new mechanism have been a hot issue in the field of drug design recent years. Further more, there is a growing awareness that governments all over the world give more and more financial support to the anti-AIDS drug design related research work.

---

* This work was supported by National Natural Science Foundation of China through projects: Research on Computational Chemistry e-SCIENCE and its Applications (Grant no. 90612016).
** Corresponding author.

F.P. Preparata and Q. Fang (Eds.): FAW 2007, LNCS 4613, pp. 74–83, 2007.

The existing anti-AIDS drugs mainly work well in the following three targets, i.e., retrovirus, protease and integrase enzyme. The key factor of designing a new drug always relies on looking for appropriate inhibitors in those targets mentioned above. Nowadays, effective anti-AIDS inhibitors, mechanism of inhibitors and HIV, reciprocity among inhibitors in human body as well as HIV resistance [1][2] have been reported in thousands of literature. Meanwhile, the correlative literature is increasing rapidly. Hence, it is very expensive and time-consuming for scientists to track the up-to-date progress of the fields they are interested in. In short, how to extract information on inhibitors from immense scientific literature is a very urgent issue for anti-AIDS drug designers.

The aim to information extraction is providing a more powerful information access tool to help researchers overcome the challenge of information overloading. Compared with information retrieval which is usually used to find the highest ranked documents to match a user query, information extraction is to extract specified information from a passage text, to produce well-formed output and to deposit the results into a database. Information extraction is the comprehensive application of a variety of natural language processing technology.

Research on ontology is becoming widespread in the computer science community. While in the past this terminology has been rather confined to the philosophical sphere, it is now gaining a specific role in Artificial Intelligence, Computational Linguistics, and Database Theory. In particular, its importance is being recognized in research fields as diverse as knowledge engineering[3][4][5], knowledge representation[6][7][8], qualitative modeling[9], information modeling[10], information retrieval and extraction[11][12] and so on. Ontology is an augmented conceptual-model instance that serves as a wrapper for a narrow domain of interest [13]. Ontology of anti-AIDS inhibitors will provide a model that can be used to form a semantic framework for information extraction.

In this paper we present the Ontology-based Information Extraction approach to extract anti-AIDS inhibitors from biological literature. While, key to this approach is the construction of anti-AIDS inhibitor ontology and the annotation of corpus. Therefore, we chiefly focus on the ontology of anti-AIDS inhibitor and the manual annotation of original corpus to fulfill this approach. The paper is organized as follows: Section 2 provides some information about related works and projects. The architecture of OIE system, the ontology of anti-AIDS inhibitor and the annotation of corpus are described in Section 3. Section 4 presents a case study to show how to annotate the corpus using the ontology we have constructed. Finally, Section 5 draws the conclusion and points out the future work.

2   Related Work

At present, the application of information extraction in the biology, medicine and chemistry research have been paid more and more attention in the international and domestic research institutions. e-Science project in Cambridge Molecular Information Center extracts the name and properties of compounds from the chemical literature [14]. The primary task of GENIA project in Tsujii laboratory of the University of Tokyo is to automatically extract useful information from texts written by scientists to help overcome the problems caused by information overloading. They are currently working on the key task of extracting event information about protein interactions [15][16][17]. The Gene Ways project of Genome Center in Colombia University [18] extracts information about molecular composition and the interaction between them. PennBioIE, the biomedical information extraction project at the University of Pennsylvania, has developed software named FABLE which provides biomedical researchers a way to more thoroughly identify MEDLINE articles relevant to their interests. Currently, FABLE assists with finding MEDLINE abstracts that mention gene(s), RNA(s) or proteins(s), and the system is optimized for human genes and gene products[19]. In a word, their work all extract useful information from scientific literature.

Some domestic research institutes also do similar work,such as the Institute of Computational Linguistics, Peking University research on Computational Linguistics & Language Information Processing which covers a wide range of areas, including Chinese syntax, language parsing, computational lexicography, semantic dictionaries, computational semantics and application systems[20]. Information Retrieval Lab, Harbin Institute of Technology mainly research information extraction technology, auto index, text classification and cluster and so on [21].

Extracting protein-protein interactions from the biological literature has been done by many institutes and researchers. However, researches about anti-AIDS inhibitor extraction have not been reported yet. In this paper, we utilize the Ontology-based Information Extraction method to extract anti-AIDS inhibitors.

3   The Proposed Approach: OIE

The feasible approach we presented for extracting anti-AIDS inhibitors from biological literature, called Ontology-based Information Extraction(OIE), consists of four main steps: (1) the construction of initial corpus, (2) building anti-AIDS inhibitors ontology, (3) annotating initial corpus and (4) extraction of annotated entity which is described in the section of Architecture of OIE. Before discussing (2) and (3) in detail, we first briefly discuss the construction of initial corpus.

Initial corpus usually composes of raw articles related to some specific domain. In our work, the construction of the initial corpus is just to search articles with keyword 'HIV' and 'inhibitor' from PubMed database.

In this section, we aim at drawing the architecture of OIE and discussing the construction of ontology and the annotation of original corpus.

3.1   Architecture of OIE

The ontology provides specific domain knowledge for understanding anti-AIDS inhibitors and annotating the original corpus. Be top of the architecture of OIE, we also draw the interaction between the ontology module and annotation module. Apparently, in order to extract inhibitor from biological literature, there are other modules besides ontology and annotation modules. Fig. 1 denotes a generic architecture which consists of seven main modules which are Interface Module,

Retrieval Module, Corpus Module, Ontology Module, Thesaurus Module, Extraction Module and Database Module. For some common modules, we collect, share and reuse existing software packages, such as Thesaurus Module etc, and we extend them to meet what scientific researchers need. There are several key features in the Fig. 1, illustrated as follows:

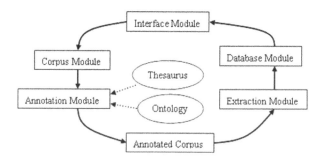

Fig. 1. Architecture of OIE Overview

Interface Module. It is a very useful client to send user request and query knowledge stored in database. Initially, users start a query with some keywords and then get the results of search. The set of results are the raw material for Corpus Module.

Corpus Module. It is regarded as the basic material for annotation module. In general, corpus is constituted of some articles searched from the literature database such as PubMed. We have known that the abstract outlines the whole article. In order to improve the performance of extraction and lessen the complexity, the corpus only contains the abstracts.

Ontology Module. This module, which provides a semantic framework for information extraction and annotation,is the key module of the overall architecture. According to the Ontology Module, domain experts manually annotate the original corpus.

Annotation Module. Annotation the original corpus is a significant, pervasive topic throughout the whole of architecture. Ontology and Thesaurus Module are regarded as the background knowledge of annotation.

Extraction Module. The aim of this module is to extract information about inhibitors from annotated corpus and interaction with database module to store the result into database. In this paper, the annotated corpus has been formatted as XML, so we write a Java application to extract annotated entity from annotated corpus.

Database Module. In order to make users search quickly and directly, the extracted information usually store into database. It provides interface to access for users.

3.2   Ontology of Anti-AIDS Inhibitors

We have known that the ontology describes basic concepts in a domain and defines relationship among them. Basic building blocks of ontology design include classes or concepts, properties of each concept describing various features and attributes of the concept, i.e., slots(sometimes called roles or properties), and restrictions on slots(facets, sometime called role restrictions). Ontology together with a set of individual instances of classes constitutes a knowledge base. At present, there is no one correct methodology for developing ontology, nor is there a single correct result. Therefore developing ontology is an iterative process.

The ontology has been developed in the biological sciences for several applications such as Gene Ontology[22] and BioCon Knowledge Base of the TAMBIS Project[23]. Such ontology include conceptual hierarchies for database covering diseases and drug names.

In this paper, we use Protégé tool to construct the ontology of anti-AIDS inhibitors. Protégé developed in the Musen Laboratory at Stanford Medical Informatics and written in Java is a free, open source ontology editor and knowledge-base framework , which also provides a Java API for independent development of plug-ins.The Protégé platform supports two main ways of modeling ontology via the Protégé-Frames and Protégé-OWL editors. The Protégé-Frames editor provides a full-fledged user interface and knowledge server to support users in constructing and frame-based domain ontology, customizing data entry forms, and entering instance data. In this model, ontology consists of a set of classes organized in a subsumption hierarchy to represent a domain's salient concepts, a set of slots associated to classes to describe their properties and relationships, and a set of instances of those classes-individual exemplars of the concepts that hold specific values for their properties. The Protégé-OWL editor is an extension of Protégé that supports the Web Ontology Language (OWL). OWL is the most recent development in standard ontology languages, endorsed by the World Wide Web Consortium (W3C)[24]. The Protégé-OWL way is chosen as the editor of anti-AIDS inhibitor ontology in order to annotate corpus based on the OWL ontology.

The anti-AIDS inhibitor ontology, a taxonomy of some entities involved in anti-AIDS inhibitors, is developed as the semantic classification used in the corpus. In biological field, anti-AIDS inhibitors are divided into two types via structure type, Nucleoside Analogs and Non-Nucleoside.But according to the different mechanism, the anti-AIDS inhibitors contain three main types, Retrovirus, Protease and Integrase Inhibitors which can be divided into more subdivision.

In Protégé, each inhibitor concept is a class, reproducing the hierarchical nature of inhibitor. The root concept, Inhibitor_Ontology_Entity, has children *mechanism* and *structure*. Fig. 2 shows the class hierarchy of inhibitor ontology and some instance of protease inhibitors.

3.3   Corpus Annotation

The lack of an extensively annotated corpus of biological literature can be seen as a major bottleneck for applying NLP techniques to bioinformatics. The

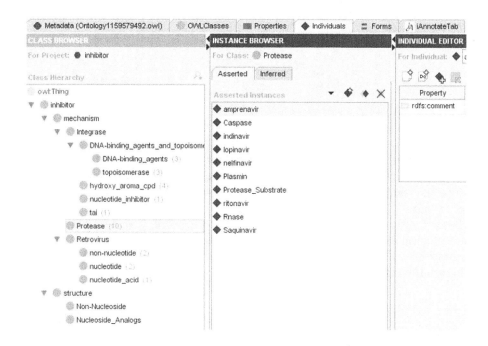

Fig. 2. Class hierarchy and Protease individuals in Protégé

annotated corpus aims at providing high quality reference materials to let NLP techniques work for bioinformatics and at providing the gold standard for the evaluation of text mining systems.

The task of annotation can be regarded as identifying and classifying the terms that appears in the texts according to a pre-defined classification which is just provided by anti-AIDS inhibitor ontology we constructed. Hence, an OWL-based Protégé plugin named iAnnotate Tab [25] is used to manually annotating text with ontology concepts. The iAnnotate Tab can be used for developing semantically annotated corpus and save the annotated document in XML with 'class' tags enclosing the annotated text.

In this paper, the original corpus consists of several abstracts from PubMed database. Since we wanted our annotation work to converge on anti-AIDS inhibitors, we selected articles with the keywords, anti-AIDS and inhibitor from PubMed database.

Fig. 3 shows the basic architecture that we are employing. The annotation tool interacts Ontology Editor (Protégé) and Inhibitor ontology. By the annotation tool, original corpus changes into annotated one.

4   Case Study

In order to examine the ontology we constructed, we have an experience with a demonstrated application.

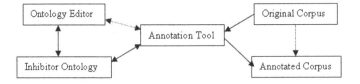

Fig. 3. The basic Annotation Architecture

Firstly, the original corpus is obtained by searching articles related to 'HIV' && 'inhibitor' from PubMed database and access 7400 abstracts. In order to reduce the complexity, our original corpus consists of 500 abstracts with plain text format.

Secondly, the original corpus is imported into Protégé, with IAnnotate Tab, the biological experts manually annotate the corpus. Fig. 4 shows the abstract annotated by IAnnotation tool. In left column, the words annotated by other color are the instances of ontology class .

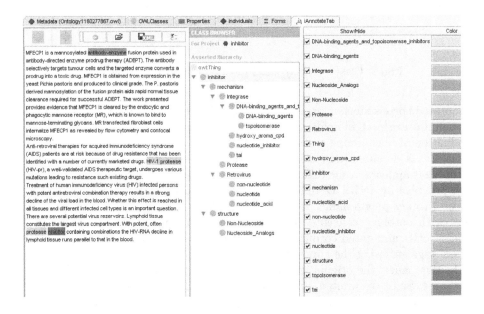

Fig. 4. Annotated Abstract by IAnnotation Tool

The annotated abstracts can be saved as XML format by IAnnotation tool. Fig. 5 shows the XML format of annotated abstract. In this XML file, the abstract is between tag IAnn and /IAnn. Other tags, such as tag 'Retrovirus', 'Protease', 'Inhibitor' denote the corresponding class of inhibitor ontology, the instance of class is between a pair of tags. Here Antlbody-enzyme is an instance of Retrovirus class.

```
<?xml version="1.0" ?>
- <IAnn>
    MFECP1 is a mannosylated
    <Retrovirus>antibody-enzyme</Retrovirus>
    fusion protein used in antibody-directed enzyme prodrug therapy (ADEPT). The antibody selectively targets tumour cells and
    the targeted enzyme converts a prodrug into a toxic drug. MFECP1 is obtained from expression in the yeast Pichia pastoris
    and produced to clinical grade. The P. pastoris derived mannosylation of the fusion protein aids rapid normal tissue clearance
    required for successful ADEPT. The work presented provides evidence that MFECP1 is cleared by the endocytic and phagocytic
    mannose receptor (MR), which is known to bind to mannose-terminating glycans. MR transfected fibroblast cells internalize
    MFECP1 as revealed by flow cytometry and confocal microscopy. Anti-retroviral therapies for acquired immunodeficiency
    syndrome (AIDS) patients are at risk because of drug resistance that has been identified with a number of currently marketed
    drugs.
    <Protease>HIV-1 protease</Protease>
    (HIV-pr), a well-validated AIDS therapeutic target, undergoes various mutations leading to resistance such existing drugs.
    Treatment of human immunodeficiency virus (HIV) infected persons with potent antiretroviral combination therapy results in
    a strong decline of the viral load in the blood. Whether this effect is reached in all tissues and different infected cell types is an
    important question. There are several potential virus reservoirs. Lymphoid tissue constitutes the largest virus compartment.
    With potent, often
    <Protease>protease</Protease>
    <inhibitor>inhibitor</inhibitor>
    containing combinations the HIV-RNA decline in lymphoid tissue runs parallel to that in the blood.
  </IAnn>
```

Fig. 5. Annotated Abstracts of XML format

The effectiveness of ontology is evaluated by annotation of original corpus. During the process of annotation, the experts can fully annotate the instances related to inhibitors depending on the ontology we constructed. Since original corpus is manually annotated by expert, the inhibitor ontology satisfied the requirements of domain experts, and annotated corpus is XML format, the precision of entity extraction is very high. Therefore, the corpus manually annotated can be regarded as the training sets of automatic annotation. However, the efficiency of manual annotation is indeed very low.

5   Conclusion and Future Work

In this paper, we mainly discussed how to build up the ontology of anti-AIDS inhibitors, and later to extract inhibitor information from the biological literature based on the ontology we had constructed. The contributions of this paper chiefly reflect on four aspects. Firstly, a feasible architecture for Ontology-based Information Extraction is presented. Secondly, an conceptual model, anti-AIDS inhibitor ontology, is set up by an ontology editor named Protégé tool. Further more, original corpus is manually annotated by experts based on the ontology. Finally, we extract the entities of inhibitors from the annotated corpus. According to the entities, we prove that the ontology is fulfilled as the background knowledge of anti-AIDS inhibitors. In the near future, in order to improve the efficiency of manual annotation, the next step of our work will focus on automatic annotation of corpus, and the manually annotated corpus with high precision and recall will be considered as the training sets of automatic annotation.

References

1. Henry, K., Melroe, H., Huebsch, J., et al.: Severe premature coronary artery disease with protease inhibitors [letter]. Lancet 351, 1328 (1998)
2. Murata, H., Hruz, P.W., Mueckler, M.: The mechanism of insulin resistance caused by HIV protease inhibitor therapy. J Biol Chem 275, 20251–20254 (2000)

3. Gruber, T.R.: A translation approach to portable ontology specifications. Knowledge Acquisition 5, 199–220 (1993)
4. Uschold, M., Gruninger, M.: Ontologies: Principles, Methods and Applications. The Knowledge Engineering Review 11(2), 93–136 (1996)
5. Gaines, B.: Editorial: Using Explicit Ontologies in Knowledge-based System Development. International Journal of Human-Computer Systems 46, 181 (1997)
6. Guarino, N.: Formal Ontology, Conceptual Analysis and Knowledge Representation. International Journal of Human and Computer Studies 43(5/6), 625–640 (1995)
7. Artale, A., Franconi, E., Guarino, N., Pazzi, L.: Part-Whole Relations in Object-Centered Systems: an Overview. Data and Knowledge Engineering 20(3), 347–383 (1996)
8. Sowa, J.F.: Knowledge Representation: Logical, Philosophical, and Computational Foundations. PWS Publishing Co., Boston (forthcoming, 1998)
9. Gotts, N.M., Gooday, J.M., Cohn, A.G.: A Connection Based Approach to Commonsense Topological Description and Reasoning. The Monist: An International Journal of General Philosophical Inquiry 79(1) (1996)
10. Ashenhurst, R.L.: Ontological Aspects of Information Modelling. Minds and Machines 6, 287–394 (1996)
11. Guarino, N.: Semantic Matching: Formal Ontological Distinctions for Information Organization,Extraction, and Integration. In: Pazienza, M.T. (ed.) Information Extraction: A Multidisciplinary Approach to an Emerging Information Technology, pp. 139–170. Springer, Heidelberg (1997)
12. Benjamins, V.R., Fensel, D.: The Ontological Engineering Initiative (KA)2. In: Guarino, N. (ed.) Formal Ontology in Information Systems, IOS Press, Amsterdam (this volume) (1998)
13. Embley, D.W.: Towards semantic understanding - an approach based on information extraction ontologies. In: Scheweand, K.-D., Williams, H.E. (eds.) ADC. CRPIT, vol. 27, Australian Computer Society (2004)
14. Cambridge eScience Centre Projects - Molecular Informatics (October 2006), http://www.escience.cam.ac.uk/projects/mi/
15. The GENIA project: (October 2006), http://www-tsujii.is.s.u-tokyo.ac.jp/genia/
16. Sekine, S., Nobata, C.: An Information Extraction System and a Customization Tool. In: Proceedings of Hitachi workshop-98 (1998)
17. Collier, N., Park, H., Ogata, N., Tateishi, Y., Nobata, C., Ohta, T., Sekimizu, T., Imai, H., Tsujii, J.: The GENIA project: corpus-based knowledge acquisition and information extraction from genome research papers. In: Proceedings of the Annual Meeting of the European chapter of the Association for Computational Linguistics (EACL'99), Bergen, Norway (1999)
18. Rzhetsky, A., Koike, T., Kalachikov, S., Gomez, S.M., Krauthammer, M., Kaplan, S.H., Kra, P., Russo, J.J., Friedman, C.: A knowledge model for analysis and simulation ofregulatory networks 16(12) 1120–1128, 2000 (October 2006), http://genome6.cpmc.columbia.edu/tkoike/ontology
19. PennBioIE (October 2006), http://bioie.ldc.upenn.edu/index.jsp
20. Institute of Computational Linguistics: Peking University (October 2006), http://www.icl.pku.edu.cn/icl_intro/default_en.asp
21. Information Retrieval Lab: Harbin Institute of Technology, http://www.ir-lab.org/english/
22. Gene Ontology Project (October 2006), http://www.geneontology.org/

23. TAMBIS Ontology (October 2006)
    http://www.ontologos.org/OML/..%5COntology%5CTAMBIS.htm
24. Protégé project (October 2006)
    http://Protege.stanford.edu/
25. iAnnotate Plugin (October 2006),
    http://www.dbmi.columbia.edu/~cop7001/iAnnotateTab/iannotate.htm

A Novel Biology-Based Reversible Data Hiding Fusion Scheme

Tianding Chen

Institute of Communications and Information Technology,
Zhejiang Gongshang University, Hangzhou 310035
chentianding@163.com

Abstract. Recently, hiding secret data in DNA becomes an important and interesting research topic. Some researchers hid data in non-transcribed DNA, non-translated RNA regions, or active coding segments. Unfortunately, these schemes either alter the functionalities or modify the original DNA sequences. As a result, how to embed the secret data into the DNA sequence without altering the functionalities and to have the original DNA sequence be able to be retrieved is worthy of investigating. This paper apply two reversible information hiding schemes on DNA sequence by using the difference expansion technique and lossless compression. The reversible property makes the secret data hidden in anywhere in DNA without altering the functionalities because the original DNA sequence can be recovered in our schemes.

1 Introduction

Nowadays, biology techniques become more and more popular, and they are applied to many kinds of applications, authentication protocols [1], biochemistry, cryptography [2][3][4][5], and so on. As we know, DNA is two twisted strands composed of four bases, adenine (A), cytosine (C), thymine (T) and guanine (G). Every DNA strand, RNA, seems both random and meaningful. Recently, hiding secret data in DNA becomes an important and interesting research topic because of this property. In [6], a simple substitution scheme is used, where three consecutive bases are treated as a character. For example, 'B' = CCA, 'E' = GGC, and so on. As a result, there are at most 64 characters can be encoded in [6]. The frequencies of characters 'E' and 'I' appeared in the text are quite high. Consequently, the simple substitution scheme is dangerous because it cannot defend against the attack by analyzing the frequency of each set consisting of three DNA bases. In [7], two methods are proposed. The first method is a simple technique to hide data in non-transcribed DNA or non-translated RNA regions, and it can be treated as a complicated version of the scheme proposed in [6]. The second method is to hide data in active coding segments without influencing the result amino acid sequence by using arithmetic coding. Because there are various specific restrictions to which codons can be used in live DNA, the second method will degrade, where the length of an mRNA codon is 3-base. In [4], a stegographic approach is shown.

F.P. Preparata and Q. Fang (Eds.): FAW 2007, LNCS 4613, pp. 84–95, 2007.

Unfortunately, these schemes either alter the functionalities or modify the original DNA sequences. As a result, how to embed the secret data into the DNA sequence without altering the functionalities and to have the original DNA sequence be able to be retrieved is worthy of investigating. Nowadays, plenty of data hiding schemes are proposed. To have the hidden data unnoticeable, the stego-medium must be meaningful. Different from the products of the data hiding schemes and the traditional encryption ones, DNA is not only random but also significant. Thus some schemes based on DNA were been presented. In this paper, two data hiding schemes are proposed. In the schemes, secret messages are hidden in a DNA sequence so that the hidden data will not be detected. Moreover, the host DNA sequence can be reconstructed after the reverse operation, which far differs from the previous schemes also based on DNA. This property not only ensures the security of the secret data but also preserves the functionality of the original DNA.

2 Preliminaries

The biological background needs to be established to have the article easier to understand. In the following, how a DNA sequence eventually leads to a protein is demonstrated. The basics are first given followed by transcription and translation [7].

DNA is two twisted strands composed of four bases, adenine (A), cytosine (C), thymine (T) and guanine (G). The four bases represent the genetic code. A bonds with the complementary T, G bonds with the complementary C, and vice versa. This is very useful for error detection. Thus one strand and the corresponding complementary strand constitute DNA. For example, one strand is AACGTC, and the other must be TTGCAG. The DNA sequence determines the arrangement of amino acids which form a protein. Proteins are responsible for almost everything that goes on in the cells. In a word, DNA determines (1) what, (2) when, and (3) how much of everything which the cellular machinery produces.

Transcription is the process to create RNA, an intermediary copy of the instructions contained in DNA. RNA is a single strand and contains nucleotide (N), uracil (U), where thymine (T) would appear in DNA. For clarity, the four bases in RNA are adenine (A), cytosine (C), uracil (U) and guanine (G). Note that RNA is actually interpreted by the cellular machinery. In other words, DNA copies are created in abundance and translated into proteins while DNA is retained safely in the nucleus of the cell. A promoter is an RNA polymerase binding to a thermodynamically favorable region of the DNA, and it can be treated as the start signal for transcription. Moreover, it is one of the methods to control how much of a protein should be produced. Thus, many deficiency diseases and cancers are linked to the problems of the promoter sites.

The RNA copy (transcript) is referred to mRNA (message RNA) by discarding the intervening sequences of RNA. Note that the intervening sequences, having structural functions and not innocuous, do not code for protein. mRNA leaves the nucleus and binds a ribosome which facilitates the translation of the mRNA sequence into protein. Two ribosomal subunits, forming a complex, are bounded by mRNA. The ribosome steps linearly along the mRNA strand. There are twenty distinct amino acids, Phe, Leu, Ile, Val, Ser, Pro, Thr, Ala, Tyr, His Gln, Asn, Lys, Asp, Glu, Cys, Trp, Arg, and

Gly, which can be chained together during protein synthesis. On an mRNA, a codon, three nucleotides, indicates which amino acid will be attached next. The codon binds a group of three nucleotides on an anticodon, a tRNA molecule. There are about forty distinct tRNA molecules, and each one of them has a binding site for one of the amino acids. tRNA can be treated as a medium to translate nucleic acid code into protein. Appropriate tRNA binds to codon on the mRNA. Translation is completed while the ribosome encounters a STOP codon, and the protein is released.

In 2003, Shimanovsky et al. [8] exploited the codon redundancy to hide data in mRNA sequence. Generally, an mRNA codon is composed of three nucleotides. The possible nucleotides are U, C, A, and G. Hence, there are 4^3 combinations to form an mRNA codon. However, there are only twenty distinct amino acids, encoded from the mRNA codon. This clearly shows that some codons might be mapped to the same amino acids. For example, the codons 'UUA', 'CUU', 'CUA', 'UUG', 'CUC', and 'CUG' are mapped to the same amino acid Leu. Shimanovsky et al. exploited this redundancy to embed information in the mRNA codon.

In their scheme, if the codon should be encoded with 'UUA' but the secret message is four, they use the codon 'UUG' to replace the original one. It is because 'UUG' is the fourth codon of the set of codons whose mapping amino acid is Leu. Although the replacement will not influence the transcription results, they modify the nucleotides of the original DNA sequence, which might potentially cause unknown effects. Therefore, we need a reversible hiding mechanism that can not only conceal information into the DNA sequence but also completely restore the original sequence.

3 The Reversible Data Hiding Schemes

The proposed schemes adopt the lossless compression approach and the difference expansion technique to hide the secret message in a DNA sequence, respectively. Different from an image composed of pixels, a DNA sequence is composed of four nucleotides, A, C, G, and T. Hence, we need to transform the representation format of the nucleotides such that the hiding techniques can be used to conceal the secret message in a DNA sequence.

First, each nucleotide symbol of the DNA sequence is converted into a binary string. A convenient strategy is to encode each nucleotide with two bits in alphabetical order. For example, the nucleotide A is encoded with '00', C is encoded with '01', G is encoded with '10', and T is encoded with '11'.

Next, several bits of the binary formatted DNA sequence are combined to form a bit string, and then the bit string is converted to a decimal integer. Each integer in the decimal formatted DNA sequence is called a word. Let $|w|$ be the length of a bit string to form a word. Let us take a DNA sequence 'AGTTCAGT' as an example. The binary format of the sequence is '0010111101001011'. Assume that $|w| = 4$, the first four bits '0010' are converted to the decimal integer 2 because $(0010)_2=(2)_{10}$. Hence, the decimal format of the DNA sequence 'AGTTCAGT' is '2 15 4 11'. After that, the decimal formatted DNA sequence can be used to conceal the secret message.

3.1 The Type-I Reversible Data Hiding Scheme

The first proposed scheme is a lossless compression-based information hiding scheme. First, the scheme compresses the decimal formatted DNA sequence by using the lossless compression method. Next, the secret message is appended to the end of the compression result to form a bit stream. Then, the scheme adds a 16-bit header before the bit stream, where the header records the size of the compression result.

After the above procedures, the proposed scheme applies least-significant-bit (LSB) substitution to hiding the bit stream into the decimal formatted DNA sequence. Finally, the scheme converts the hidden result from the decimal format to nucleotides. The diagram of the hiding scheme is shown in Fig. 1.

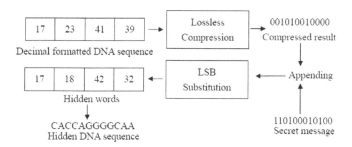

Fig. 1. Diagram of the lossless compression-based information hiding scheme

Nowadays, many lossless compression techniques have been proposed such as Huffman coding, Lemple-Ziv coding, run-length coding, arithmetic coding, and so on [7][9][10]. The proposed scheme adopts the arithmetic coding technique to compress the decimal formatted DNA sequence. It is because the arithmetic coding technique possesses the great compression performance and needs not to maintain complex data structure in the compression process. In addition, the technique is suitable for bit streams compression.

In the embedding phase, the secret bits of the bit stream are embedded in the rightmost k bits of each word, which is represented with w bits. In order to increase the security of the hidden message, the scheme assigns different k's to different words. The value of k is determined by using a pseudo-random number generator with a secret key. Only the legal users who own the secret key can extract the secret message from the hidden DNA sequence.

The total hiding payload of the lossless compression-based scheme is

$$P_{LC} = \frac{2 \times N}{|w|} \times k - \|DNA\|$$ (1)

where N is the number of nucleotides in a DNA sequence, k is the number of bits concealed in a word, and $\|DNA\|$ is the length of the compressed bit stream of the DNA sequence.

In the decoding process, the hidden DNA sequence is converted from nucleotides to a binary string. The scheme collects rightmost k bits from each word to form a bit stream. Because the first 16 bits are the header which records the size of the compression result, we divide the compression result and the secret message from the bit stream. Then, the compression result is used to restore the decimal formatted DNA sequence. Finally, the scheme converts the decimal formatted DNA sequence from the decimal format to nucleotides to obtain the original DNA sequence.

3.2 The Type-II Reversible Data Hiding Scheme

The second proposed scheme is a Type-II information hiding scheme which adopts the difference expansion technique to conceal a secret bit in two neighboring words. However, the difference expansion technique suffers from the overflow and underflow problems. The value of a word ranges between 0 and $2^{|w|} - 1$. The overflow or underflow problems occur when the difference between two neighboring words is great such that the hidden result is greater than the upper bound $2^{|w|} - 1$ (or smaller than the lower bound 0). For example, assume that $|w| = 8$, the value of a word is ranged from 0 to 255. Suppose that there are two neighboring words $A = 178$ and $B = 239$, and the secret bit $h = 1$. The difference between A and B is 61. The embedded results of A and B are $A' = 208 + \left\lfloor \dfrac{(2 \times 61 + 1) + 1}{2} \right\rfloor = 270$ and $B' = 208 - \left\lfloor \dfrac{(2 \times 61 + 1) + 1}{2} \right\rfloor = 147$, respectively. Obviously, A' is greater than the upper bound 255.

To prevent the underflow or overflow problems, the hidden results must satisfy the fowling constraints:

$$0 \le A' = \ell + \left\lfloor \frac{d' + 1}{2} \right\rfloor \le 2^{|w|} - 1 \text{ and}$$

$$0 \le B' = \ell - \left\lfloor \frac{d'}{2} \right\rfloor \le 2^{|w|} - 1 \tag{2}$$

Hence, if two neighboring words, namely a word pair, satisfy the first condition

$$|2 \times d + h| \le \min\{2 \times (2^{|w|} - 1 - \ell), 2 \times \ell + 1\} \tag{3}$$

The word pair can be used to conceal a secret bit. If the word pair satisfies the first condition, we call this pair expandable. According to the first condition, we can observe that the word pair $A = 178$ and $B = 239$ is not expandable since $|2 \times 61 + 1| \ge \min\{2 \times (255 - 208), 2 \times 208 + 1\}$.

If the word pair does not satisfy the first condition, we conceal the secret bit on the difference value of the pair without expansion. The new difference value is obtained by

$$d' = 2 \times \left\lfloor \frac{d}{2} \right\rfloor + h \tag{4}$$

Hence, the hiding results must satisfy the second condition

$$\left| 2 \times \left\lfloor \frac{d}{2} \right\rfloor + h \right| \leq \min\{2 \times (2^{|w|} - 1 - \ell), 2 \times \ell + 1\} \tag{5}$$

If the pair satisfies the second condition, we call the pair changeable. With the same example, the word pair $A = 178$ and $B = 239$ is changeable because $\left| 2 \times \left\lfloor \frac{61}{2} \right\rfloor + 1 \right| \leq \min\{2 \times (255 - 208), 2 \times 208 + 1\}$. The hidden results of A and B are

$$A' = 208 + \left\lfloor \frac{(2 \times \lfloor 61/2 \rfloor + 1) + 1}{2} \right\rfloor = 239 \quad \text{and} \quad B' = 208 - \left\lfloor \frac{(2 \times \lfloor 61/2 \rfloor + 1) + 1}{2} \right\rfloor = 178,$$

respectively.

The scheme uses a location map to indicate whether the pair is expandable or not. If the pair is expandable, the indicator of the pair is 1; otherwise, the indicator of the pair is 0. The location map is used to restore the original words. Hence, the location map must be kept for restoring the DNA sequence. The scheme losslessly compresses the location map by using arithmetic coding. In addition, the scheme collects the original LSBs of the changeable pairs. The collected LSBs are used to reconstruct the original words of the changeable pairs in the decoding process.

The scheme then concatenates the compressed bit stream of the location map, the collected LSBs, and the secret message to form a bit stream. The pre-process procedure is summarized as shown in Algorithm 1. Following that, the scheme performs the hiding procedure shown in Algorithm 2 on each word pair to obtain the hidden words. Finally, the scheme converts the hidden words from the decimal format to nucleotides.

The total hiding payload of the difference expansion-based scheme is

$$P_{DE} = \frac{N}{|w|} - \|LSB + Map\| \tag{6}$$

where $\|LSB + Map\|$ is the length of the compressed bit stream of the LSBs in the changeable pairs concatenated with the location map.

Algorithm 1. The pre-process procedure
Input: the decimal formatted DNA sequence and the secret message
Output: the bit stream
1) Partition the words of the sequence into word pairs.
2) Classify the set of the word pairs into three subsets, S_1, S_2, and S_3, where the pairs in S_1 are expandable, the pairs in S_2 are changeable, and the non-changeable pairs are in S_3.
3) Create a location map to indicate whether the word pair is expandable or not.
4) Compress the location map by using arithmetic coding.
5) Collect the original LSBs of the word pairs in S_2 except the pairs whose difference value is equal to zero or -1.
6) Append the collected LSBs and the secret message to the end of the compression result to form a bit stream.
7) Return the bit stream.

Algorithm 2. The hiding procedure
Input: the word pair, A and B , and the secret bit b of the bit stream
Output: the hidden results A' and B'
1) Compute the difference d between A and B by $d = A-B$.
2) Calculate the least integer value ℓ of the average of A and B by

$$\ell = \left\lfloor \frac{A+B}{2} \right\rfloor.$$

3) If the word pair belongs to S_1 then
$d' = 2 \times d + h$
else

$$d' = 2 \times \left\lfloor \frac{d}{2} \right\rfloor + h.$$

4) Obtain the hidden results A' and B' by

$$A' = \ell + \left\lfloor \frac{d'+1}{2} \right\rfloor \text{ and } B' = \ell - \left\lfloor \frac{d'}{2} \right\rfloor, \text{ respectively.}$$

5) Return A' and B'.

In the decoding process, the scheme extracts the hidden secret message and recovers the original DNA sequence from the hidden DNA sequence. First of all, the scheme converts the hidden DNA sequence from nucleotides to a binary string, and composes the words from the binary string.

Next, the scheme partitions the words into word pairs and checks whether the pairs are changeable or not. For the changeable pair, the scheme collects the LSBs of the difference to form a bit stream. Then, we can identify the compressed bit stream of the location map, the original LSBs of the changeable pairs, and the secret message from the produced bit stream.

Finally, the scheme decodes the location map from the compressed bit stream and uses the map to recover the original DNA sequence. The recovering procedure for each word pair is stated in Algorithm 3. For the changeable word pair with the difference value d' such that $-2 \le d' \le 1$, the scheme uses its original LSBs to restore the original values.

Algorithm 3. The recovering procedure
Input: the word pair, A' and B'
Output: the original words A and B
1) Compute the difference d' between A' and B' by $d' = A' - B'$.
2) Calculate the least integer value ℓ of the average of A' and B' by

$$\ell = \left\lfloor \frac{A'+B'}{2} \right\rfloor$$

3) If the word pair is changeable then
If the indicator of the pair in the location map is 1 then

$$d = \left\lfloor \frac{d'}{2} \right\rfloor \quad //\text{for expandable word pair}$$

else if ($0 \leq d' \leq 10$) then

$d = 1$

else if ($-2 \leq d' \leq -1$) then

$d = -2$

else $d = 2 \times \left\lfloor \dfrac{d'}{2} \right\rfloor + h$.

4) If the word pair is expandable or is changeable with the difference value d', which satisfies $-2 \leq d' \leq 1$, then

$$A = \ell + \left\lfloor \frac{d+1}{2} \right\rfloor \text{ and } B = \ell - \left\lfloor \frac{d}{2} \right\rfloor$$

else $A = \ell + \left\lfloor \dfrac{d+1}{2} \right\rfloor$ +the original LSB and $B = \ell - \left\lfloor \dfrac{d}{2} \right\rfloor$ + the original LSB.

5) Return A and B.

4   Experiments and Results

Experiments were carried out to evaluate the performance of the proposed schemes. The proposed schemes were tested on a Pentium IV 2.4 GHz personal computer with 512 RAM. As shown in Table 8-2, eight DNA sequences, AC153526, AC166252, AC167221, AC168874, AC168897, AC168901, AC168907, and AC168908, were used as the test DNA sequences [11]. In Table 1, 'Mus musculus' is the scientific name of house mice, and 'Bos taurus' is the scientific name of cows. The system uses the Visual C++ function random() to generate pseudo-random numbers and a secret message.

Table 1. Twelve tested DNA sequences

Locus	Number of nucleotides	Definition
AC153526	257,731	Mus musculus clone RP23-383C2, WORKING DRAFT SEQUENCE, 11 unordered pieces
AC166252	169,736	Mus musculus clone RP23-100G10, WORKING DRAFT SEQUENCE, 9 unordered pieces
AC167221	207,591	Mus musculus clone RP23-3P24, WORKING DRAFT SEQUENCE
AC168874	30,868	Bos taurus clone CH240-209N9, 11 unordered pieces
AC168897	50,145	Bos taurus clone CH240-190B15, 18 unordered pieces
AC168901	21,286	Bos taurus clone CH240-185I1, 7 unordered pieces
AC168907	61,902	Bos taurus clone CH240-195I7, 17 unordered pieces
AC168908	61,438	Bos taurus clone CH240-195K23, 18 unordered pieces

First, the proposed lossless compression-based scheme was tested on the sequence AC166252 with different $|w|$'s to determine the proper length of a bit string to form a word. In this experiment, we only hide a secret bit in the rightmost bit of each word. The experimental results are given in Table 2. The symbol $|w|$ in the table is the length of a bit string to form a word. The column 'Capacity' is the total number of bits embedded in the sequence, 'Compressed results' is the length of the compressed bit stream of the DNA sequence, and 'Payload' is the length of the secret message which we conceal in the sequence. Obviously, the lossless compression-based scheme has the best performance for AC166252 with $|w| = 2$, where the scheme is able to hide 135,936 bits.

Table 2. Experimental results of the lossless compression-based scheme on AC166252 with different $|w|$'s

lwl	Number of words	Capacity	Compressed results	Payload
2	168, 936	168, 936	33,000	135,936
4	84,468	84,468	18,160	66,308
6	56,312	56,312	12,816	43,496
8	42,234	42,234	10,016	32,218
10	33,787	33,787	8,016	25,771
12	28,156	28,156	6,880	21,276

Next, we performed the lossless compression-based scheme on the tested sequences with $|w| = 2$ to examine the hiding performance. The experimental results are shown in Table 3. The column 'bpn' is the abbreviation of bit-per-nucleotide that is the measurement to estimate the hiding ability of each nucleotide. The bpn is computed by $bpn = \dfrac{Payload}{Number\_of\_words}$.

We can observe that the lossless compression-based scheme has the best performance for AC153526, where the scheme is able to hide 208,753 bits (0.41 bpn). As to AC168901, the scheme is able to hide 15,372 bits (0.37 bpn).

Table 3. Experimental results of the lossless compression-based scheme on the tested sequences with $|w| = 2$

Locus	Number of words	Capacity	Compressed results	Payload	bpn
AC153526	256,729	256,729	47,976	208,753	0.81
AC166252	168,936	168,936	33,000	135,936	0.80
AC167221	168,936	168,936	33,000	135,936	0.80
AC168874	29,153	29,153	7,120	22,033	0.76
AC168897	48,434	48,434	11,168	37,266	0.77
AC168901	20,580	20,580	5,208	15,372	0.75
AC168907	60,302	60,302	13,536	46,766	0.78
AC168908	59,736	59,736	13,440	46,296	0.78

The difference expansion-based hiding scheme was performed on the sequence AC166252 with different $|w|$'s to determine the proper $|w|$. From Table 4, it is obvious that the scheme has the better result on the sequence while $|w| = 2$.

Table 4. Experimental results of the difference expansion-based scheme on AC166252 with different $|w|$'s

|w|	Number of words	Capacity	LSBs	Compressed results	Payload
2	168,936	40,810	10,184	18,176	12,450
4	84,468	20,037	5,424	10,024	4,589
6	56,312	12,849	3,936	6,704	2,209
8	42,234	9,990	2,904	5,272	1,814
10	33,787	7,561	2,440	4,176	945
12	28,156	6,668	2,048	3,688	932

After that, the difference expansion-based scheme is performed on the tested sequences with $|w|$ =2 to examine the hiding performance. Table 5 shows that the difference expansion-based scheme has the best performance on AC166252, where the scheme is able to hide 22,634 bits (0.13 bpn). As to AC168901, the scheme can hide 1,819 bits (0.09 bpn).

In addition, we tested the performance of the difference expansion-based scheme on two human chromosomes, Hs1_34565 Homo sapiens chromosome 1 (chromosome 1) and Hs5_23304 Homo sapiens chromosome 5 (chromosome 5) [12]. The size of chromosome 1 is 226,051,537, and that of chromosome 5 is 177,846,507. The capacity of chromosome 1 is 104,399,592, and that of chromosome 5 is 80,466,351.

Table 5. Experimental results of the difference expansion-based scheme on the tested sequences with $|w|$ =2

Locus	Number of words	Capacity	Compressed results	Payload	bpn
AC153526	256,729	57,128	26,080	31,048	0.12
AC166252	168,936	40,810	18,176	22,634	0.13
AC167221	168,936	40,838	18,272	22,566	0.13
AC168874	29,153	6,550	3,728	2,822	0.10
AC168897	48,434	10,821	5,968	4,853	0.10
AC168901	20,580	4,515	2,696	1,819	0.09
AC168907	60,302	14,443	7,376	7,067	0.12
AC168908	59,736	12,931	7,120	5,811	0.10

The proposed scheme not only can conceal secret data in the gray-level images, but also can widely hide secret data in the color images. In this subsection, we compare the performance of the proposed scheme with that of Alattar's scheme applying spatial triplet algorithm and quad algorithm using Lena and Baboon color images [13]. The results are shown in Fig. 2 and Fig. 3. In Fig. 2, the proposed scheme outperforms Alattar's scheme with quad algorithm at low PSNR, and Alattar's scheme with triplet algorithm outperforms our scheme at low PSNR. However, the proposed scheme outperforms Alattar's scheme with triplet algorithm at high PSNR, and Alattar's scheme with quad algorithm outperforms our scheme at high PSNR. In Fig. 3, the proposed scheme outperforms Alattar's schemes with quad algorithm and triplet algorithm at low PSNR. For PSNR=30 dB, the payload capacity of the proposed scheme is higher than that of Alattar's scheme.

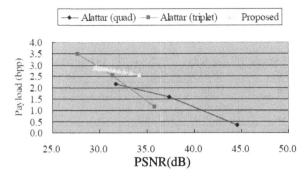

Fig. 2. Capacity-distortion performance comparisons between Alattar's scheme and the proposed scheme on Lena

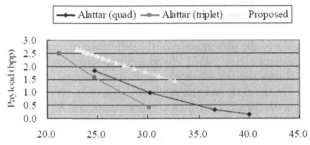

Fig. 3. Capacity-distortion performance comparisons between Alattar's scheme and the prop osed scheme on Baboon

5 Conclusion

This paper introduced two reversible information hiding schemes for DNA sequence based on lossless compression and difference expansion. The lossless compression-based scheme uses the compressed bit stream to achieve reversibility while the difference expansion–based scheme explores the redundancy in the word to perform reversibility. Both of the schemes are implemented and performed on different DNA sequences, including homo, house mice and cows. For $|w|=2$, the average payload of the lossless compression-based scheme is 0.78 (bpn) while that of the difference expansion-based scheme is 0.11(bpn). The proposed schemes can not only hide secret information in the DNA sequence but also recover the original DNA sequence from the hidden results without loss.

Reversible information hiding is a new technique that can be broadly applied in covert communication, digital rights management, and content authentication, especially, useful for sensitive military, legal, and medical data. Hence, we would like to develop some reversible information hiding techniques for various multimedia, such as video, audio, animation, text, graphic, and so on. A DNA sequence appears almost like a random sequence. However, each nucleotide in the sequence has its own meaning. Therefore, how to explore the features from the sequence, and stand on the characteristics of the features to design applicable data embedding scheme is another target of our future work.

Acknowledgements

The author gratefully acknowledge the support of this work by Youth Fund Team Project in Zhejiang Gongshang University.

References

1. Leier, A., Richter, C., Banzhaf, W., Rauhe, H.: Cryptography with DNA Binary Strands. BioSystems 57, 13–22 (2000)
2. Data Encryption Standard: NIST FIPS PUB 46-3 (October 25, 1999)
3. Hwang, R.J., Chang, C.C.: Hiding a Picture in Two Pictures. Optical Engineering. 40(3), 342–351 (2001)
4. Lin, M.H., Hu, Y.C., Chang, C.C.: Both Color and Gray Scale Secret Images Hiding in a Color Image. International Journal on Pattern Recognition and Artificial Intelligence 16(6), 697–713 (2002)
5. Rivest, R.L., Shamir, A., Adelman, L.: A Method for Obtaining Digital Signature and Public Key Cryptosystem. Communications of the ACM 21(2), 120–126 (1978)
6. Peterson, I.: Hiding in DNA, Muse, 22 (2001)
7. Shimanovsky, B., Feng, J., Potkonjak, M.: Hiding Data in DNA. In: Petitcolas, F.A.P. (ed.) IH 2002. LNCS, vol. 2578, pp. 373–386. Springer, London, UK (2003)
8. The, DNA sequences are downloaded from the NCBI database: The webpage is http://www.ncbi.nlm.nih.gov/
9. Chan, C.S., Chang, C.C.: An Image Hiding Scheme Based on Multi-bit-reference Substitution Table Using Dynamic Programming Strategy. Fundamenta Infor maticae 65(4), 291–305 (2005)
10. Wu, H.C., Chang, C.C.: Hiding Digital Watermarks Using Fractal Compression Technique. Fundament Informaticae 58(2), 189–202 (2003)
11. The chromosomes are downloaded from NCBI FTP: The webpage of the FTP are ftp://ftp.ncbi.nih.gov/genomes/H_sapiens/CHR_01~05/
12. Tsai, P.Y., Hu, Y.C., Chang, C.C.: A Color Image Watermarking Scheme Based on Color Quantization. Signal Processing 84(4), 95–106 (2004)
13. Alatar, A.M.: Reversible Watermark Using the Difference Expansion of a Generalized Integer Transform. IEEE Transactions on Image Processing 13(8), 1147–1156 (2004)

On the Approximation and Smoothed Complexity of Leontief Market Equilibria

Li-Sha Huang[1,*] and Shang-Hua Teng[2,**]

[1] State Key Laboratory of Intelligent Technology and Systems, Department of
Computer Science and Technology, Tsinghua University, Beijing, China
[2] Computer Science Department, Boston University, Boston, Massachusetts, USA
Also affiliated with Akamai Technologies Inc., Cambridge, Massachusetts, USA

Abstract. In this paper, we resolve two open questions on the computation and approximation of an Arrow-Debreu equilibrium in a Leontief exchange economy:

- We prove that the Leontief market exchange problem does not have a fully polynomial-time approximation scheme, unless **PPAD** ⊆ **P**.
- We show that the smoothed complexity of any algorithm for computing a market equilibrium in a Leontief economy, is not polynomial, unless **PPAD** ⊆ **RP**.

1 Introduction

The computation of a market equilibrium is a fundamental problem in modern economics [13] as the equilibria might provide useful information for the prediction of market trends, in the decision for future investments, and in the development of economic policies. In this paper, we consider the complexity of a classic model, the Leontief exchange economy. We resolve two open questions concerning the computation and approximation of market equilibria in this model.

A *Leontief economy* with m divisible goods (or commodities) and n traders is specified by two $m \times n$ matrices [1] $\mathbf{E} = (e_{i,j})$ and $\mathbf{D} = (d_{i,j})$: $\mathbf{E}$ is the *endowment matrix* of the traders where $e_{i,j}$ specifies the amount of commodity i that trader j initially has. We assume, without loss of generality, that there is exactly one unit

---

* Supported by National Key Foundation R&D Plan of China (2004CB318108, 2004CB318110, 2003CB317007) and NSF of China (60321002).
** Supported by the NSF grants CCR-0311430 and ITR CCR-0325630.

[1] We use bold lower-case Roman letters such as $\mathbf{x}$, $\mathbf{a}$, $\mathbf{b}_j$ to denote vectors. Whenever a vector, say $\mathbf{a} \in \mathbb{R}^n$ is present, its components is denoted by $a_1, \ldots, a_n$. Matrices are denoted by bold upper-case Roman letters such as $\mathbf{A}$. The $(i,j)^{th}$ entry of a matrix $\mathbf{A}$ is denoted by $a_{i,j}$. We use $\mathbf{a}_i$ to denote the ith column of $\mathbf{A}$. We also use the following notations: (1) $\mathbb{R}_+^m$: the set of m-dimensional vectors with non-negative real entries. (2) $\mathbb{R}_{[a:b]}^{m \times n}$: the set of all $m \times n$ matrices with real entries between a and b. (3) $\mathbb{P}^n$: the set of all vectors $\mathbf{x}$ in n dimensions such that $\sum_{i=1}^{n} x_i = 1$ and $x_i \geq 0$ for all $1 \leq i \leq n$. (4) $\langle \mathbf{a} | \mathbf{b} \rangle$: the dot-product of two vectors in the same dimension. (5) $\|\mathbf{x}\|_p$: the p-norm of vector $\mathbf{x}$, that is, $\left(\sum |x_i^p| \right)^{1/p}$ and $\|\mathbf{x}\|_\infty = \max_i |x_i|$.

F.P. Preparata and Q. Fang (Eds.): FAW 2007, LNCS 4613, pp. 96–107, 2007.
© Springer-Verlag Berlin Heidelberg 2007

of each type of good. With this assumption, the sum of every row of $\mathbf{E}$ is equal to 1. $\mathbf{D}$ is the *utility matrix*. It defines n utility functions $u_1, \ldots, u_n$, one for each trader. The exchange of goods can also be expressed by a non-negative $m \times n$ matrix $\mathbf{X} = (x_{i,j})$. Let $\mathbf{x}_j$ be the jth column of $\mathbf{X}$. The utility of trader j in $\mathbf{X}$ is $u_j(\mathbf{x}_j) = \min_i \{(x_{i,j})/(d_{i,j})\}$.

The *individual objective* of each trader is to maximize his or her utility. However, the utilities that these traders can achieve depend on the initial endowments, the individual utilities, and potentially, the (complex) process that they perform their exchanges.

In Walras' pure view of economics [16], the individual objectives of traders and their initial endowments enable the market to establish a price vector $\mathbf{p}$ of the goods in the market. Then the whole exchange can be conceptually characterized as: the traders sell their endowments – to obtain money or budgets – and individually optimize their objectives by buying the bundles of goods that maximize their utilities. By selling the initial endowment $\mathbf{e}_j$, trader j obtains a budget of $\langle \mathbf{e}_j | \mathbf{p} \rangle$ amount.

The *optimal bundle* for u_j is a solution to the following mathematical program:

$$\max u_j(\mathbf{x}_j) \quad \text{subject to } \langle \mathbf{x}_j | \mathbf{p} \rangle \leq \langle \mathbf{e}_j | \mathbf{p} \rangle , \mathbf{x}_j \geq \mathbf{0} \quad (A1)$$

A solution to (A1) is referred to as an *optimal demand* of trader j under prices $\mathbf{p}$. The price vector $\mathbf{p}$ is an *Arrow-Debreu equilibrium price* or simply an *equilibrium* of the Leontief economy $(\mathbf{E}, \mathbf{D})$ if there exists optimal solution $\mathbf{X} = (\mathbf{x}_1, \ldots, \mathbf{x}_n)$ to (A1) such that (A2) $\mathbf{x}_i = \mathbf{0}$ if $\langle \mathbf{e}_i | \mathbf{p} \rangle = 0$ and (A3) $\sum_j \mathbf{x}_j \leq \mathbf{1}$, where $\mathbf{1}$ is the m-dimensional column vector with all ones. This last constraint states that the traders' optimal demands can be met by the market. In other words, an equilibrium price vector essentially allows each trader to make individual decision without considering others' utilities nor how they achieve their objectives.

A central complexity question in the Leontief market exchange problem is: *Is the problem of finding an equilibrium of a Leontief economy in $\mathbf{P}$?* So far no polynomial-time algorithm has been found for this problem. The combination of two recent results greatly dashed the hope for a positive answer to this question. Codenotti, Saberi, Varadarajan, and Ye [4] gave a polynomial-time reduction from two-person games to a special case of the Leontief economy. In a remarkable breakthrough, Chen and Deng [2] proved that the problem of finding a Nash equilibrium of a two-person game is $\mathbf{PPAD}$-complete[2].

1.1   Our Result

In practice, one may be willing to relax the condition of equilibria and considers the computation of approximate market equilibria. Deng, Papadimitriou,

---

[2] We refer the readers who are not familiar with the complexity class $\mathbf{PPAD}$ to the paper by Papadimitriou [12] and also to the subsequent papers on the $\mathbf{PPAD}$-completeness of normal games [5,1,6,2,3]. This class includes several important search problems such as discrete fixed-point problems and the problem of finding a Nash equilibrium.

Safra [7] proposed a notion of an approximate market equilibrium. A price $\mathbf{p}$ is an ϵ-*approximate equilibrium* of a Leontief economy $(\mathbf{E}, \mathbf{D})$ if there exists an ϵ-approximately optimal demand $\mathbf{X} = (\mathbf{x}_1, \ldots, \mathbf{x}_n)$ such that $\sum_j \mathbf{x}_j \le (1 + \epsilon)\mathbf{1}$.

Question 1 (Fully Polynomial Approximate Leontief?). Can an ϵ-approximate equilibrium of a Leontief economy with m goods and n traders be computed in time polynomial in m, n, and $1/\epsilon$?

In this paper, we provide an answer to question 1: We show that there is no fully polynomial-time approximation scheme for Leontief economies unless $\mathbf{PPAD} \subseteq \mathbf{P}$. The main ingredient of our result is a numerical stability analysis of the reduction introduced by Condenotti *et al* [4]. We prove that the problem of finding an ϵ-approximate Nash equilibrium of an $n \times n$ two-person games can be reduced to the computation of an $(\epsilon/n)^2$-approximate equilibrium of a Leontief economy with $2n$ goods and $2n$ traders. This polynomial relationship between approximate Nash equilibria of bimatrix games and approximate Arrow-Debreu equilibria of Leontief economies is significant because it enables us to apply the result of Chen, Deng, and Teng [3] to show that finding an approximate market equilibrium is as hard as finding an exact market equilibrium, and is as hard as finding a Nash equilibrium of a two-player game or a discrete Brouwer fixed point in the most general settings.

We also consider the smoothed complexity of the Leontief market exchange problem. In the smoothed model introduced by Spielman and Teng [14], an algorithm receives and solves a perturbed instance. The smoothed complexity of an algorithm is the maximum over its inputs of the expected running time of the algorithm under slight random perturbations of that input. The smoothed complexity is then measured as a function of both the input length and the magnitude σ of the perturbations. An algorithm has *smoothed polynomial-time complexity* if its smoothed measure is polynomial in n, the problem size, and in $1/\sigma$ [14,15].

In the smoothed analysis of Leontief economies, we consider an arbitrary input $(\bar{\mathbf{E}}, \bar{\mathbf{D}})$ with $0 \le \bar{d}_{i,j} \le 2$ and $0 \le \bar{e}_{i,j} \le 1$. Suppose $\mathbf{E} = (e_{i,j})$ and $\mathbf{D} = (d_{i,j})$ are perturbations of $\bar{\mathbf{E}}$ and $\bar{\mathbf{D}}$.

The *smoothed complexity* of the Leontief exchange problem $(\bar{\mathbf{E}}, \bar{\mathbf{D}})$ is measured by the expected complexity of finding an equilibrium of the Leontief economy $(\mathbf{E}, \mathbf{D})$.

Question 2 (Smoothed Polynomial Leontief?). Can an equilibrium of a Leontief economy be computed in smoothed time polynomial in m, n, and $1/\sigma$? Can an ϵ-equilibrium of a Leontief economy be computed in smoothed time polynomial in m, n, $1/\epsilon$ and $1/\sigma$?

By refining our analysis of the reduction from the two-person games to Leontief economies, we show the statement is unlikely true. We prove that the problem of finding an (approximate) equilibrium of a Leontief economy is not in smoothed polynomial time, unless $\mathbf{PPAD} \subseteq \mathbf{RP}$.

2   Approximate Nash Equilibria

The *two-person game* or the *bimatrix game* is a non-cooperative game between two players [11,9], the *row player* and the *column player*. If the row player has m pure strategies and the column player has n pure strategies, then their payoffs are given by a pair of $m \times n$ matrices $(\mathbf{A}, \mathbf{B})$.

A mixed row strategy is a vector $\mathbf{x} \in \mathbb{P}^m$ and a mixed column strategy is a vector $\mathbf{y} \in \mathbb{P}^n$. The expected payoffs to these two players are respectively $\mathbf{x}^\top \mathbf{A} \mathbf{y}$ and $\mathbf{x}^\top \mathbf{B} \mathbf{y}$. A *Nash equilibrium* is then a pair of vectors $(\mathbf{x}^* \in \mathbb{P}^m, \mathbf{y}^* \in \mathbb{P}^n)$ such that for all pairs of vectors $\mathbf{x} \in \mathbb{P}^m$ and $\mathbf{y} \in \mathbb{P}^n$,

$$(\mathbf{x}^*)^\top \mathbf{A} \mathbf{y}^* \geq \mathbf{x}^\top \mathbf{A} \mathbf{y}^* \quad \text{and} \quad (\mathbf{x}^*)^\top \mathbf{B} \mathbf{y}^* \geq (\mathbf{x}^*)^\top \mathbf{B} \mathbf{y}.$$

Every two-person game has at least one Nash equilibrium [11]. But in a recent breakthrough, Chen and Deng [2] proved that the problem of computing a Nash equilibrium of a two-person game is **PPAD**-complete.

One can relax the condition of Nash equilibria and considers approximate Nash equilibria. There are two notions of approximation. An ϵ-*approximate Nash equilibrium* of game $(\mathbf{A}, \mathbf{B})$ is a pair of mixed strategies $(\mathbf{x}^*, \mathbf{y}^*)$, such that for all $\mathbf{x}, \mathbf{y} \in \mathbb{P}^n$,

$$(\mathbf{x}^*)^\top \mathbf{A} \mathbf{y}^* \geq \mathbf{x}^\top \mathbf{A} \mathbf{y}^* - \epsilon \text{ and } (\mathbf{x}^*)^\top \mathbf{B} \mathbf{y}^* \geq (\mathbf{x}^*)^\top \mathbf{B} \mathbf{y} - \epsilon.$$

An ϵ-*relatively-approximate Nash equilibrium* of game $(\mathbf{A}, \mathbf{B})$ is a pair of mixed strategies $(\mathbf{x}^*, \mathbf{y}^*)$, such that for all $\mathbf{x}, \mathbf{y} \in \mathbb{P}^n$,

$$(\mathbf{x}^*)^\top \mathbf{A} \mathbf{y}^* \geq (1 - \epsilon) \mathbf{x}^\top \mathbf{A} \mathbf{y}^* \text{ and } (\mathbf{x}^*)^\top \mathbf{B} \mathbf{y}^* \geq (1 - \epsilon)(\mathbf{x}^*)^\top \mathbf{B} \mathbf{y}.$$

Note that the Nash equilibria and the relatively-approximate Nash equilibria of a two-person game $(\mathbf{A}, \mathbf{B})$ are invariant under positive scalings, i.e., the bimatrix game $(c_1 \mathbf{A}, c_2 \mathbf{B})$ has the same set of Nash equilibria and relatively-approximate Nash equilibria as the bimatrix game $(\mathbf{A}, \mathbf{B})$, as long as $c_1, c_2 > 0$. However, each ϵ-approximate Nash equilibrium $(\mathbf{x}, \mathbf{y})$ of $(\mathbf{A}, \mathbf{B})$ becomes a $c \cdot \epsilon$-approximate Nash equilibrium of the bimatrix game $(c\mathbf{A}, c\mathbf{B})$ for $c > 0$.

Thus, we often normalize the matrices $\mathbf{A}$ and $\mathbf{B}$ so that all their entries are between 0 and 1, or between -1 and 1, in order to study the complexity of approximate Nash equilibria [10,3]. Recently, Chen, Deng, and Teng [3] proved the following result.

Theorem 1 (Chen-Deng-Teng). *For any constant $c > 0$, the problem of computing a $1/n^c$-approximate Nash equilibrium of a normalized $n \times n$ two-person game is **PPAD**-complete. It remains **PPAD**-complete to compute a $1/n^c$-relatively-approximate Nash equilibrium of an $n \times n$ bimatrix game.*

3   Market Equilibria: Approximation and Smoothed Complexity

In this and the next sections, we analyze a reduction π that transforms a two-person game $(\bar{\mathbf{A}}, \bar{\mathbf{B}})$ into a Leontief economy $(\bar{\mathbf{E}}, \bar{\mathbf{D}}) = \pi(\bar{\mathbf{A}}, \bar{\mathbf{B}})$ such that from each $(\epsilon/n)^2$-approximate Walrasian equilibrium of $(\bar{\mathbf{E}}, \bar{\mathbf{D}})$ we can construct an ϵ-relatively-approximate Nash equilibrium of $(\bar{\mathbf{A}}, \bar{\mathbf{B}})$.

We will also consider the smoothed complexity of Leontief economies. To establish a hardness result for computing an (approximate) market equilibrium in the smoothed model, we will examine the relationship of Walrasian equilibria of a perturbed instance $(\mathbf{E}, \mathbf{D})$ of $(\bar{\mathbf{E}}, \bar{\mathbf{D}})$ and approximate Nash equilibria of $(\bar{\mathbf{A}}, \bar{\mathbf{B}})$. In particular, we show that if the magnitude of the perturbation is σ, then we can construct an $(\epsilon + n^{1.5}\sqrt{\sigma})$-relatively-approximate Nash equilibrium of $(\bar{\mathbf{A}}, \bar{\mathbf{B}})$ from each $(\epsilon/n)^2$-approximate equilibria of $(\mathbf{E}, \mathbf{D})$.

3.1 Approximate Market Equilibria of Leontief Economy

We first introduce the explicit definition of market equilibria and approximate market equilibria in a Leontief economy.

Let $\mathbf{D}$ be the demand matrix and $\mathbf{E}$ be the endowment matrix of a Leontief economy with m goods and n traders. Given a price vector $\mathbf{p}$, trader j can obtain a budget of $\langle \mathbf{e}_j | \mathbf{p} \rangle$ by selling the endowment. By a simple variational argument, one can show that the optimal demand $\mathbf{x}_j$ with budget $\langle \mathbf{e}_j | \mathbf{p} \rangle$ satisfies $x_{i,j}/d_{i,j} = x_{i',j}/d_{i',j}$ for all i and i' with $d_{i,j} > 0$ and $d_{i',j} > 0$. Thus, under the price vector $\mathbf{p}$, the maximum utility that trader j can achieve is 0 if $\langle \mathbf{e}_j | \mathbf{p} \rangle = 0$, and $\langle \mathbf{e}_j | \mathbf{p} \rangle / \langle \mathbf{d}_j | \mathbf{p} \rangle$ otherwise. Moreover, in the latter case, $x_{i,j} = d_{i,j} \left(\langle \mathbf{e}_j | \mathbf{p} \rangle / \langle \mathbf{d}_j | \mathbf{p} \rangle \right)$. Let $\mathbf{u} = (u_1, \ldots, u_n)^\top$ denote the vector of utilities of the traders. Then $\mathbf{p}$ is an equilibrium price if

$$\mathbf{p} \geq \mathbf{0}, \quad u_i = \langle \mathbf{e}_i | \mathbf{p} \rangle / \langle \mathbf{d}_i | \mathbf{p} \rangle, \quad \text{and} \quad \mathbf{Du} \leq \mathbf{1}. \qquad (1)$$

In the remainder of this paper, we will refer to a pair of vectors $(\mathbf{u}, \mathbf{p})$ that satisfies Equation (1) as an *equilibrium* of the Leontief economy $(\mathbf{E}, \mathbf{D})$. Then, an ϵ-*approximate equilibrium* of the Leontief economy $(\mathbf{E}, \mathbf{D})$ is a pair of utility and price vectors $(\mathbf{u}, \mathbf{p})$ satisfying:

$$\begin{cases} u_i \geq (1 - \epsilon) \langle \mathbf{e}_i | \mathbf{p} \rangle / \langle \mathbf{d}_i | \mathbf{p} \rangle, \forall i & \text{(C1)} \\ u_i \leq (1 + \epsilon) \langle \mathbf{e}_i | \mathbf{p} \rangle / \langle \mathbf{d}_i | \mathbf{p} \rangle, \forall i. & \text{(C2)} \\ \mathbf{Du} \leq (1 + \epsilon) \cdot \mathbf{1}. & \text{(C3)} \end{cases}$$

(C1) states that all traders are approximately satisfied. (C2) states that budget constraints approximately hold. (C3) states that the demands approximately meet the supply.

If $(\mathbf{u}, \mathbf{p})$ is an equilibrium of $(\mathbf{E}, \mathbf{D})$, so is $(\mathbf{u}, \alpha\mathbf{p})$ for every $\alpha > 0$. Similarly, if $(\mathbf{u}, \mathbf{p})$ is an ϵ-equilibrium of $(\mathbf{E}, \mathbf{D})$, so is $(\mathbf{u}, \alpha\mathbf{p})$ for every $\alpha > 0$. So we can assume $\|\mathbf{p}\|_1 = 1$. For approximate equilibria, we assume without loss of generality that $\mathbf{u}$ and $\mathbf{p}$ are strictly positive to avoid division-by-zero — a small perturbation of an approximate equilibrium is still a good approximate equilibrium.

3.2 Reduction from NASH to LEONTIEF

Let $(\bar{\mathbf{A}}, \bar{\mathbf{B}})$ be a two-person game in which each player has n strategies. Below we assume $\bar{\mathbf{A}} \in \mathbb{R}^{n \times n}_{[1,2]}$ and $\bar{\mathbf{B}} \in \mathbb{R}^{n \times n}_{[1,2]}$. We use the reduction introduced by

Codenotti, Saberi, Varadarajan, and Ye [4] to map a bimatrix game to a Leontief economy. This reduction constructs a Leontief economy with $(\bar{\mathbf{E}}, \bar{\mathbf{D}}) = \pi(\bar{\mathbf{A}}, \bar{\mathbf{B}})$ where the endowment matrix is simply $\bar{\mathbf{E}} = \mathbf{I}_{2n}$, the $(2n) \times (2n)$ identity matrix and the utility matrix is given by

$$\bar{\mathbf{D}} = \begin{pmatrix} 0 & \bar{\mathbf{A}} \\ \bar{\mathbf{B}} & 0 \end{pmatrix}.$$

$(\bar{\mathbf{E}}, \bar{\mathbf{D}})$ is a special form of Leontief exchange economies [17,4]. It has $2n$ goods and $2n$ traders. The jth trader comes to the market with one unit of the jth-good. In addition, the traders are divided into two groups $\mathcal{M} = \{1, 2, ..., n\}$ and $\mathcal{N} = \{n+1, ..., 2n\}$. Traders in $\mathcal{M}$ only interests in the goods associated with traders in $\mathcal{N}$ and vice versa.

Codenotti *et al* [4] prove that there is a one-to-one correspondence between Nash equilibria of the two person game $(\mathbf{A}, \mathbf{B})$ and market equilibria of Leontief economy $(\bar{\mathbf{E}}, \bar{\mathbf{D}})$. It thus follows from the theorem of Nash [11], that the Leontief economy $(\bar{\mathbf{E}}, \bar{\mathbf{D}})$ has at least one equilibrium.

We will prove the following extension of their result in the next section.

Lemma 1 (Approximation of Games and Markets). *For any bimatrix game $(\bar{\mathbf{A}}, \bar{\mathbf{B}})$, let $(\bar{\mathbf{E}}, \bar{\mathbf{D}}) = \pi(\bar{\mathbf{A}}, \bar{\mathbf{B}})$. Let $(\mathbf{u}, \mathbf{w})$ be an ϵ-approximate equilibrium of $(\bar{\mathbf{E}}, \bar{\mathbf{D}})$ and assume $\mathbf{u} = (\mathbf{x}^\top, \mathbf{y}^\top)^\top$ and $\mathbf{w} = (\mathbf{p}^\top, \mathbf{q}^\top)^\top$. Then, $(\mathbf{x}, \mathbf{y})$ is an $O(n\sqrt{\epsilon})$-relatively-approximate Nash equilibrium for $(\bar{\mathbf{A}}, \bar{\mathbf{B}})$.*

Apply Lemma 1 with $\epsilon = n^{-h}$ for a sufficiently large constant h and Theorem1, we can prove one of our main results of this paper.

Theorem 2 (Market Approximation is Likely Hard). *The problem of finding a $1/n^{\Theta(1)}$-approximate equilibrium of a Leontief economy with n goods and n traders is **PPAD**-hard. Thus, if **PPAD** is not in **P**, then there is no algorithm for finding an ϵ-equilibrium of Leontief economies in time $\text{poly}(n, 1/\epsilon)$.*

3.3 Smoothed Market

In the smoothed analysis of the Leontief market exchange problem, we assume that entries of the endowment and utility matrices is subject to slight random perturbations.

Consider an economy with $\left(\bar{\mathbf{E}} \in \mathbb{R}^{n \times n}_{[0,2]}, \bar{\mathbf{D}} \in \mathbb{R}^{n \times n}_{[0,1]} \right)$. For a $\sigma > 0$, a perturbed economy is defined by a pair of random matrices $(\mathbf{\Delta}^E, \mathbf{\Delta}^D)$ where $\delta^E_{i,j}$ and $\delta^D_{i,j}$ are independent random variables of magnitude σ. Two common models are the uniform perturbation and Gaussian perturbation. In the *uniform perturbation* with magnitude σ, a random variable is chosen uniformly from the interval $[-\sigma, \sigma]$. In the *Gaussian perturbation* with variance σ^2, a random variable δ is chosen with density $e^{-\delta^2/2\sigma^2}/(\sqrt{2\pi}\sigma)$.

Let $\mathbf{D} = \max(\bar{\mathbf{D}} + \mathbf{\Delta}^D, \mathbf{0})$ and let $\mathbf{E} = \max(\bar{\mathbf{E}} + \mathbf{\Delta}^E, \mathbf{0})$. Although we can re-normalize $\mathbf{E}$ so that the sum of each row is equal to 1, we choose not to do so in favor of a simpler presentation. The perturbed game is then given by $(\mathbf{E}, \mathbf{D})$.

Following Spielman and Teng [14], the smoothed complexity of an algorithm W for the Leontief economy is defined as following: Let $T_W(\mathbf{E}, \mathbf{D})$ be the complexity of algorithm W for solving a market economy defined by $(\mathbf{E}, \mathbf{D})$. Then, the smoothed complexity of algorithm W under perturbations $N_\sigma()$ of magnitude σ is

$$\text{Smoothed}_W[n, \sigma] = \max_{\bar{\mathbf{D}} \in \mathbb{R}_{[0,2]}^{n \times n}, \bar{\mathbf{E}} \in \mathbb{R}_{[0,1]}^{n \times n}} \mathbf{E}_{\mathbf{E} \leftarrow N_\sigma(\bar{\mathbf{E}}), \mathbf{D} \leftarrow N_\sigma(\bar{\mathbf{D}})} [T_W(\mathbf{E}, \mathbf{D})],$$

where we use $\mathbf{E} \leftarrow N_\sigma(\bar{\mathbf{E}})$ to denote that $\mathbf{E}$ is a perturbation of $\bar{\mathbf{E}}$ according to $N_\sigma(\bar{\mathbf{E}})$.

An algorithm W for computing Walrasian equilibria has *polynomial smoothed time complexity* if for all $0 < \sigma < 1$ and for all positive integer n, there exist positive constants c, k_1 and k_2 such that

$$\text{Smoothed}_W[n, \sigma] \le c \cdot n^{k_1} \sigma^{-k_2}.$$

The Leontief exchange economy is in *smoothed polynomial time* if there exists an algorithm W with polynomial smoothed time-complexity for computing a Walrasian equilibrium.

To relate the complexity of finding an approximate Nash equilibrium of two-person games with the smoothed complexity of Leontief economies, we examine the equilibria of perturbations of the reduction presented in the last subsection. In the remainder of this subsection, we will focus on the smoothed complexity under uniform perturbations with magnitude σ. One can similarly extend the results to Gaussian perturbation with standard deviation σ.

Let $(\bar{\mathbf{A}}, \bar{\mathbf{B}})$ be a two-person game in which each player has n strategies. Let $(\bar{\mathbf{E}}, \bar{\mathbf{D}}) = \pi(\bar{\mathbf{A}}, \bar{\mathbf{B}})$. Let $(\boldsymbol{\Delta}^E, \boldsymbol{\Delta}^D)$ be a pair of perturbation matrices with entries drawn uniformly at random from $[-\sigma, \sigma]$. The perturbed game is then given by $\mathbf{E} = \max(\bar{\mathbf{E}} + \boldsymbol{\Delta}^E, \mathbf{0})$ and $\mathbf{D} = \max(\bar{\mathbf{D}} + \boldsymbol{\Delta}^D, \mathbf{0})$.

Let $\Pi_\sigma(\bar{\mathbf{A}}, \bar{\mathbf{B}})$ be the set of all $(\mathbf{E}, \mathbf{D})$ that can be obtained by perturbing $\pi(\bar{\mathbf{A}}, \bar{\mathbf{B}})$ with magnitude σ. Note that the off-diagonal entries of $\mathbf{E}$ are between 0 and σ, while the diagonal entries are between $1 - \sigma$ and $1 + \sigma$. In the next section, we will prove the following lemma.

Lemma 2 (Approximation of Games and Perturbed Markets). *Let $(\bar{\mathbf{A}}, \bar{\mathbf{B}})$ be a bimatrix game with $\bar{\mathbf{A}}, \bar{\mathbf{B}} \in \mathbb{R}_{[1,2]}^{n \times n}$. For any $0 < \sigma < 1/(8n)$, let $(\mathbf{E}, \mathbf{D}) \in \Pi_\sigma(\bar{\mathbf{A}}, \bar{\mathbf{B}})$. Let $(\mathbf{u}, \mathbf{w})$ be an ϵ-approximate equilibrium of $(\mathbf{E}, \mathbf{D})$ and assume $\mathbf{u} = (\mathbf{x}^\top, \mathbf{y}^\top)^\top$ and $\mathbf{w} = (\mathbf{p}^\top, \mathbf{q}^\top)^\top$. Then, $(\mathbf{x}, \mathbf{y})$ is an $O(n\sqrt{\epsilon} + n^{1.5}\sqrt{\sigma})$-relatively-approximate Nash equilibrium for $(\bar{\mathbf{A}}, \bar{\mathbf{B}})$.*

We now follow the scheme outlined in [15] and used in [3] to use perturbations as a probabilistic polynomial reduction from the approximation problem of two-person games to market equilibrium problem over perturbed Leontief economies.

Lemma 3 (Smoothed Leontief and Approximate Nash). *If the problem of computing an equilibrium of a Leontief economy is in smoothed polynomial time under uniform perturbations, then for any $0 < \epsilon' < 1$, there exists a randomized* *algorithm for computing an ϵ'-approximate Nash equilibrium in expected time polynomial in n and $1/\epsilon'$.*

Proof. Suppose W is an algorithm with polynomial smoothed complexity for computing a equilibrium of a Leontief economy. Let $T_W(\mathbf{E}, \mathbf{D})$ be the complexity of algorithm W for solving the market problem defined by $(\mathbf{E}, \mathbf{D})$. Let $N_\sigma()$ denotes the uniform perturbation with magnitude σ. Then there exists constants c, k_1 and k_2 such that for all $0 < \sigma < 1$,

$$\max_{\bar{\mathbf{E}} \in \mathbb{R}^{n \times n}_{[0,1]}, \bar{\mathbf{D}} \in \mathbb{R}^{n \times n}_{[0,2]},} \mathbf{E}_{\mathbf{E} \leftarrow N_\sigma(\bar{\mathbf{E}}), \mathbf{D} \leftarrow N_\sigma(\bar{\mathbf{D}})} [T_W(\mathbf{E}, \mathbf{D})] \le c \cdot n^{k_1} \sigma^{-k_2}.$$

Consider a bimatrix game $(\bar{\mathbf{A}}, \bar{\mathbf{B}})$ with $\bar{\mathbf{A}}, \bar{\mathbf{B}} \in \mathbb{R}^{n \times n}_{[1,2]}$. For each $(\mathbf{E}, \mathbf{D}) \in \Pi_\sigma(\bar{\mathbf{A}}, \bar{\mathbf{B}})$, by Lemma 2, by setting $\epsilon = 0$ and $\sigma = O(\epsilon'/n^3)$, we can obtain an ϵ'-approximate Nash equilibrium of $(\bar{\mathbf{A}}, \bar{\mathbf{B}})$ in polynomial time from an equilibrium of $(\mathbf{E}, \mathbf{D})$. Now given the algorithm W with polynomial smoothed time-complexity, we can apply the following randomized algorithm to find an ϵ-approximate Nash equilibrium of game $(\bar{\mathbf{A}}, \bar{\mathbf{B}})$:

`ApproximateNashFromSmoothedLeontief`$(\bar{\mathbf{A}}, \bar{\mathbf{B}})$

1. Let $(\bar{\mathbf{E}}, \bar{\mathbf{D}}) = \pi(\bar{\mathbf{A}}, \bar{\mathbf{B}})$.
2. Randomly choose two perturbation matrices $(\mathbf{\Delta}^E, \mathbf{\Delta}^D)$ of magnitude σ.
3. Let $\mathbf{D} = \max(\bar{\mathbf{D}} + \mathbf{\Delta}^D, 0)$ and let $\mathbf{E} = \max(\bar{\mathbf{E}} + \mathbf{\Delta}^E, 0)$.
4. Apply algorithm W to find an equilibrium $(\mathbf{u}, \mathbf{w})$ of $(\mathbf{E}, \mathbf{D})$.
5. Apply Lemma 2 to get an approximate equilibrium $(\mathbf{x}, \mathbf{y})$ of $(\bar{\mathbf{A}}, \bar{\mathbf{B}})$.

The expected time complexity of `ApproximateNashFromSmoothedLeontief` is bounded from above by the smoothed complexity of W when the magnitude perturbations is ϵ'/n^3 and hence is at most $c \cdot n^{k_1 + 3k_2} (\epsilon')^{-k_2}$. ∎

We can use this randomized reduction to prove the second main result of this paper.

Theorem 3 (Hardness of Smoothed Leontief Economies). *Unless* **PPAD** $\subseteq$ **RP**, *the problem of computing an equilibrium of a Leontief economy is not in smoothed polynomial time, under uniform or Gaussian perturbations.*

4   The Approximation Analysis

In this section, we prove Lemma 2. Let us first recall all the matrices that will be involved: We start with two matrices $(\bar{\mathbf{A}}, \bar{\mathbf{B}})$ of the bimatrix game. We then obtain the two matrices $(\bar{\mathbf{E}}, \bar{\mathbf{D}}) = \pi(\bar{\mathbf{A}}, \bar{\mathbf{B}})$ of the associated Leontief economy, and then perturb $(\bar{\mathbf{E}}, \bar{\mathbf{D}})$ to obtain $(\mathbf{E}, \mathbf{D})$. Note that $\bar{\mathbf{E}} = \mathbf{I}_{2n}$ and we can write $\bar{\mathbf{D}}$ and $\mathbf{D}$ as:

$$\bar{\mathbf{D}} = \begin{pmatrix} 0 & \bar{\mathbf{A}} \\ \bar{\mathbf{B}} & 0 \end{pmatrix} \quad \text{and} \quad \mathbf{D} = \begin{pmatrix} \mathbf{Z} & \mathbf{A} \\ \mathbf{B} & \mathbf{N} \end{pmatrix}$$

where for all $\forall i, j$, $z_{i,j}, n_{i,j} \in [0, \sigma]$ and $a_{i,j} - \bar{a}_{ij}, b_{i,j} - \bar{b}_{ij} \in [-\sigma, \sigma]$, Note also because $\bar{\mathbf{A}}, \bar{\mathbf{B}} \in \mathbb{R}^{n \times n}_{[1,2]}$ and $0 < \sigma < 1$, $\mathbf{A}$ and $\mathbf{B}$ are uniform perturbations with magnitude σ of $\bar{\mathbf{A}}$ and $\bar{\mathbf{B}}$, respectively. Moreover, $z_{i,j}$ and $n_{i,j}$ are 0 with probability 1/2 and otherwise, they are uniformly chosen from $[0, \sigma]$.

Let $(\mathbf{u}, \mathbf{w})$ be an ϵ-approximate equilibrium of $(\mathbf{E}, \mathbf{D})$. Assume $\mathbf{u} = (\mathbf{x}^\top, \mathbf{y}^\top)^\top$ and $\mathbf{w} = (\mathbf{p}^\top, \mathbf{q}^\top)^\top$, where all vectors are column vectors. By the definition of ϵ-approximate equilibrium, we have:

$$\begin{cases} (1 - \epsilon)\mathbf{E}^\top \mathbf{w} \le \text{diag}(\mathbf{u})\mathbf{D}^\top \mathbf{w} \le (1 + \epsilon)\mathbf{E}^\top \mathbf{w}, \\ \mathbf{Zx} + \mathbf{Ay} \le (1 + \epsilon) \cdot \mathbf{1} \\ \mathbf{Bx} + \mathbf{Ny} \le (1 + \epsilon) \cdot \mathbf{1} \end{cases} \tag{2}$$

where $\text{diag}(\mathbf{u})$ is the diagonal matrix whose diagonal is $\mathbf{u}$. Since the demand functions are homogeneous with respect to the price vector $\mathbf{w}$, we assume without loss of generality that $\|\mathbf{w}\|_1 = \|\mathbf{p}\|_1 + \|\mathbf{q}\|_1 = 1$.

We will prove $(\mathbf{x}, \mathbf{y})$ is an $O\left(n\sqrt{\epsilon} + n^{1.5}\sqrt{\sigma}\right)$-relatively-approximate Nash equilibrium of the two-person game $(\bar{\mathbf{A}}, \bar{\mathbf{B}})$. To this end, we first list the following three properties of the approximate equilibrium $(\mathbf{u}, \mathbf{w})$.

Property 1 (Approximate Price Symmetry). If $\|\mathbf{w}\|_1 = 1$, $0 < \epsilon < 1/2$, and $0 < \sigma < 1/(2n)$, then

$$\frac{1 - \epsilon - 4n\sigma}{2 - 4n\sigma} \le \|\mathbf{p}\|_1, \|\mathbf{q}\|_1 \le \frac{1 + \epsilon}{2 - 4n\sigma}.$$

Proof. Recall $\mathbf{u} = (\mathbf{x}^\top, \mathbf{y}^\top)^\top$ and $\mathbf{w} = (\mathbf{p}^\top, \mathbf{q}^\top)^\top$. By (2) and the fact that the diagonal entries of $\mathbf{E}$ are at least $1 - \sigma$, we have

$$(1 - \epsilon)(1 - \sigma)\|\mathbf{p}\|_1 \le \mathbf{1}^\top\left(\text{diag}(\mathbf{x})\,\mathbf{Z}^\top \mathbf{p} + \text{diag}(\mathbf{x})\,\mathbf{B}^\top \mathbf{q}\right) = (\mathbf{Zx})^\top \mathbf{p} + (\mathbf{Bx})^\top \mathbf{q}$$
$$\le (1 + \epsilon)\|\mathbf{q}\|_1 + \langle \mathbf{Zx}|\mathbf{p}\rangle \le 3n\sigma\|\mathbf{p}\|_1 + (1 + \epsilon)\|\mathbf{q}\|_1,$$

Applying, $\|\mathbf{q}\|_1 = \|\mathbf{w}\|_1 - \|\mathbf{p}\|_1 = 1 - \|\mathbf{p}\|_1$ to the inequality, we have

$$\|\mathbf{p}\|_1 \le (1 + \epsilon)/\left[(1 - \epsilon)(1 - \sigma) + (1 + \epsilon) - 3n\sigma\right] \le (1 + \epsilon)/(2 - 4n\sigma)$$

Thus, $\|\mathbf{q}\|_1 = 1 - \|\mathbf{p}\|_1 \ge (1 - \epsilon - 4n\sigma)/(2 - 4n\sigma)$. We can similarly prove the other direction. ∎

Property 2 (Approximate Utility Symmetry). If $\|\mathbf{w}\|_1 = 1$, $0 < \epsilon < 1/2$, and $0 < \sigma < 1/(8n)$, then

$$\frac{(1 - \epsilon)(1 - \sigma)(1 - \epsilon - 4n\sigma)}{(1 + \epsilon)(2 + 2\sigma)} \le \|\mathbf{x}\|_1, \|\mathbf{y}\|_1 \le \frac{(1 + \epsilon)^2 + n\sigma(1 + \epsilon)(2 - 4n\sigma)}{(1 - \sigma)(1 - \epsilon - 4n\sigma)}.$$

Proof. By our assumption on the payoff matrices of the two-person games, $1 \le \bar{a}_{ij}, \bar{b}_{ij} \le 2$, for all $1 \le i, j, \le n$. Thus, $1 - \sigma \le a_{ij}, b_{ij} \le 2 + \sigma$. By (2) and the fact the diagonal entries of $\mathbf{E}$ is at least $1 - \sigma$, we have

$$x_i \ge \frac{(1 - \epsilon)(1 - \sigma)p_i}{\langle \mathbf{b}_i|\mathbf{q}\rangle + \langle \mathbf{z}_i|\mathbf{p}\rangle} \ge \frac{(1 - \epsilon)(1 - \sigma)p_i}{(2 + \sigma)\|\mathbf{q}\|_1 + \sigma\|\mathbf{p}\|_1} \ge \frac{(1 - \epsilon)(1 - \sigma)(2 - 4n\sigma)p_i}{(2 + 2\sigma)(1 + \epsilon)},$$

where the last inequality follows from Property 1. Summing it up, we obtain,

$$\|\mathbf{x}\|_1 \geq \frac{(1-\epsilon)(1-\sigma)(2-4n\sigma)}{(2+2\sigma)(1+\epsilon)}\|\mathbf{p}\|_1 \geq \frac{(1-\epsilon)(1-\sigma)(1-\epsilon-4n\sigma)}{(1+\epsilon)(2+2\sigma)},$$

where again, we use Property 1 in the last inequality. On the other hand, from (2) we have

$$x_i \leq \frac{(1+\epsilon)\langle \mathbf{e}_i | \mathbf{w} \rangle}{\langle \mathbf{b}_i | \mathbf{q} \rangle + \langle \mathbf{z}_i | \mathbf{p} \rangle} \leq \frac{(1+\epsilon)(p_i + \sigma)}{\langle \mathbf{b}_i | \mathbf{q} \rangle} \tag{3}$$

$$\leq \frac{(1+\epsilon)(p_i+\sigma)}{(1-\sigma)\|\mathbf{q}\|_1} \leq \frac{(1+\epsilon)(p_i+\sigma)(2-4n\sigma)}{(1-\sigma)(1-\epsilon-4n\sigma)}. \tag{4}$$

Summing it up, we obtain,

$$\|\mathbf{x}\|_1 \leq \frac{(1+\epsilon)(2-4n\sigma)}{(1-\sigma)(1-\epsilon-4n\sigma)}(\|\mathbf{p}\|_1 + n\sigma) \leq \frac{(1+\epsilon)^2 + n\sigma(1+\epsilon)(2-4n\sigma)}{(1-\sigma)(1-\epsilon-4n\sigma)}.$$

We can similarly prove the bound for $\|\mathbf{y}\|_1$.   ∎

Property 3 (Utility Upper Bound). Let $\mathbf{s} = \mathbf{Zx} + \mathbf{Ay}$ and $\mathbf{t} = \mathbf{Bp} + \mathbf{Ny}$. Let $\lambda = \max\{\epsilon, n\sigma\}$. Under the assumption of Property 2, if $s_i \leq (1+\epsilon)(1-\sigma) - \sqrt{\lambda}$, (similarly $t_i \leq (1+\epsilon)(1-\sigma) - \sqrt{\lambda}$), then

$$x_i \leq \frac{(1+\epsilon)(2-4n\sigma)(5\sqrt{\lambda}+\sigma)}{(1-\sigma)(1-\epsilon-4n\sigma)}, \quad y_i \leq \frac{(1+\epsilon)(2-4n\sigma)(5\sqrt{\lambda}+\sigma)}{(1-\sigma)(1-\epsilon-4n\sigma)}.$$

Property 3 is a direct corollary of Property 2. We now use these three properties to prove Lemma 2.

Proof. [of Lemma 2] In order to prove that $(\mathbf{x}, \mathbf{y})$ is a δ-relatively approximate Nash equilibrium for $(\bar{\mathbf{A}}, \bar{\mathbf{B}})$, it is sufficient to establish:

$$\begin{cases} \mathbf{x}^\top \bar{\mathbf{A}} \mathbf{y} \geq (1-\delta) \max_{\|\tilde{\mathbf{x}}\|_1 = \|\mathbf{x}\|_1} \tilde{\mathbf{x}}^\top \bar{\mathbf{A}} \mathbf{y} \\ \mathbf{y}^\top \bar{\mathbf{B}}^\top \mathbf{x} \geq (1-\delta) \max_{\|\tilde{\mathbf{y}}\|_1 = \|\mathbf{y}\|_1} \tilde{\mathbf{y}}^\top \bar{\mathbf{B}}^\top \mathbf{x}. \end{cases} \tag{5}$$

Let $\mathbf{s} = \mathbf{Zx} + \mathbf{Ay}$. We observe,

$$\begin{aligned} \mathbf{x}^\top \bar{\mathbf{A}} \mathbf{y} &= \mathbf{x}^\top (\mathbf{Ay} + \mathbf{Zx} - \mathbf{Zx} + (\bar{\mathbf{A}} - \mathbf{A})\mathbf{y}) = \mathbf{x}^\top \mathbf{s} - \mathbf{x}^\top (\mathbf{Zx} + (\mathbf{A} - \bar{\mathbf{A}})\mathbf{y}) \\ &\geq \mathbf{x}^\top \mathbf{s} - \|\mathbf{x}\|_1 (\|\mathbf{Zx}\|_\infty + \|(\mathbf{A} - \bar{\mathbf{A}})\mathbf{y}\|_\infty) \geq \mathbf{x}^\top \mathbf{s} - \sigma \|\mathbf{x}\|_1 (\|\mathbf{x}\|_1 + \|\mathbf{y}\|_1) \\ &\geq \mathbf{x}^\top \mathbf{s} - \frac{2\sigma(1+\epsilon)^2 + 2n\sigma^2(1+\epsilon)(2-4n\sigma)}{(1-\sigma)(1-\epsilon-4n\sigma)} \|\mathbf{x}\|_1 \\ &= \mathbf{x}^\top \mathbf{s} - O(\sigma) \|\mathbf{x}\|_1, \end{aligned}$$

where the last inequality follows from Property 2. Let $\lambda = \max(\epsilon, n\sigma)$. By Property 3, we can estimate the lower bound of $\langle \mathbf{x} | \mathbf{s} \rangle$:

$$\langle \mathbf{x} | \mathbf{s} \rangle = \sum_{i=1}^{n} x_i s_i \geq \sum_{i:s_i > (1+\epsilon)(1-\sigma) - \sqrt{\lambda}} x_i s_i$$

$$\geq [(1+\epsilon)(1-\sigma) - \sqrt{\lambda}](\|\mathbf{x}\|_1 - n\frac{(1+\epsilon)(2-4n\sigma)(5\sqrt{\lambda}+\sigma)}{(1-\sigma)(1-\epsilon-4n\sigma)})$$

$$\geq \|\mathbf{x}\|_1 (1 - O(\sqrt{\lambda}))(1 - n\frac{(1+\epsilon)(2-4n\sigma)(5\sqrt{\lambda}+\sigma)}{(1-\sigma)(1-\epsilon-4n\sigma)\|\mathbf{x}\|_1})$$

$$= \|\mathbf{x}\|_1 [1 - O(n\sqrt{\lambda} + n\sigma)].$$

On the other hand, by (2), we have $\mathbf{Ay} \leq (1+\epsilon)\mathbf{1}$ and hence

$$\max_{\|\tilde{\mathbf{x}}\|_1 = \|\mathbf{x}\|_1} \tilde{\mathbf{x}}^\top \bar{\mathbf{A}} \mathbf{y} = \|\mathbf{x}\|_1 \|\bar{\mathbf{A}}\mathbf{y}\|_\infty \leq (1+\epsilon+\sigma\|\mathbf{y}\|_1)\|\mathbf{x}\|_1 \leq (1+\epsilon+O(\sigma))\|\mathbf{x}\|_1.$$

Therefore,

$$\mathbf{x}^\top \bar{\mathbf{A}} \mathbf{y} \geq \mathbf{x}^\top \mathbf{s} - O(\sigma)\|\mathbf{x}\|_1 \geq \|\mathbf{x}\|_1[(1 - O(n\sqrt{\lambda}+n\sigma)) - O(\sigma)]$$

$$\geq \frac{1}{1+\epsilon+O(\sigma)}[1 - O(n\sqrt{\lambda}+n\sigma)] \max_{\|\tilde{\mathbf{x}}\|_1 = \|\mathbf{x}\|_1} \tilde{\mathbf{x}}^\top \bar{\mathbf{A}} \mathbf{y}$$

$$= (1 - O(n\sqrt{\lambda}+n\sigma)) \max_{\|\tilde{\mathbf{x}}\|_1 = \|\mathbf{x}\|_1} \tilde{\mathbf{x}}^\top \bar{\mathbf{A}} \mathbf{y}.$$

We can similarly prove

$$\mathbf{y}^\top \bar{\mathbf{B}}^\top \mathbf{x} = (1 - O(n\sqrt{\lambda}+n\sigma)) \max_{\|\tilde{\mathbf{y}}\|_1 = \|\mathbf{y}\|_1} \tilde{\mathbf{y}}^\top \bar{\mathbf{B}}^\top \mathbf{x}.$$

We then use the inequalities $\sqrt{\lambda} = \sqrt{\max(\epsilon, n\sigma)} \leq \sqrt{\epsilon + n\sigma} \leq \sqrt{\epsilon} + \sqrt{n\sigma}$ and $\sigma \leq \sqrt{\sigma}$ to complete the proof. ■

5   Discussions

Our results as well as the combination of Codenotti, Saberi, Varadarajan, and Ye [4] and Chen and Deng [2] demonstrate that exchange economies with Leontief utility functions are fundamentally different from economies with linear utility functions. In Leontief economies, not only finding an exact equilibrium is likely hard, but finding an approximate equilibrium is just as hard.

Although, we prove that the computation of an $O(1/n^{\Theta(1)})$-approximate equilibrium of Leontief economies is **PPAD**-hard. Our hardness result does not cover the case when ϵ is a constant between 0 and 1. The following are two optimistic conjectures.

Conjecture 1 (PTAS Approximate LEONTIEF). There is an algorithm to find an ϵ-approximate equilibrium of a Leontief economy in time $O(n^{k+\epsilon^{-c}})$ for some positive constants c and k.

Conjecture 2 (Smoothed LEONTIEF: Constant Perturbations). There is an algorithm to find an equilibrium of a Leontief economy with smoothed time complexity $O(n^{k+\sigma^{-c}})$ under perturbations with magnitude σ, for some positive constants c and k.

References

1. Chen, X., Deng, X.: 3-Nash is PPAD-complete. ECCC, TR05-134 (2005)
2. Chen, X., Deng, X.: Settling the Complexity of 2-Player Nash-Equilibrium. In: The Proceedings of FOCS 2006, pp. 261–272 (2006)
3. Chen, X., Deng, X., Teng, S.-H.: Computing Nash Equilibria: Approximation and smoothed complexity. In: The Proceedings of FOCS 2006, pp. 603–612 (2006)
4. Codenotti, B., Saberi, A., Varadarajan, K., Ye, Y.: Leontief economies encode nonzero sum two-player games. In: The Proceedings of SODA 2006, pp. 659–667 (2006)
5. Daskalakis, C., Goldberg, P.W., Papadimitriou, C.H.: The Complexity of Computing a Nash Equilibrium. In: The Proceedings of STOC 2006, pp. 71–78 (2006)
6. Daskalakis, C., Papadimitriou, C.H.: Three-player games are hard. ECCC, TR05-139 (2005)
7. Deng, X., Papadimitriou, C., Safra, S.: On the complexity of price equilibria. Journal of Computer and System Sciences 67(2), 311–324 (2003)
8. Deng, X., Huang, L.-S.: Approximate Economic Equilibrium Algorithms. In: Gonzalez, T. (ed.) Approximation Algorithms and Metaheuristics (2005)
9. Lemke, C.E., Howson Jr., J.T.: Equilibrium points of bimatrix games. J. Soc. Indust. Appl. Math. 12, 413–423 (1964)
10. Lipton, R.J., Markakis, E., Mehta, A.: Playing large games using simple strategies. In: The Proceedings of EC 2003, pp. 36–41 (2003)
11. Nash, J.: Noncooperative games. Annals of Mathematics 54, 289–295 (1951)
12. Papadimitriou, C.H.: On the complexity of the parity argument and other inefficient proofs of existence. Journal of Computer and System Sciences 48(3), 498–532 (1994)
13. Scarf, H.: The Computation of Economic Equilibria. Yale University Press, New Haven, CT (1973)
14. Spielman, D.A., Teng, S.-H.: Smoothed analysis of algorithms: Why the simplex algorithm usually takes polynomial time. J. ACM 51(3), 385–463 (2004)
15. Spielman, D.A., Teng, S.-H.: Smoothed analysis of algorithms and heuristics: Progress and open questions. In: The Proceedings of Foundations of Computational Mathematics, pp. 274–342 (2006)
16. Walras, L.: Elements of Pure Economics, or the Theory of Social Wealth (1874)
17. Ye, Y.: On exchange market equilibria with Leontief's utility: Freedom of pricing leads to rationality. In: Deng, X., Ye, Y. (eds.) WINE 2005. LNCS, vol. 3828, pp. 14–23. Springer, Heidelberg (2005)

On Coordination Among Multiple Auctions*

Peter Chuntao Liu[1] and Aries Wei Sun[2]

[1] Software Institute, Nanjing University, China
`peterliu@software.nju.edu.cn`
[2] Department of Computer Science, City University of Hong Kong
`sunwei@cs.cityu.edu.hk`

Abstract. Our research is motivated by finding auction protocols to elegantly coordinate the sellers and buyers when there are multiple auctions. In our model, there are multiple sellers selling different items that are substitute to each other, multiple buyers each demanding exactly one item, and a market that is *monopolistic competitive*. We implement our auction coordination protocol by polynomial running time algorithms and establishes its incentive compatibleness.

1 Introduction

In incentive compatible auctions, a buyer's utility is maximized when he[1] reports his true valuation, without information of other buyers' strategies. Thus incentive compatible auctions can be regarded as coordination protocols in which a player (buyer) can find his best strategy without any prediction or expectation of other players. This kind of coordination protocol is elegant, and can usually guarantee an optimal allocation result that maximizes social efficiency.

However, it becomes less obvious how to elegantly coordinate the buyers when there are multiple auctions available, each managed by a distinct seller. It will be a difficult task for a buyer to decide how many and which auctions to attend, because he can not predict which auctions other buyers will attend, a piece of information now does matter. What if he attended an auction and win nothing, while he could have won something if he had attended some other auction. What if he attended multiple auctions and won multiple items, while he demands only one of them?

The problem of finding a protocol to elegantly coordinate multiple sellers and buyers has **motivated** our research presented in this paper.

1.1 The Model

We model the market as *monopolistic competitive* where there are m sellers and n buyers. Each seller (auctionner) a_j can supply at most s_j units of identical items. The items provided by different sellers are *substitute* to each other. However, *the items sold by different sellers is allowed to be different*. Each buyer $b^{(i)}$ demands

---

* This research is supported by SRG grant (7001989) of City University of Hong Kong.
[1] For convinience, we refer to buyers as males and sellers as females.

F.P. Preparata and Q. Fang (Eds.): FAW 2007, LNCS 4613, pp. 108–116, 2007.

exactly one item from any of the sellers. However, each buyer may have different valuations for the item from different sellers. Buyer $b^{(i)}$ *privately* values the item from seller a_j at $V_j^{(i)}$.

At first, each buyer $b^{(i)}$ *declares* $V_j^{(i)}$ to be $v_j^{(i)}$, and submits $(v_j^{(i)}, i = 1, 2, \ldots, n, j = 1, 2, \ldots, m)$ to the *market coordinator* as its bid.

After that, the market coordinator announces the list of winners, from which seller each winner have the item, and the price the winner should pay to that seller for the item. If buyer $b^{(i)}$ have won an item from seller a_j, then $w_j^{(i)} = 1$ and $W^{(i)} = j$, $b^{(i)}$ should pay a price $p^{(j)}$ to seller a_j, and he is called a **winner**. If buyer $b^{(i)}$ has not won any item from seller a_j, then $w_j^{(i)} = 0$. If buyer $b^{(i)}$ has not won any item from any seller, then $W^{(i)} = 0$ and it is called a **loser**. For convinience, we define $v_0^{(i)} = 0$, $i = 1, 2, \ldots, n$.

We name our auction coordination protocol as **Coordination Protocol For Auctions With Unit Demand**, or **CPA-U** for short.

1.2   Our Work

In this paper, we present a coordination protocol among multiple auctions. Our model has been formallized in §1.1.

The preliminary part of our paper is in §2. §2.1 gives some basic definitions. Since the mechanism design part of our protocol is based on the VCG mechanism, we briefly introduce the VCG mechanism in §2.2 to facilitate the incentive compatible analysis of our protocol. We briefly introduce the minimum cost flow problem in §2.3, whose solution will be used by our algorithm to produce an optimal allocation result.

The algorithmic solution presented in §3 is divided in to two steps. The first step is to determine the allocation, with the aim to maximize the social efficiency. The second step is to determine the price each buyer should pay, with the aim to make the protocol incentive compatible. All of our algorithm are in *polynomial* running time.

We prove the correctness of our approach in §4 by two steps. The first step in §4.1 proves that the allocation result *maximizes* social efficiency. The second step in §4.2 establishes the *incentive compatibleness* of our protocol.

§5 concludes the paper, discusses its application in $C2C^2$ e-commerce websites, and gives an open problem as our future works.

1.3   Related Works

We reduce the auction problem into an equivalent **minimum cost flow** problem in order to get an optimal allocation. The minimum cost flow problem is one of the most important problems within network optimization research. [3] was the first to solve the minimum cost flow problem in polynomial time. Their *Edmonds-Karp Scaling Technique*, which reduces the minimum cost flow problem to a sequence of shortest path problems, has been followed by many researchers. In

---

[2] Customer to customer.

this paper, we take the algorithm from [4] to solve the shortest path problem in $O(|E| + |V| \cdot log(|V|))$ time and apply the scaling technique from [9] to reduce the minimum cost flow problem to a sequence of $O((|E| + |V|) \cdot log(|V|))$ shortest path problems.

The mechanism design part of our protocol is based on the **Vickrey-Clark-Groves** (VCG [1,5]) mechanism, which generalizes the Vickrey[11] auction to the case of heterogeneous items, and is *"arguably the most important positive result in mechanism design"*[8].

A similar but different way to coordinate the market when there are multiple buyers and sellers is known as *double auction*. Some of the most representative works in this field are [7,10,12,6,2]. In double auctions, the market is *completely competitive*, i.e. the items provided by the sellers are all the same. Whereas in our model, the market is *monopolistic competitive*, the sellers sell different items that are substitute to each other, and the buyers may have different valuations for different items.

2 Preliminaries

This section is the preliminary part of the paper. §2.1 adapts some definitions from the mechanism design literature to our model for analytical convinience in later sections. §2.2 briefly introduce the VCG mechanism to facilitate the incentive compatible analysis of our protocol. §2.3 briefly introduce the minimum cost flow problem, whose solution will be used by our algorithm to produce an optimal allocation result.

2.1 Definitions

Definition 1 (Feasible). *An allocation is feasible, if and only if it satisfies both the demand and supply constraints:*

1. *Demand Constraint: Each buyer is allocated at most one item.*
2. *Supply Constraint: The number of items each seller a_j should provide according to the allocation result does not exceed its supply ability s_j.*

Definition 2 (Social Efficiency). *Given an allocation result, the **social efficiency** e is defined as: $e = \sum_{i=1}^{n} v_{W^{(i)}}^{(i)}$, where $v_j^{(i)}$ is buyer $b^{(i)}$'s reported valuation (or bid) on an item from seller a_j, $w_j^{(i)}$ is the number of items buyer $b^{(i)}$ has won from seller a_j.*

Definition 3 (Incentive Compatible, Truthful). *An auction is **incentive compatible**, or **truthful**, for buyers, if and only if sincere bidding, i.e. reporting $v_j^{(i)} = V_j^{(i)}$, is each buyer $b^{(i)}$'s dominant strategy.*

Definition 4 (Optimal). *We say an allocation result is **optimal** if it is feasible and it maximizes the social efficiency.*

2.2   The VCG Mechanism

Generalizing the Vickrey[11] auction to the case of heterogeneous items, the celebrated Vickrey-Clark-Groves (VCG [1,5]) mechanism is arguably the most important positive result in mechanism design. Our auction is partially based on the VCG mechanism. In this subsection, we briefly introduce the VCG mechanism to facilitate our analysis.

Definition 5. *We say that an auction is a* **VCG Mechanism** *[1,5] if*

1. *The allocation result maximizes the social efficiency e.*
2. *Each buyer $b^{(i)}$'s price $p^{(i)}$ is determined by the VCG payment formula:*
   $p^{(i)} = ewo_i - (e - e^{(i)})$.

where e and ewo_i are the maximum social efficiency with and without participation of $b^{(i)}$, respectively. $e^{(i)}$ is buyer $b^{(i)}$'s reported valuation of the allocation result.

Notice that in our model $e^{(i)} = v_{W^{(i)}}^{(i)}$.

Theorem 1 (VCG Incentive Compatibleness [1,5]). *Assume that the bidders have pure private values for an arbitrary set of items, then sincere bidding is a weakly dominant strategy for every bidder in the Vickrey-Clarke-Groves mechanism, yielding an efficient outcome.*

2.3   The Minimum Cost Flow Problem

The minimum cost flow problem is concerned with finding a network flow of a designated size on a given network while observing the capacity constraints. In the graph $G = (V, E)$ representing the flow network, there are a source vertex s, a termination vertex t, and some other vertices. Each edge has a maximum allowed capacity which can not be exceeded. Each edge also has a value representing the cost of each unit of flow. The cost of sending x units of flow through an edge where the cost value per unit is v equals to $x \times v$. The problem is to find a flow to send at least f units from source vertex to termination vertex such that the cost is minimized while observing the capacity constraint of every edge.

[3] was the first to solve the minimum cost flow problem in polynomial time. They have used a technique called "Edmonds-Karp Scaling" to reduce the minimum cost flow problem to a sequence of shortest path problems. The technique has been followed by many researchers.

In our algorithm presented in Algorithm 1, we take the algorithm from [4] to solve the shortest path problem in $O(|E| + |V| \cdot log(|V|))$ time and apply the scaling technique from [9] to reduce the minimum cost flow problem to a sequence of $O((|E| + |V|) \cdot log(|V|))$ shortest path problems.

Algorithm 1. Minimum Cost Flow

Input: f the designated flow size; $G = (V, E)$ representing the flow
network.
Output: A solution to the minimum cost flow problem

1 Apply the scaling technique from [9] to reduce the minimum cost flow
   problem to a sequence of shortest path problems;
2 Solve the shortest path problems by algorithm in [4];
3 Form a solution to the minimum cost flow problem;

Proposition 1. *The running time of Algorithm 1 is*

$$O((|E| + |V|) \cdot log(|V|) \cdot (|E| + |V| \cdot log(|V|)))$$

where V is the set of vertices and E is the set of edges.

3   Algorithm For *CPA-U* Protocol

In this section, we present three algorithms related to the *CPA-U* protocol. Algorithm 2 reduces the social efficiency optimization problem to the minimum cost flow problem introduced in §2.3. Algorithm 3 determines the allocation among buyers and sellers, and calculates the resulting social efficiency. Finally, Algorithm 4 solves the incentive compatible pricing problem and presents a complete solution to the auction coordination problem under our model introduced in §1.1. We later establish the optimality and incentive compatibleness of *CPA-U* in §4. The running times analysis of the three algorithms are quite straight forward. The detailed analysis are omitted to save space and are certainly available upon our readers' requests.

Algorithm 2. *CPA-U* To **Min-Cost-Flow** Reduction

Input: $s_j, v_j^{(i)}$; $j = 1, 2, \ldots, m$; $i = 1, 2, \ldots, n$.
Output: A graph $G = (V, E)$ and a desired flow f.
1 $f \leftarrow min(\sum_{j=1}^{m} s_j, n)$;
2 $V \leftarrow \{s, t\} \cup (\cup_{i=1}^{n} \{b^{(i)}\}) \cup (\cup_{j=1}^{m} \{a_j\})$;
3 $E \leftarrow (\cup_{i=1, j=1}^{n, m} \{(b^{(i)}, a_j)\}) \cup (\cup_{i=1}^{n} \{(s, b^{(i)})\}) \cup (\cup_{j=1}^{m} \{(a_j, t)\})$;
4 **for** *i=1* **to** *n* **and** *j=1* **to** *m* **do**
5 | Set $(b^{(i)}, a_j)$ with cost $-v_j^{(i)}$ and capacity 1;
6 | Set $(s, b^{(i)})$ with cost 0 and capacity 1;
7 | Set (a_j, t) with cost 0 and capacity s_j;
8 **end**

Proposition 2. *The running time of the Algorithm 2 is $O(m \times n)$, where m is the number of time sellers and n is the number of buyers.*

Algorithm 3. *CPA-U* Allocation

Input: s_i, $v_j^{(i)}$; $j = 1, 2, \ldots, m$; $i = 1, 2, \ldots, n$.
Output: Social Efficiency e; $W^{(i)}$, $j = 1, 2, \ldots, m$; .

1 Run Algorithm 2 to convert the input of an allocation problem to the input of an minimum cost flow problem;
2 Run Algorithm 1 to solve the minimum cost flow problem;
3 **for** i=1 **to** n **and** j=1 **to** m **do**
4 $\quad \mid \quad W^{(i)} \leftarrow j \times (b^{(i)}, a_j).flow$;
5 **end**
6 $e \leftarrow \sum_{i=1}^{n} v_{W^{(i)}}^{(i)}$;

Proposition 3. *The running time of Algorithm 3 is:*

$$O(m \cdot n \cdot log(m + n) \cdot (m \cdot n + (m + n) \cdot log(m + n)))$$

where m is the number of sellers and n is the number of buyers.

The pricing algorithm is based on the VCG mechanism. The intuition is that each buyer pays the difference between the sum of the individual efficiency of all other buyers when it is not present and when it is present, or the decrement in all other buyers' efficiency caused by its participation in the auction.

Algorithm 4. *CPA-U* Protocol

Input: $v_j^{(i)}$; $j = 1, 2, \ldots, m$; $i = 1, 2, \ldots, n$.
Output: Allocations $W^{(i)}$; Prices $p^{(i)}$; $j = 1, 2, \ldots, m$; $i = 1, 2, \ldots, n$.

1 $(e, (W^{(i)})_n) \leftarrow$ Run Algorithm 3;
2 **for** i=1 **to** n **do**
3 $\quad \mid \quad ewo_i \leftarrow$ Run Algorithm 3 to obtain the maximum social efficiency without the existence of buyer $b^{(i)}$;
4 $\quad \mid \quad p^{(i)} \leftarrow e - v_{W^{(i)}}^{(i)} - ewo_i$;
5 **end**

Proposition 4. *The running time of Algorithm 4 is:*

$$O(m \cdot n^2 \cdot log(m + n) \cdot (m \cdot n + (m + n) \cdot log(m + n)))$$

where m is the number of sellers and n is the number of buyers.

4   Economic Analysis

In this section, we establish the optimality and incentive compatibleness of our *CPA-U* protocol.

4.1 Optimality

Lemma 1. *There exists an social efficiency maximizing allocation in which the total number of sold units equals to $f = min(\sum_{i=1}^{m} s_j, n)$.*

Proof. We observe that in a feasible allocation, allocating one more unit of item to a buyer will never make the social efficiency decrease. Thus for any feasible allocation that maximizes social efficiency and the total sold item is less than f, sell one more unit will still result in an allocation that is both feasible and maximizes the social efficiency.

On the other hand, the maximum possible number of sold items is f. Thus all feasible allocation must sell no more than f units.

Thus the argument is done.   □

Lemma 2. *An allocation is feasible if and only if the corresponding network flow resulted from Algorithm 2 is feasible.*

Proof. By Definition 1, an allocation is feasible if and only if it satisfies both the demand and supply constraints, i.e.:

1. Demand Constraint: Each buyer is allocated at most one item.
2. Supply Constraint: The number of items each seller a_j should provide according to the allocation result does not exceed its supply ability s_j.

If the above constraints are *not* satisfied, then the corresponding flow in network resulted from Algorithm 2 is *not* fesible, since there must exist an edge starting from source vertex s or ending at termination vertex t whose flow is larger than its capacity.

On the other hand, if the above constraints are satisfied, then the capacity constraints of edges starting from source vertex s or ending at termination vertex t are satisfied. Further, since no buyer can get more than 1 item and the capacity of every edge starting from a buyer vertex and ending at a seller vertex equals to 1, it is not possible that any of these edges' capacity constraint is broken. Thus the network flow is feasible.

The above arguments complete the proof.   □

Lemma 3. *Consider an input into and its corresponding flow network graph resulted from Algorithm 2, if a flow of size f minimizes the cost in the network, then it maximizes the social efficiency in the auction.*
Proof.

$$\text{flow cost} = \sum_{i=1}^{n} \sum_{j=1}^{m} (-v_j^{(i)}) \times w_j^{(i)}$$

$$= -\sum_{i=1}^{n} \sum_{j=1}^{m} v_j^{(i)} \times w_j^{(i)}$$

$$= -\text{efficiency}$$

Thus when the cost of the flow is minimized, the social efficiency is maximized.   □

Theorem 2. *The allocation resulted from Algorithm 3 maximizes the social efficiency.*

Proof. Algorithm 3 coverts an input of the coordination protocol into an input of a corresponding minimum cost flow problem, solve the minimum cost flow problem by Algorithm 1, and uses the resulting flow network to give the solution to the auction coordination problem.

By Lemma 1, the converted minimum cost flow problem has a feasible solution. By Lemma 2, the corresponding allocation is also feasible. And finally, by Lemma 3, the corresponding allocation maximizes the social efficiency.   □

4.2   Incentive Compatibleness

Theorem 3. CPA-U *is a VCG mechanism.*

Proof. By Theorem 2, the allocation result maximizes the social efficiency.

From Algorithm 4 we know that the price function exhibits the form of the VCG payment formula.

Thus, **CPA-U** is a VCG mechanism.   □

Corollary 1. CPA-U *is incentive compatible.*

Proof. The corollary follows from the theorem that that a VCG mechanism is incentive compatible (Theorem 1) and the fact that **CPA-U** is a VCG mechanism (Theorem 3).   □

5   Conclusions, Discussions and Future Works

In this paper, we have studied a coordination protocol among multiple auctions when the market is *monopolistic competitive*. Unlike double auction [7,10,12,6,2] where all sellers sell identical items, our protocol allows the sellers to sell different items that are substitute to each other, and allows buyers to have different willingness-to-pay for different items. Analysis show that our protocol is *incentive compatible* for buyers, *optimal*, and implementable in *polynomial time*.

Our protocol is of practical value to the C2C e-commerce websites, such as eBay, Taobao, and Yahoo auctions. Buyers will benefit a lot from our protocol. Becuase with the help of our protocol, the *complexity of decision making* is dramatically decreased due to its incentive compatibleness, the *risk of winning no items* is reduced, and the *risk of winning redundant items* is eliminated completely. Thus buyers are encouraged to participate in C2C e-commerce more often. And it is the active participation of buyers that determines the value of a C2C e-commerce website to the sellers.

The **CPA-U** protocol is not *yet* a generalization of incentive compatible double auction protocols, because such protocols usually address the incentive compatibleness for both buyers and sellers, while our protocol only address the incentive compatibleness for buyers. It remains an interesting **open problem** whether both seller and buyer incentive compatibleness can be achieved in the *monopolistic competitive* market studied in our paper. We plan to solve this problem as our future works.

Acknowledgments. We would like to thank Prof. Xiaotie Deng from City University of Hong Kong for stimulating discussions and invaluable comments.

References

1. Clarke, E.H.: Multipart pricing of public goods. Public Choice 11, 17–33 (1971)
2. Deshmukh, K., Goldberg, A.V., Hartline, J.D., Karlin, A.R.: Truthful and competitive double auctions. In: Möhring, R.H., Raman, R. (eds.) ESA 2002. LNCS, vol. 2461, pp. 361–373. Springer, Heidelberg (2002)
3. Edmonds, J., Karp, R.M.: Theoretical improvements in algorithmic efficiency for network flow problems. Journal of the ACM 19(2), 248–264 (1972)
4. Fredman, M.L., Tarjan, R.E.: Fibonacci heaps and their uses in improved network optimization algorithms. Journal of the ACM 34(3), 596–615 (1987)
5. Groves, T.: Incentives in teams. Econometrica 41(4), 617–631 (1973)
6. Huang, P., Scheller-Wolf, A., Sycara, K.: Design of a multi-unit double auction e-market. Computational Intelligence 18(4), 596–617 (2002)
7. Myerson, R.B., Satterthwaite, M.A.: Efficient mechanisms for bilateral trading. Journal of Economic Theory 29(2), 265–281 (1983)
8. Nisan, N.: Algorithms for selfish agents. In: Meinel, C., Tison, S. (eds.) STACS 99. LNCS, vol. 1563, Springer, Heidelberg (1999)
9. Orlin, J.B: A faster strongly polynomial minimum cost flow algorithm. Operations Research 41(2), 338–350 (1993)
10. Satterthwaite, M.A., Williams, S.R.: Bilateral trade with the sealed bid k-double auction: Existence and efficiency. Journal of Economic Theory 48(1), 107–133 (1989)
11. Vickrey, W.: Counterspeculation, auctions and sealed tenders. Journal of Finance 16(1), 8–37 (1961)
12. Yoon, K.: The modified vickrey double auction. Journal of Economic Theory 101, 572–584 (2001)

The On-Line Rental Problem with Risk and Probabilistic Forecast*

Yucheng Dong[1], Yinfeng Xu[1,2], and Weijun Xu[3]

[1] School of Management, Xi'an Jiaotong University, Xi'an, 710049, China
[2] The State Key Lab for Manufacturing Systems Engineering,
Xi'an, 710049, China
{ycdong, yfxu}@mail.xjtu.edu.cn
[3] School of Business Administration, South China University of Technology,
Guangzhou, 510641, China
xuwj@scut.edu.cn

Abstract. This paper proposes a generalized on-line risk-reward model, by introducing the notion of the probabilistic forecast. Using this model, we investigate the on-line rental problem. We design the risk rental algorithms under the basic probability forecast and the geometric distribution probability forecast, respectively. In contrast to the existing competitive analyses of the on-line rental problem, our results are more flexible and can help the investor choosing the optimal algorithm according to his/her own risk tolerance level and probabilistic forecast. Moreover, we also show that this model has a good linkage to the stochastic competitive ratio analysis.

1 Introduction

In Karp's ski-rental problem [1], we wish to acquire equipment for skiing. However, since we do not know how many times we will use this equipment, we do not know if it is cheaper to rent or to buy. Let c be the rental price every time, and p the buying price. For simplicity, assume that $c \mid p$ and $c, p > 0$. It is easy to prove that the algorithm that achieves the optimal competitive ratio, $2 - c/p$, is to rent for the first $p/c - 1$ times, and then buy the equipment in the p/c time.

Considering the real-life situation of the rental problem, many researchers expanded Karp's ski-rental problem. Irani and Ramanathan [2] studied the rental algorithm under the condition that the buying price is fluctuated but the rental price remains unchanged. Xu [3] further discussed the rental algorithm under the circumstance that the buying price and rental price both fluctuate. EI-Yaniv et al [4] introduced the interest rate factors to the on-line rental problem. Xu [3] considered the discount factors in the on-line rental study. Some more complicated versions based on the ski-rental problem also have been presented, such as EI-Yaniv and Karp's replacement problem [5] and Fleischer's Bahncard problem [6].

---

* This research is supported by NSF of China (No. 70525004, 70121001 and 70471035) and PSF of China (No. 20060400221).

F.P. Preparata and Q. Fang (Eds.): FAW 2007, LNCS 4613, pp. 117–123, 2007.
© Springer-Verlag Berlin Heidelberg 2007

Moreover, some researchers also focus on constructing more flexible competitive analysis frameworks to study on-line rental algorithm. al-Binali [7] proposed the notable on-line risk-reward model. Fujiwara and Iwama [8], and Xu et al [9] integrated probability distribution into the classical competitive analysis [10] to study the rental problem. The main purpose of this paper is to propose a generalized on-line risk-reward model under the probability forecast. We also show a good linkage of our model to the existing competitive analysis frameworks. The rest of this paper is organized as follows. In Section 2, we proposes a generalized on-line risk-reward model, by introducing the notion of the probabilistic forecast. In Section 3, we design the risk rental algorithm under the basic probability forecast. In Section 4, we study the risk rental algorithm under the geometric distribution probability forecast. Concluding remarks and future research are included in Section 5.

2   On-Line Risk-Reward Model Under Probabilistic Forecast

In 1985, Sleator and Tarjian [10] proposed the concept of the competitive ratio to study on-line problems, by comparing the performance of on-line algorithms to a benchmark (optimal off-line) algorithm. During this classical competitive analysis, there are an algorithm set S for the on-line decision-maker and a uncertain information set I dominated by the off-line opponent. The on-line decision-maker's goal is to design a good algorithm $A \in S$ to deal with the uncertainty input sequence $\sigma \in I$ of the off-line rival. For a known sequence σ, let $C_{opt}(\sigma)$ be the total cost of the optimal off-line algorithm to complete σ. For an on-line algorithm A, if there are constants λ_A and ζ satisfying

$$C_A(\sigma) \leq \lambda_A C_{opt}(\sigma) + \zeta$$

for any $\sigma \in I$, then A is called a λ_A-competitive algorithm and λ_A is called the competitive ratio of A, where $C_A(\sigma)$ is the total cost taken with algorithm A to complete σ. That is to say, $\lambda_A = \sup_{\sigma \in I} \frac{C_A(\sigma)}{C_{opt}(\sigma)}$. We denote $\lambda^* = \inf_{A \in S}(\lambda_A)$ as the optimal competitive ratio for the on-line problem. If $\lambda_{A^*} = \lambda^*$, then A^* is called the optimal on-line algorithm.

The above competitive analysis is the most fundamental and significant approach, yet it is not very flexible, especially in the economic management issues, many investors want to manage their risk. Al-Binali [9] first defined the concepts of risk and reward for on-line financial problems. Al-Binali defined the risk of an algorithm A to be $r_A = \frac{\lambda_A}{\lambda^*}$. The greater the value of r_A, the higher the risk of A. Let $F \subset I$ be a forecast, then denote $\overline{\lambda_A} = \sup_{\sigma \in F} \frac{C_A(\sigma)}{C_{opt}(\sigma)}$ as the restricted competitive ratio of A restricted to cases when the forecast is correct. The optimal restricted competitive ratio under the forecast F is $\overline{\lambda^*} = \inf_{A \in S}(\overline{\lambda_A})$. When the forecast is correct, Al-Binali defined the reward of the algorithm A to be $f_A = \frac{\lambda^*}{\overline{\lambda_A}}$.

The above reward definition is based on the certain forecast that is described to be a subset of I. When the forecast selected is correct, it will bring reward; otherwise bring risk. This paper extends the certain forecast to the probability forecast. Let $F_1, F_2, ..., F_m$ be a group of subsets of I, where $\bigcup F_i = I$ and $F_i \cap F_j = \phi$ for $i \neq j$. Denote P_i as the probability that the on-line decision maker anticipates that $\sigma \in F_i$, where $\sum_{i=1}^{m} P_i = 1$. We call $\{(F_i, P_i)|i = 1, 2, ..., m\}$ a probability forecast. Let $\overline{\lambda_{A,i}} = \sup\limits_{\sigma \in F_i} \frac{C_A(\sigma)}{C_{opt}(\sigma)}$ be the restricted competitive ratio under the forecast F_i. Let $R_{A,i} = \frac{\lambda^*}{\overline{\lambda_{A,i}}}$ be the reward after the success of the forecast F_i. Based on this, we define $\widetilde{\lambda_A} = \sum_{i=1}^{m} P_i \overline{\lambda_{A,i}}$ as the restricted competitive ratio under the probability forecast $\{(F_i, P_i)|i = 1, 2, ..., m\}$, and define $\widetilde{R_A} = \frac{\lambda^*}{\widetilde{\lambda_A}}$ as the reward under the probability forecast.

The reward definition based on the probability forecast has some desired properties.

Theorem 1. For any $A \in S$, $\min\limits_{i}\{R_{A,i}\} \leq \widetilde{R_A} \leq \max\limits_{i}\{R_{A,i}\}$.

Proof. Since $\widetilde{\lambda_A} = \sum_{i=1}^{m} P_i \overline{\lambda_{A,i}}$, $\min\limits_{i}\{\overline{\lambda_{A,i}}\} \leq \widetilde{\lambda_A} \leq \max\limits_{i}\{\overline{\lambda_{A,i}}\}$. Consequently, $\min\limits_{i}\{\lambda^*/\overline{\lambda_{A,i}}\} \leq \lambda^*/\widetilde{\lambda_A} \leq \max\limits_{i}\{\lambda^*/\overline{\lambda_{A,i}}\}$, that is $\min\limits_{i}\{R_{A,i}\} \leq \widetilde{R_A} \leq \max\limits_{i}\{R_{A,i}\}$.

Let $\{(F_i, P_i)|i = 1, 2, ..., m\}$ be a probability forecast. We divide F_i into $F_{i,1}$ and $F_{i,2}$, where $F_{i,1} \cup F_{i,2} = F_i$ and $F_{i,1} \cap F_{i,2} = \phi$. We also divide P_i into $P_{i,1}$ and $P_{i,2}$, where $P_{i,1} + P_{i,2} = P_i$. In this way, we can construct a more detailed probability forecast based on $\{(F_i, P_i)|i = 1, 2, ..., m\}$, that is $\{(F_1, P_1), (F_2, P_2), ..., (F_{i-1}, P_{i-1}), (F_{i,1}, P_{i,1}), (F_{i,2}, P_{i,2}), (F_{i+1}, P_{i+1})..., (F_m, P_m)\}$. Denote $\widetilde{\widetilde{R_A}} = \frac{\lambda^*}{\widetilde{\widetilde{\lambda_A}}}$ as the reward under the newly constructed probability forecast.

Theorem 2. For any $A \in S$, $\widetilde{R_A} \leq \widetilde{\widetilde{R_A}}$.

Proof. From the definition of the restricted competitive ratio, we know that $\overline{\lambda_{A,i1}} \leq \overline{\lambda_{A,i}}$ and $\overline{\lambda_{A,i2}} \leq \overline{\lambda_{A,i}}$. Besides $P_{i,1} + P_{i,2} = P_i$, thus $\frac{\lambda^*}{\widetilde{R_A}} - \frac{\lambda^*}{\widetilde{\widetilde{R_A}}} = P_i \overline{\lambda_{A,i}} - P_{i,1} \overline{\lambda_{A,i1}} - P_{i,2} \overline{\lambda_{A,i2}} \geq 0$, that is $\widetilde{R_A} \leq \widetilde{\widetilde{R_A}}$.

Theorem 2 shows that if a probability forecast can be described more detailedly, the reward under the probability forecast will be greater.

Based on these newly introduced concepts, we propose a generalized risk-reward model under the probability forecast. If r is the risk tolerance level of the on-line decision maker (where $r \geq 1$ and higher values of r denote a higher risk tolerance), then denote $S_r = \{A|\lambda_A \leq r\lambda^*\}$ by the set of all algorithms with the risk tolerance level r. Our main aim is to look for an optimal risk algorithm $A^* \in S_r$ that maximizes the reward under the probability forecast $\{(F_i, P_i)|i = 1, 2, ..., m\}$, that is $\widetilde{R_{A^*}} = \sup\limits_{A \in S_r} \frac{\lambda^*}{\widetilde{\lambda_A}}$. The mathematic model can be described as follows:

$$\begin{cases} \max\limits_{A} \ \widetilde{R_A} = \frac{\lambda^*}{\widetilde{\lambda_A}} \\ s.t \ \ \lambda_A \leq r\lambda^* \end{cases}. \tag{1}$$

The steps to use this model can be described as follows.

Step 1: Divide I into $F_1, F_2, ..., F_m$, where $\bigcup F_i = I$ and $F_i \cap F_j = \phi$ for $i \neq j$;

Step 2: Denote P_i as the probability that the on-line decision maker anticipates that $\sigma \in F_i$, where $\sum_{i=1}^m P_i = 1$;

Step 3: According to definitions, compute the risk and reward under the probability forecast $\{(F_i, P_i) | i = 1, 2, ..., m\}$;

Step 4: Set the risk tolerance level to be r;

Step 5: Solve the model (1) to obtain the optimal risk algorithm A^*.

In the following two sections, we will demonstrate the model that we have just introduced using Karp's ski-rental problem.

3   Risk Rental Algorithm Under Basic Probability Forecast

We consider the following deterministic on-line rental algorithm T: rent up to $T - 1$ times and buy in T. Let $Cost_T(t)$ and $Cost_{opt}(t)$ denote the cost of the on-line algorithm T and the cost of the optimal off-line algorithm, respectively, where t is the total number of the actual leases.

For the off-line rental problem, if $t \geq p/c$, then buy; otherwise rent. So we have that

$$Cost_{opt}\{t\} = \begin{cases} ct & 0 \leq t < p/c \\ p & p/c \leq t \end{cases}. \tag{2}$$

For the on-line problem, if $t < T$, then always lease. According to on-line algorithm T ($T = 0, 1, 2, ...$), then it is not difficult to see that

$$Cost_T\{t\} = \begin{cases} ct & 0 \leq t \leq T \\ cT + p & T < t \end{cases}. \tag{3}$$

According to the off-line optimal rental algorithm, we construct a basic probability forecast, $\{(F_1, P_1), (F_2, P_2)\}$, as follows.

Forecast F_1: $F_1 = \{t : t < p/c\}$. The probability when F_1 appears is P_1.
Forecast F_2: $F_2 = \{t : t \geq p/c\}$. The probability when F_2 appears is P_2.

Theorem 3. When setting the risk tolerance level $r \geq \frac{2p}{2p-c}$, the optimal risk rental algorithm under the probability forecast $\{(F_1, P_1), (F_2, P_2)\}$ is

$$T^* = \begin{cases} p/c, & P_1 > 1/2 \\ \frac{p}{c}\sqrt{\frac{P_1}{1-P_1}}, & p^2/((2pr - cr - p)^2 + p^2) \leq P_1 \leq 1/2 \\ p^2/(2prc - c^2r - pc), & P_1 \leq p^2/((2pr - cr - p)^2 + p^2) \end{cases} ; \tag{4}$$

otherwise, when $1 \leq r \leq \frac{2p}{2p-c}$,

$$T^* = \begin{cases} p/c, & P_1 > 1/2 \\ p^2/(2prc - c^2r - pc), & P_1 \leq 1/2 \end{cases}. \tag{5}$$

Proof. According to the definition of the competitive ratio, we know that $\lambda_T = \frac{cT+p}{min\{cT,p\}}$. Since the risk tolerance level set is r, we have that $\lambda_T \leq r(2 - c/p)$.
When $cT \leq p$, we have that $\lambda_T = \frac{cT+p}{cT} \leq r(2 - c/p)$, that is $T \geq \frac{p^2}{2prc-c^2r-pc}$.
When $cT > p$, we have that $\lambda_T = \frac{cT+p}{p} \leq r(2 - c/p)$, that is $T \leq \frac{2pr-cr-p}{c}$.
Consequently,

$$\frac{p^2}{2prc - c^2r - pc} \leq T \leq \frac{2pr - cr - p}{c} \tag{6}$$

Denote $S_r = [\frac{p^2}{2prc-c^2r-pc}, \frac{2pr-cr-p}{c}]$ as the algorithm set with risk level r. From the definition of the restricted competitive ratio under the probability forecast, we have that $\widetilde{\lambda_T} = \sum_{i=1}^2 P_i \overline{\lambda_{T,i}}$, where $\overline{\lambda_{T,i}} = \sup_{\sigma \in F_i} \frac{C_T(\sigma)}{C_{opt}(\sigma)}$. Consequently,

$$\widetilde{\lambda_T} = \begin{cases} P_1 \frac{cT+p}{cT} + \frac{cT+p}{p}(1 - P_1), & T < p/c \\ P_1 + \frac{cT+p}{p}(1 - P_1), & T \geq p/c \end{cases} \tag{7}$$

Solving $\frac{\partial \widetilde{\lambda_T}}{\partial T}$, we find that:

(1) when $P_1 > 1/2$, $\widetilde{\lambda_T}$ is monotony decreasing at $T < p/c$, and monotony increasing at $T \geq p/c$;

(2) when $P_1 \leq 1/2$, $\widetilde{\lambda_T}$ is monotony decreasing at $T < \frac{p}{c}\sqrt{\frac{P_1}{1-P_1}}$, and monotony increasing at $T \geq \frac{p}{c}\sqrt{\frac{P_1}{1-P_1}}$.

From the above monotony properties of $\widetilde{\lambda_T}$ and equations (6), we can look for the optimal risk rental algorithm T^* that makes $\widetilde{\lambda_T}$ minimum, that is equations (4) and (5).

Corollary 1. When $P_1 = 1$, $T^* = p/c$; when $P_1 = 0$, $T^* = p^2/(2prc - c^2r - pc)$. Corollary 1 shows that our model is a generalized risk-reward framework, compared with one presented in Al-Binali [7].

4   Risk Rental Algorithm Under Geometric Distribution Forecast

By dividing F_i into $F_{i,1}$ and $F_{i,2}$ (where $F_{i,1} \cup F_{i,2} = F_i$ and $F_{i,1} \cap F_{i,2} = \phi$), and dividing P_i into $P_{i,1}$ and $P_{i,2}$ (where $P_{i,1} + P_{i,2} = P_i$), we construct a more detailed probability forecast based on $\{(F_i, P_i)|i = 1, 2, ..., m\}$, that is $\{(F_1, P_1), (F_2, P_2), ..., (F_{i-1}, P_{i-1}), (F_{i,1}, P_{i,1}), (F_{i,2}, P_{i,2}), (F_{i+1}, P_{i+1})..., (F_m, P_m)\}$.
For the rental problem, we can obtain the probability distribution of t, when repeatedly dividing $\{(F_1, P_1), (F_2, P_2)\}$ in the above way. Fujiwara and Iwama [8], and Xu et al [9] integrated probability distribution into the classical competitive analysis, and introduced the concept of the stochastic competitive ratio (Definition 1) for the on-line rental problem.

Definition 1. Let the number of leases be a stochastic variable X subject to some type of probability distribution function $P(X = t)$. The discrete stochastic competitive ratio is then defined as

$$\widetilde{\widetilde{\lambda_T}} = Ex \frac{Cost_T(X)}{Cost_{opt}(X)} = \sum_{t=0}^{\infty} \frac{Cost_T(t)}{Cost_{opt}(t)} P(X = t), \tag{8}$$

where $P(X = t)$ is a probability function that is used by the on-line decision maker to approximate the input structures.

Note. It is easy to find that the definition of the discrete stochastic competitive ratio is consistency of one of the restricted competitive ratio under the corresponding probability distribution forecast.

For the rental problem, Xu ea tal [9] consider the geometric distribution function $P(X = t) = \theta^{t-1}(1 - \theta)$, $(t = 0, 1, 2, 3, \cdots)$, where θ is the hazard rate of continuous leasing in every period, and $1 - \theta$ is the hazard rate of immediately purchasing in every period. In this paper, we only discuss the situation that $\frac{1}{1-\theta} < p/c$ to illustrate our model.

Let $s = p/c$. According to equations (2), (3), and (8), we have, for T=0, 1, 2, 3, $\cdots$, s, that

$$\widetilde{\widetilde{\lambda_T}} = (1 - \theta^T) + (T + s)(1 - \theta) \sum_{t=T+1}^{s} \frac{\theta^{t-1}}{t} + \frac{T + s}{s} \theta^s, \tag{9}$$

and for $k = s + 1, s + 2, s + 3, \cdots$,

$$\widetilde{\widetilde{\lambda_T}} = (1 - \theta^s) + \frac{(1 - \theta)}{s} \Sigma_{t=s+1}^{T} t \theta^{t-1} + \frac{T + s}{s} \theta^T. \tag{10}$$

Then we obtain the following result (Theorem 4).

Theorem 4. When setting the risk tolerance level to be r, the optimal risk algorithm under the geometric distribution forecast is $T^{**} = \frac{2pr - cr - p}{c}$ (we only discuss the situation that $\frac{1}{1-\theta} < s$).

Proof. For $t < s - 1$, we have that

$$\widetilde{\widetilde{\lambda_{T+1}}} - \widetilde{\widetilde{\lambda_T}} = -\frac{s(1 - \theta)}{T + 1} \theta^T + \frac{1}{s} \theta^s + (1 - \theta) \Sigma_{t=T+1}^{s} \frac{\theta^{t-1}}{t}$$

$$\leq -\frac{s(1 - \theta)}{T + 1} \theta^T + \frac{1}{s} \theta^s + \frac{1 - \theta}{T + 1} \frac{\theta^T - \theta^s}{1 - \theta}$$

$$= \theta^T \left(\frac{1}{T + 1} - \frac{s(1 - \theta)}{T + 1} \right) + \theta^s \left(\frac{1}{s} - \frac{1}{T + 1} \right) < 0$$

For $t \geq s - 1$, we also have that

$$\widetilde{\widetilde{\lambda_{T+1}}} - \widetilde{\widetilde{\lambda_T}} = \left(\frac{1}{s} - 1 + \theta \right) \theta^T < 0$$

Therefore, we have that $\lambda_{T+1} - \lambda_T < 0$ for any T. Besides, because $\widetilde{\widetilde{R_T}} = \frac{\lambda^*}{\widetilde{\widetilde{\lambda_T}}}$ and $T \in S_r$, we have that $T^{**} = \frac{2pr - cr - p}{c}$.

5 Conclusions

The classical competitive ratio analysis is the most fundamental and important framework to study online problems. But it is not very flexible, particularly in the financial and investment issues (such as on-line rental problem, on-line currency conversion, on-line auctions problem). Many investors hope to manage the risk. Sometimes for more reward, they are willing to take certain risk. Therefore, the notable online risk-reward idea has been proposed by Al-Binali. However, the existing concept of risk-reward is mainly based on the certainty forecast. In this paper, we further puts forward the online risk- reward model under the probability forecast. The probability forecast will not make the simple judgment about whether the forecast is correct or not, but estimate the probability that the forecast is correct. The newly introduced model makes the risk-reward idea more flexible. Moreover, some researchers presented the concept of the stochastic competitive ratio to improve the performance measure of competitive analysis, by integrating probability distribution into the classical competitive ratio analysis. This paper shows that our model has a good linkage to the stochastic competitive ratio analysis. We also argue that our model is the generalized stochastic competitive ratio analysis. By using Karp's ski-rental problem, we demonstrate the on-line risk-reward model under the probability forecast. In general, it is hard for an on-line decision maker to accurately estimate the probability forecast of the future inputs. Therefore, in our future research, we will explore the on-line risk-reward model in linguistic environments, by introducing the notations and operational laws of the linguistic variables.

References

1. Karp, R.: Online algorithms versus offline algorithms: How much is it worth to know the future? In: Proc. IFIP 12th World Computer Congress, pp. 416–429 (1992)
2. Irani, S., Ramanathan, D.: The problem of renting versus buying. Personal communication (1998)
3. Xu, W.J.: Investment algorithm design and competitive analysis for on-line rental problem. PhD thesis, Xi'an Jiaotong University, Xi'an, China (2005)
4. El-Yaniv, R., Kaniel, R., Linial, N.: Competitive optimal on-line leasing. Algorithmica 25, 116–140 (1999)
5. El-Yaniv, R., Karp, R.: Nearly optmal competitive online replacement policies. Mathematics of Operations Research 22, 814–839 (1997)
6. Fleischer, R.: On the Bahncard problem. In: Hsu, W.-L., Kao, M.-Y. (eds.) COCOON 1998. LNCS, vol. 1449, pp. 65–74. Springer, Heidelberg (1998)
7. Al-Binali, S.: A risk-reward framework for the competitive analysis of financial games. Algorithmica 25, 99–115 (1999)
8. Fujiwara, H., Iwama, K.: Average-case competitive analyses for ski-rental problems. In: The 13th Annual International Symposium on Algorithms and Computation, pp. 476–488 (2002)
9. Xu, Y.F., Xu, W.J., Li, H.Y.: On the on-line rent-or-buy problem in probabilistic environments. Journal of Global Optimization 2006 (in press)
10. Sleator, D.D., Tarjan, R.E.: Amortized efficiency of list update and paging rules. Communications of the ACM 28, 202–208 (1985)

Distributed Planning for the On-Line Dial-a-Ride Problem

Claudio Cubillos, Broderick Crawford, and Nibaldo Rodríguez

Pontificia Universidad Católica de Valparaíso, Escuela de Ingeniería Informática,
Av. Brasil 2241, Valparaíso, Chile
{claudio.cubillos,broderick.crawford,nibaldo.rodriguez}@ucv.cl

Abstract. This paper describes the experiments and results obtained from distributing an improved insertion heuristic for the scheduling of passengers' trip requests over a fleet of vehicles. The distribution has been obtained by means of an agent architecture implemented over Jade. Agents make use of the contract-net protocol as base coordination mechanism for the planning and scheduling of passenger trips. In particular, this paper focuses on the insertion heuristic implementation details within the agent-based architecture and its performance in diverse distributed scenarios when varying the number of hosts.

1 Introduction

The online Dial-a-Ride problem refers to transport services capable of satisfying personal transportation requests at relatively low costs, thanks to an integrated planning and by using the latest IT infrastructure. Furthermore, the research in the field of passenger transportation planning has received an increasing attention in the last decades due to diverse factors. On one side, traffic jams and pollution are frequent problems. On the other, the mobility patterns of citizens have changed in the last years. Therefore, traditional public-transport planning systems are no longer adequate to tackle these newer challenges. Therefore, more flexible transportation alternatives are required [1], with planning methodologies capable of considering dynamic and distributed information in the routing and scheduling process.

Under this scenario, the present work has been focused into taking a greedy insertion heuristic for the planning and scheduling of trip requests and embedding it on a software architecture that adopts the agent paradigm as a way to provide the required distribution by means of using the contract-net protocol for coordinating the agents through the planning procedure.

The work is structured as follows. Section 2 starts presenting other works in the field for then at section 3 explaining the online Dial-a-Ride problem considered in this work. Section 4 outlines the agent architecture and Section 5 explains the planning algorithm to be distributed. Then, experimental tests are presented at Section 6 for then drawing some conclusions at Section 7.

2 Related Work

The Dial-a-Ride Problem (DARP) is a sub-type of the Travel Salesman Problem (TSP) and more specifically, the Pickup and Delivery Problem (PDP) often devoted to

F.P. Preparata and Q. Fang (Eds.): FAW 2007, LNCS 4613, pp. 124–135, 2007.

goods transport. The dial-a-ride problem regards the planning of passenger trips. In this field, an important research effort has been devoted to greedy insertion heuristics (see [9][17]) which provides good solutions in small time.

Coslovich et al. [4] have addressed a dynamic dial-a-ride where people might unexpectedly ask a driver for a trip at a given stop by using a two-phase method and a neighborhood of solutions generated off-line. A software system for D-DARP was proposed by Horn [8]. The optimization capabilities of the system are based on least-cost insertions of new requests and periodic re-optimization of the planned routes.

Newer research tackling the dynamic problem tends to use a distributed market-based philosophy based in the Contract-Net Protocol (CNP) ((see [3][7] and [14]). In the soft computing field, we can find the use of an ant-colony based system [15], genetic algorithms (GA) for the optimization of the assignment ([10] [18]) and the use of fuzzy logic for the travel times ([11] [19]).

On the other hand, agent technology applied to transportation has been widely researched in literature [7],[12],[13],[16] most of them focused in the transportation of goods (e.g. vehicle routing problem, pickup & delivery problem). Although the multi-agent paradigm [20] appears as a promising technology, capable of providing a flexible assignment and service, it is hard to find in literature agent architectures devoted to the transportation of passengers (e.g. dial-a-ride problem, demand-responsive transport). [12] and [16] present agent-based systems for goods transportation. [12] uses the Contract-Net Protocol (CNP) plus a stochastic post-optimization phase to improve the result initially obtained. In [16] is presented the Provisional Agreement Protocol (PAP), based on the Extended CNP and de-commitment.

Finally, although the DARP problem is well-known in the research community, there is a lack of benchmark data for this specific problem. For the static case, benchmarks for the Vehicle Routing Problem (VRP) and PDP are usually adapted to fit DARP. This situation does no get better for the online or dynamic case in which requests arrive following a certain distribution rather than being known in advance. In fact no widely used datasets seem to be available.

3 The On-Line DARP

The problem we are treating consists of transport requests coming from a set of clients which should be satisfied by a heterogeneous fleet. From a mathematical point of view the problem corresponds to the on-line (dynamic) version of the Dial-a-ride Problem (D-DARP). It consists of a set C of geographically distributed transportation requests, coming from customers that should be served by a set of vehicles V. In addition, this dynamic variant of the DARP problem implies that requests arrive online. That is, an immediate-request approach is used, in which the system should process the trips as they come and provide an answer back (bus number, pickup and delivery times) in a timely way.

Service requests have to be assigned to vehicles and scheduled according to time restrictions. A restriction exists about the maximum number of passengers to carry (capacity). Transport requests commonly specify a pick-up and delivery place. They also indicate *time windows*, that is, time intervals within which the client has to be picked-up at the origin node and delivered at the destination node. Moreover, the requests can include further descriptions of the desired service like type and number of places, shared or exclusive use of the vehicle, wheelchair place use and any of the complementary services described before.

In practice, the dial-a-ride system we are tackling considers transport requests coming from different types of clients that should be satisfied by a homogeneous fleet of capacitated vehicles.

4 The Agent Architecture

In the following the agent architecture used for distributing the algorithm is presented. Figure 1 shows a lower layer with the Jade agent platform [2], which provides a full environment for agents to work: an agent management system (AMS), the possibility of agent containers in different hosts (distribution), a directory facilitator (DF) providing yellow-page services, a message transport system (MTS) for supporting communication between agents and mobility services between containers.

On top, the agent architecture is built, with Client and Vehicle agents that are interfaces for communicating with the involved actors. The Broker provides a matchmaking service and the Map gives support with distances and paths between points.

The Trip-Request agent is responsible of having the client's request fulfilled and of communicating him about the result and possible changes in the original plan (e.g.

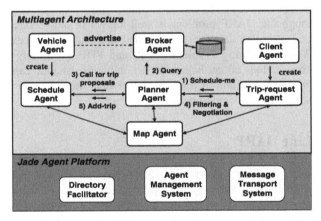

Fig. 1. The agent architecture for DARP shows the diverse agents involved, all over the Jade agent platform

delays, trip cancellations). The Planner agent is the agent in charge of executing a mediation role in the layer. It processes all the client's requests coming through the Trip-request agents. For more details on the agent architecture please refer to [5] and for the its design using the agent oriented software engineering (AOSE) methodology PASSI please refer to [6].

Schedule agents manage the trip plans (work-schedule) of their corresponding vehicles. In practical terms, the agent will have to make proposals upon request and in case of winning will have to include the trip into its actual plan. Upon changes informed either by the Vehicle agent or Planner agent, the Schedule agent will update the plan and reschedule the remaining requests. The schedule agent implements a greedy insertion heuristic for routing and scheduling which is further explained in Section 5.

The routing and scheduling functionality provided by the agent architecture is based on the contract-net protocol (CNP). The interaction among the agents is as follows (see Fig. 1): First, each transportation request coming from a Client is received by the corresponding Trip-request agent of the couple, which asks the Planner to process it. Next, the Planner processes the request first by obtaining from the Broker agent the vehicles that match the required profile, and then by making a call for trip-proposals to all the corresponding Schedule agents (call for bids in contract-net) that represent the different vehicles of the considered fleet. Each schedule agent searches for possible trip alternatives within their respective schedule by using the algorithm (insertion heuristic) detailed in Section 5. in this way, they send back their proposals and the Planner forwards them to the client (through its Trip-request agent) for him to select the best alternative. After choosing, the Planner tells the Schedule agent that won the proposal to add the trip to its actual schedule and tells the others their proposal rejection.

5 Algorithm Implementation

As stated before, during the planning process schedule agents make proposals of trip insertions which are managed by the planner. Therefore, each of these agents contains a scheduling heuristic to search in the state space for suitable alternatives. The main algorithm's implementation details are explained in the following.

5.1 Time-Windows

The model considers a *pickup time window [ept, lpt]* and a *delivery time window [edt$_i$, ldt]* (see Fig. 2). It is considered the specification of a *delivery time* for each customer. This time is assigned to the upper bound of the delivery time window (*ldt*). The parameter *WS* sets the width of all the delivery time windows. In this a way, a vehicle serving the customer i must reach the destination node nd_i, neither not before the edt_i time, nor after the ldt_i time.

The function *DRT(NxN)* $\rightarrow \Re$ defines the *direct ride time* (optimistic time) which corresponds to the time spent travelling from ns to nd through the shortest existing path. The function *MRT(NxN)* $\rightarrow \Re$, defines the *maximum ride time* (pessimistic time) which corresponds to the maximum time that can be spent by a client in reaching the destination node nd_j from the origin node ns. Delivery times define a *time window for pick-up*, the pair *(ept$_i$, lpt$_f$)*, where *ept$_i$=edt$_i$-MRT(Ns$_i$,Nd$_i$)* $\wedge$ *lpt$_i$=edt$_i$-DRTe(Ns$_i$,Nd$_i$)*.

5.2 Work-Schedule

The model used for the vehicles' work-schedules considers that along the day a vehicle can be in any of these three states: at a depot, in travel or inactive. When the vehicle is at

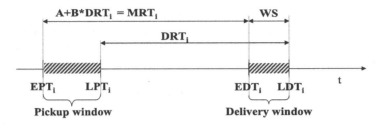

Fig. 2. Time-Windows model for clients pickup and delivery intervals

a depot means that it has not started its service period or has just finished it. When the vehicle is in travel, means that it is actually going to pickup or delivery passengers generating schedule blocks. As [9] shows, a schedule block corresponds to a sequence of pickups and deliveries for serving one or more trip requests. A schedule block always begins with the vehicle starting on its way to pick-up a customer and ends when the last on-board customer is discharged.

The third state is when the vehicle is inactive or idle generating a slack time. In this case the vehicle is parked and waiting to serve a next customer and then begin another schedule block.

Therefore, a complete vehicle's work-schedule will have periods of vehicle's utilization (schedule blocks) and inactive periods (slacks times) in which the vehicle is available and waiting.

5.3 Routing Algorithm

The routing algorithm used by vehicles (schedule agents) for finding the "optimal" sequence of pickups and deliveries is based on the ADARTW algorithm [9], a constructive greedy heuristic. ADARTW finds all the feasible ways in which a new customer can be inserted into the actual work schedule of a vehicle, choosing the one that offers the maximum additional utility according to a certain objective function.

The search must include all the schedule blocks contained in the vehicle's work-schedule. In a block with already d stops (2 per customer) there are $(d+1)(d+2)/2$ possible insertions, considering that the customer's pickup must always precede his

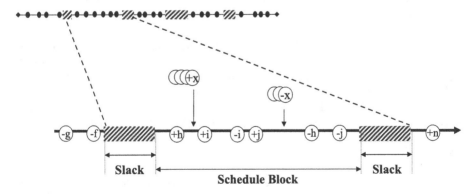

Fig. 3. Work-schedule used by vehicles, consisting in sequences of schedule blocks and slacks

delivery and that is not possible to pickup a client in one block and deliver him in another (because of the block's definition).

Figure 3 shows a schedule block that serves 3 customers (h, i, j) while evaluating the insertion of a fourth one (customer x). Each of them has their pick-up (+) and delivery (-) stops respectively.

Once we have found that a possibility of insertion is feasible, it is necessary to define the actual times for those events, that is, the Actual Pick-up Time and the Actual Delivery Time. This problem is often mentioned in literature as the scheduling problem, as once the sequence of trips (route) has been fixed the following step is to define the exact position where the sequence will be placed in time.

Commonly, there will be a time interval in which can be inserted, meaning that the sequence can be scheduled more early or late in time within that interval. Several authors program the actual times as soon as possible for reducing the travel and waiting times of the customers, reason why our implementation does it in this way.

5.4 Solution Feasibility Processing

The solution feasibility processing is tightly coupled to the work-schedule model. Within the checking algorithm, different restrictions need to be checked for a given potential solution. The most important ones are the time windows, the capacity constraints (on number and type) and the bounds on the duration of clients' ride and of vehicle route.

This represented a challenging aspect of the work, as in general is difficult to find in literature the used mechanism for tackling this point. In Jaw et al. 9 is described only in general terms and most research papers state a change from the previous work but not its specific implementation.

For a block X with w events representing either a pickup or a delivery of passengers, Jaw's work presents the following calculations representing how much the events can be anticipated/posticipated in time.

$$BUP(X_i) = Min (Min (AT(X_i) - ET(X_i)), SLK_0)$$

$$BDOWN(X_i) = Min (LT(X_i) - AT(X_i))$$
$$AUP(X_i) = Min (AT(X_i) - ET(X_i))$$
$$ADOWN(X_i) = Min (Min (LT (X_i) - AT(X_i)), SLK_{w+1})$$

With $0 < i < w+1$, SLK_0 and SLK_{w+1} being the (possible) slack periods immediately preceding and following the block respectively. $ET(X_i)$, $AT(X_i)$ and $LT(X_i)$ represent the early, actual and late times of the event X_i respectively.

Our developed model is based on the Jaw's calculations on BUP and ADOWN but adds the important idea of intersecting the time windows restrictions along a piece of route, allowing to simplify the processing of the time windows feasibility check and making it possible to evaluate the insertion of the whole client (pickup and delivery) at the same time.

Therefore, in the case of the implemented insertion heuristic the starting point is the schedule block under which to evaluate the insertion of the new client. The Fig. 4 (a) shows a detailed view when evaluating the insertion of the pickup (X^+) and delivery (X^-) of a client. Between the pickup and the delivery are one or more events separating them and at the beginning (or ending) of the block is a slack or the bus depot. The approach is to divide the schedule block in three sub-blocks A, B and C for the

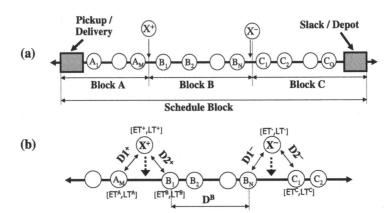

Fig. 4. Feasibility-check model

events before the pickup, in between and after the delivery respectively. A special case is when both events are consecutive meaning that the block B includes only the distance from the pickup to the delivery of the new client.

The Fig. 4 (b) shows the time windows and distances needed for the evaluation and intersection. The interval [ET^A, LT^A] represents the earliest and latest times to which the event A_M can be shifted (anticipated / posticipated) without violating the time window constraints of all the events within its block. A similar thing happens with intervals [ET^B, LT^B] and [ET^C, LT^C] on the events B_1 and C_1 for the blocks B and C respectively. Therefore, is needed to identify the feasible shift up and shift down for each of the three blocks. For the block A are used the BUP and BDOWN of the event A_M as they consider the previous events, while for the block C the AUP and ADOWN of C_1 are needed. For block B is needed the AUP and ADOWN for B_1 but considering only until B_N and not the events on block C as the normal calculations would. Then, for interval [ET^A, LT^A] we have: ET^A = $AT(A_M)$ - $BUP(A_M)$ and LT^A = $AT(A_M)$ + $BDOWN(A_M)$.

A similar thing happens with [ET^B, LT^B] and [ET^C, LT^C]. Distance $D1^+$, $D2^+$, $D1^-$ and $D2^-$ correspond to the distances between the nodes indicated by the respective arrows in the figure. The next step is intersecting the time intervals of the three blocks and the two time windows coming from the new client's pickup and delivery events. This intersection needs to consider the distances separating each of the five intervals. For this reason, a point in the schedule is used as reference and all the intervals are translated to that reference obtaining a single time interval [ET, LT]. By using the pickup event (X^+) of the new client as reference point and following Fig. 4 (b) are obtained:

ET = Max (ET^A + $D1^+$; ET^+; ET^B - $D2^+$; ET^- - $D1^-$ - D^B - $D2^+$; ET^C - $D2^-$ - $D1^-$ - D^B - $D2^+$)

LT = Min (LT^A + $D1^+$; LT^+; LT^B - $D2^+$; LT^- - $D1^-$ - D^B - $D2^+$; LT^C - $D2^-$ - $D1^-$ - D^B - $D2^+$)

This [ET, LT] interval represents the feasibility area in which to set the new schedule with respect to the reference point. The actual time for the reference point (X^+ in this case) must be set and hence the actual times for the whole schedule block can be calculated as they depend on the fixed distances between one event and another. Defining the optimal place within this interval corresponds to the scheduling problem mentioned before.

6 Experiment Setup

As mentioned earlier, the original architecture's planning approach is based on the contract-net protocol (CNP) under self-interested agents. Therefore, Vehicle agents pursue the optimization of travelling costs (utility function with total slack time and total travel time) and Client agents were oriented towards the maximization of the perceived service quality (utility function with excess travel time and waiting time). In [5] is presented a comparison of this architecture with a traditional centralized system based on a well-known insertion heuristic developed by Jaw et al. [9]. That analysis only compares the quality of the solutions obtained but not the processing time involved when distributing among diverse hosts.

All the tests considered the same geographical net which models a small mountain community near Ivrea – Italy. 20 demand scenarios were generated, labeled from U1.txt to U20.txt, each considering 50 trip requests distributed uniformly in a two-hour horizon. For each demand scenario 25 runs were carried out.

Regarding the considered distributed environment, the hosts were PCs with Intel Pentium 4 of 2 GHz. with 256 MB Ram, connected through a 10/100 Mb. Router.

The following operational decisions were adopted: 1) the same utility function and scheduling algorithm have been used for all the vehicles, 2) all the clients share the same utility function, 3) the available fleet is of 30 identical vehicles with capacity 20, 4) one depot is used for all the vehicles and 5) in all cases the effectiveness measures (utility variables) were weighted with the same value.

For the simulations were considered an agent devoted to the generation of the Trip-request agents and another devoted to generating the Schedule agents. In addition, a Main agent was in charge of managing all the aspects related to the simulation control, specifically centered on the generation of the agents, request of output data and deletion operations along the diverse runs and scenarios.

The generation of Trip-request agents (and hence the arrival of trip-requests) to the system follows a Poisson distribution. Then, the time between arrivals distributes Exponential, $E(\lambda)$, with lambda in terms of requests per second.

The agents involved in the simulations were the three involved in the planning (Trip-request, Planner and Schedule agents) plus two service providers (Map and Broker agents) as Table 1 details below.

Table 1. Distribution of agents among hosts over the 3 scenarios

2 hosts	3 hosts	5 hosts	Agents
1	1	1	Map agent
1	2	2	Trip-request agents
2	3	3	Schedule agents
1	1	4	Planner agent
1	1	5	Broker agent

6.1 Hosts Sensibility Experiment

This first experiment considered evaluating how the way agents distribute among several hosts has an impact on performance. More precisely, the focus was settled not on schedule agents (which contain the planning heuristic) but on the diverse types of agents in order to analyze how the topology of hosts and agents under a contract-net-protocol-based communication affected the planning response time for each request.

Figure 5 shows different curves for 2, 3 and 5 hosts' configuration. At first sight it is possible to see that a big improvement exists when changing from 2 hosts to 3 hosts, while little improvement is obtained when changing from 3 to 5 hosts. A closer look on how agents were distributed – see table 1 – helps to infer that separating Trip-request and Schedule agents from the rest of agents has a big impact on performance, but separating the Planner from the Broker and Map agents on diverse host gets only a small improvement in terms of processing time.

6.2 Lambda Sensibility Experiment

A second experiment focused in contrasting the effect of changing the lambda (λ) coefficient in the overall performance of the planning system in processing a request. In Fig. 6 are compared 2 diverse arrival rates; $\lambda=3$ and $\lambda=5$ requests per second for the 3-host and 5-host scenarios. The two curves in the lover part correspond to $\lambda=3$ scenarios while the other two at the top, to $\lambda=5$.

All curves present an increasing trend as expected with a pick around the 45th arrival. This is explained by the fact that with each new arrival vehicles increase their schedule plans, increasing also the time required to process feasible solutions (trip-insertion possibilities). The interesting point is in the increasing rate, while with

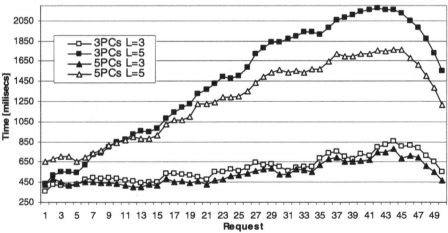

Fig. 5. Processing time for trip requests ordered by arrival at Exp(3) and Exp(5) rates, over 3 and 5 hosts. The mean times are over the 20 scenarios and their 25 runs each.

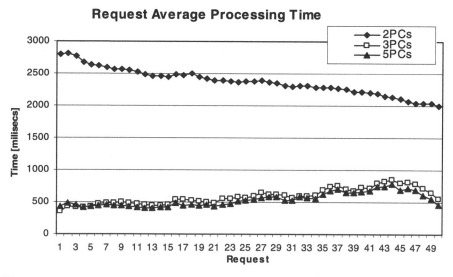

Fig. 6. Processing time for trip requests ordered by arrival at a rate of Exponential(3), varying the number of hosts to 2, and 5. The mean times are over the 20 scenarios and their 25 runs.

arrivals at $\lambda=3$ the processing time of a request increases at an average rate of around 10 milliseconds, at $\lambda=5$ it increases at a 32 milliseconds rate.

Besides, it is important to mention that in all cases exists a descending slope by the end explained by the fact that requests stop to arrive, lowering the systems congestion. Corresponds to an ending transient phase, reason why this data must not be considered in the analysis.

Finally, it is important to highlight that the average quality of the results for the two arrival rates and the three hosts' configurations does not vary that much. In fact, for all the 20 demand scenarios, the number of vehicles used and the cost of the solutions provided were not significantly different when varying arrival rates and number of hosts.

7 Conclusions

A distributed planning of trip request based on a greedy insertion heuristic plus the contract net protocol has been implemented using an agent architecture. Its performance was tested with diverse scenarios of 50 requests, diverse request arrival rates (λ) following an exponential distribution and with different number of hosts. Results shows that comparable results are obtained and how the average time for processing a single request increases through time with a higher rate as the λ parameter gets bigger.

Further work considers extending the present experiments in order to consider probability distributions for other events such as delays, no-show of clients and vehicles breakdowns, all of which imply a re-planning process.

Acknowledgement

This work is part of Project No. 209.746/2007 entitled "Coordinación en una sociedad multiagentededicada a la programación y control bajo ambiente dinámico", funded by the Pontifical Catholic University of Valparaíso (www.pucv.cl).

References

1 Ambrosino, G., et al.: EBusiness Applications to Flexible Transport and Mobility Services (2001), Available online at: http://citeseer.nj.nec.com/ ambrosino01ebusiness.html
2 Bellifemine, F. et al.: JADE - A FIPA Compliant Agent Framework. C SELT Internal Technical Report (1999)
3 Bürckert, H., Fischer, K., et al.: TeleTruck: A Holonic Fleet Management System. In: 14th European Meeting on Cybernetics and Systems Research, pp. 695–700 (1998)
4 Coslovich, L., Pesenti, R., Ukovich, W.: A Two-Phase Insertion Technique of Unexpected Customers for a Dynamic Dial-a-Ride Problem. Technical Report Working paper, Universita di Trieste, Italy (2003)
5 Cubillos, C., Guidi-Polanco, F., Demartini, C.: MADARP: Multi-Agent Architecture for Passenger Transportation Systems. In: Proceedings of the 8th IEEE International Conference on Intelligent Transportation Systems (ITSC'05), Vienna, Austria, IEEE Compurter Society Press, Los Alamitos (2005)
6 Cubillos, C., Gaete, S.: Design of an Agent-Based System for Passenger Transportation using PASSI. In: Mira, J., Alvarez, J.R. (eds.) IWINAC 2007. LNCS, vol. 4528, pp. 531–540. Springer, Heidelberg (2007)
7 Fischer, K., Müller, J.P., Pischel, M.: Cooperative Transportation Scheduling: An application Domain for DAI. Journal of Applied Artificial Intelligence 10 (1996)
8 Horn, M.E.T.: Fleet Scheduling and Dispatching for Demand-Responsive Passenger Services. Transportation Research C 10C, 35–63 (2002)
9 Jaw, J., et al.: A heuristic algorithm for the multiple-vehicle advance request dial-a-ride problem with time windows. Transportation Research 20B(3), 243–257 (1986)
10 Jih, W., Hsu, J.: Dynamic Vehicle Routing using Hybrid Genetic Algorithms. In: IEEE Int. Conf. on robotics & Automation, Detroit, IEEE Computer Society Press, Los Alamitos (May 1999)
11 Kikuchi, S., Donnelly, R.A.: Scheduling Demand-Responsive Transportation Vehicles using Fuzzy-Set Theory. Journal of Transportation Engineering 118(3), 391–409 (1992)
12 Kohout, R., Erol, K., Robert, C.: In-Time Agent-Based Vehicle Routing with a Stochastic Improvement Heuristic. In: AAAI/IAAI Int. Conf., Orlando, Florida, pp. 864–869 (1999)
13 Li, H., Lim, A.: A Metaheuristic for the Pickup and Delivery problem with Time Windows. In: 13th IEEE International Conference on Tools with Artificial Intelligence (ICTAI'01), Texas, IEEE Computer Society Press, Los Alamitos (November 2001)
14 Miyamoto, T., Nakatyou, K., Kumagai, S.: Route planning method for a dial -a-ride problem. IEEE Int. Conf. on SMC 4, 4002–4007 (2003)
15 Montemanni, R., Gambardella, et al.: A new algorithm for a dynamic vehicle routing problem based on ant colony system. In: 2nd Int. Workshop on Freight Transportation and Logistics (2003)
16 Perugini, D., Lambert, D., et al.: A distributed agent approach to global transportation scheduling. In: IEEE/ WIC Int. Conf. on Intelligent Agent Technology, pp. 18–24. IEEE Computer Society Press, Los Alamitos (2003)

17 Psaraftis, N.H.: Dynamic vehicle routing: Status and prospects. Annals of Operation Research 61, 143–164 (1995)
18 Uchimura, K., et al.: Demand responsive services in hierarchical public transportation system. IEEE Trans. on Vehicular Technology 51(4), 760–766 (2002)
19 Teodorovic, D., Radivojevic, G.A: Fuzzy Logic Approach to Dynamic Dial-A-Ride Problem. Fuzzy Sets and Systems 116, 23–33 (2000)
20 Weiss, G., Multiagent Systems, A.: Modern Approach to Distributed Artificial Intelligence. MIT Press, Massachusetts, USA (1999)

Maximizing the Number of Independent Labels in the Plane*

Kuen-Lin Yu[1], Chung-Shou Liao[2,**], and D. T. Lee[2,**]

[1] Department of Computer Science and Information Engineering,
National Chiao Tung University, HsinChu, Taiwan
[2] Department of Computer Science and Information Engineering,
National Taiwan University, Taipei, Taiwan
dtlee@ieee.org

Abstract. In this paper, we consider a map labeling problem to maximize the number of independent labels in the plane. We first investigate the point labeling model that each label can be placed on a given set of anchors on a horizontal line. It is known that most of the map labeling decision models on a single line (horizontal or slope line) can be easily solved. However, the label number maximization models are more difficult (like 2SAT vs. MAX-2SAT). We present an $O(n \log \Delta)$ time algorithm for the four position label model on a horizontal line based on dynamic programming and a particular analysis, where n is the number of the anchors and Δ is the maximum number of labels whose intersection is nonempty. As a contrast to Agarwal et al.'s result [Comput. Geom. Theory Appl. 11 (1998) 209-218] and Chan's result [Inform. Process. Letters 89(2004) 19-23] in which they provide $(1 + 1/k)$-factor PTAS algorithms that run in $O(n \log n + n^{2k-1})$ time and $O(n \log n + n\Delta^{k-1})$ time respectively for the fixed-height rectangle label placement model in the plane, we extend our method to improve their algorithms and present a $(1 + 1/k)$-factor PTAS algorithm that runs in $O(n \log n + kn \log^4 \Delta + \Delta^{k-1})$ time using $O(k\Delta^3 \log^4 \Delta + k\Delta^{k-1})$ storage.

1 Introduction

In cartographic literature, the main approach to conveying information concerning what is on the map is to attach texts or labels to geographic features on the map. Automated label placement subject to the constraint that the labels are pairwise disjoint is a well-known important problem in geographic information systems (GIS). In the ACM Computational Geometry Impact Task Force

---

* This work was supported in part by the National Science Council under the Grants NSC95-2221-E-001-016-MY3, NSC-94-2422-H-001-0001, and NSC-95-2752-E-002-005-PAE, and by the Taiwan Information Security Center (TWISC) under the Grants NSC NSC95-2218-E-001-001, NSC95-3114-P-001-002-Y, NSC94-3114-P-001-003-Y and NSC 94-3114-P-011-001.
** Also with the Institute of Information Science, Academia Sinica, Nankang, Taipei, Taiwan.

F.P. Preparata and Q. Fang (Eds.): FAW 2007, LNCS 4613, pp. 136–147, 2007.
© Springer-Verlag Berlin Heidelberg 2007

report[4] the map label placement is listed as an important research area. Since this problem in general is known to be NP-complete, many heuristics or special cases for which polynomial time algorithms are given have been presented. For instance, there are many algorithms that have been developed for labeling points that are on lines[5,9,11,12,19,22] or in a region[7,8,13,14,15,16,17,18,20,21].

Let $\mathcal{A}$ denote a set of points $\{A_1, A_2, \ldots, A_n\}$ in the plane, called *anchors*. Associated with each anchor there is an axis-parallel rectangle, called *label*. The *point-feature label placement problem* or simply *point labeling problem*, is to determine a placement of these labels such that the anchors coincide with one of the corners of their associated labels and no two labels overlap. The point labeling problem for labeling an arbitrary set of points has been shown to be NP-complete[8,14,15,18] and there were some heuristic algorithms[6,8,21].

There are many variations of the point labeling problem, including shapes of the labels, locations of the anchors to be labeled and where the labels are placed. Consider the case that the placement of the labels are restricted. For instance, one is *fixed-position model*, denoted 4P model, in which a label must be placed so that the anchor coincides with one of its four corners; and another is *slider model*, denoted 4S model, in which a label can be placed so that the anchor lies on one of the four boundary edges of the label. The coordinate positions $\{1, 2, 3, 4\}$ in 4P model denote the corner positions of labels coincident with the anchor, and the arrows in 4S model indicate the directions along which the label can slide, maintaining contact with the anchor.

In this paper we consider the case when the anchors lie on a line and are to be labeled with rectangular labels. This problem has been studied previously [9,16,19,5]. The prefix 1d or Slope refers to the problem in which the anchors lie on a horizontal or a sloping line, respectively. Garrido et al.[9] gave linear time algorithms for 1d4P rectangle label, 1d4S square label, and Slope4P square label models, and a quadratic time algorithm for Slope4S square label model as well. They also showed 1d4S rectangle label is NP-complete and consider the maximization version to maximize the size of labels. Chen et al.[5] further provided linear time algorithms for the decision version of Slope4P fixed-height(or width) rectangle label and elastic rectangular label (of a given area) models. They also presented a lower bound $\Omega(n \log n)$ time and a different method to maximize the label size for 1d4S square label model. Maximizing the number of labels that can be placed or the so-called maximum independent set problem, is yet another common problem. Although the label size maximization model is as easy as the decision model, the label number maximization model has been considered to be harder. Most of cases where we can tackle the decision models in polynomial time are more or less comparable to 2SAT, and yet the label number maximization model is relatively more difficult (like MAX-2SAT). In 1998, Agarwal et al.[1] provided a $(1 + 1/k)$-factor algorithm that runs in $O(n \log n + n^{2k-1})$ time, for any integer $k \geq 1$, for fixed-height rectangle label placement model in the plane and an $O(\log n)$-factor approximation algorithm that runs in $O(n \log n)$ time for arbitrary rectangle labels. Poon et al.[19] further considered the weighted case in which each label is associated with a given weight and provided the same approximation result for 4P fixed-height weighted rectangle model. They also gave a $(2+\epsilon)$-factor approximation algorithm that runs in $O(n^2/\epsilon)$ time for 1d4S weighted rectangle label. As for arbitrary rectangle label, Berman et al.[2] presented a $\lceil O(\log_k n)\rceil$-factor approximation algorithm that runs in $O(n^{k+1})$ time, for any integer $k \geq 2$. In 2004, Chan[3] improved the previous results an gave a $(1+1/k)$-factor algorithm that runs in $O(n\log n + n\Delta^{k-1})$ time, where $\Delta \leq n$ denotes the maximum number of rectangles whose intersection is nonempty, for fixed-height rectangle label model and a $\lceil O(\log_k n)\rceil$-factor approximation algorithm that runs in $n^{O(k/\log k)}$ time for arbitrary rectangle label.

We first investigate the maximization version of the feasible number of labels when the anchors lie on a horizontal line. That is, we want to maximize the number of labels whose associated anchors lie on a horizontal line for which a *feasible* placement exists that no two labels overlap. In other words, these labels form an independent set. We refer to this model as *Max-1d4P rectangle label model*, or *Max-1d4P* for short. Since most of the decision model of map labeling problems on a single line are easily solved in a greedy manner, we are looking for an almost linear time algorithm for the Max-1d4P model. As a contrast to previous related results[1,3] in which the maximum independent set of label placement problem in the plane was considered and polynomial time approximation schemes (PTAS) were provided using the *line stabbing* technique and the shifting idea, we present a faster approach based on a different form of dynamic programming strategy and a particular analysis to solving this Max-1d4P model in $O(n\log \Delta)$ time which improves previously known results that run in $O(n^2)$ and $O(n\Delta)$ time in the worse case. We also point out an implicit difference between point labeling problem and label placement problem, mentioned in the intuitive proof of the reduction[19]. In addition, we further extend our method to solve the fixed-height rectangle label placement model in the plane and present a $(1+1/k)$-factor polynomial time approximation scheme (PTAS) algorithm that runs in $O(n\log n + kn\log^4 \Delta + \Delta^{k-1})$ time, using $O(k\Delta^3 \log^4 \Delta + k\Delta^{k-1})$ storage.

This paper is organized as follows. In Section 2, we introduce some definitions. Then we present in Section 3 an $O(n\log \Delta)$ time algorithm for the Max-1d4P model. In Section 4 we specify an implicit difference of point labeling problem and label placement problem and give a $(1+1/k)$-factor PTAS algorithm for the fixed-height rectangle label placement model in the plane. Finally we conclude in Section 5 with some discussions of future work.

2   Preliminaries

Consider a set of anchors $\mathcal{A} = \{A_1, A_2, \ldots, A_n\}$ on a horizontal line, and each anchor A_k is associated with its position (in x-coordinate) x_k and label size l_k. The aim is to maximize the number of *feasible* labels so that they do not overlap with each other. A feasible solution to the point labeling problem is called a *realization*.

Since we consider the problem on a horizontal line and put the label either *above* or *below* the line, we can simply associate a 2-tuple, namely (a, b), to represent the current labeling state of a realization R, with $R.a = a$ and $R.b = b$, representing respectively the coordinates of the right edge of the rightmost label *above* the line and *below* the line. A realization R will also contain a specification of the label placement at feasible positions associated with a subset of anchors. To be more precise, we can use $A_i.\ell \in R$, where $A_i.\ell \in \{0, 1, 2, 3, 4\}$ indicates the label position for anchor A_i included in R, with $A_i.\ell = 0$ representing anchor A_i is not labeled. If R contains k-feasible labels, i.e., it contains k non-zero $A_i.\ell$'s, then R is called a k-realization, and we use $R.c$ to denote the cardinality of the subset of feasible labels. We shall use the notation R to not only represent a realization R of $\mathcal{A}$, which corresponds to a subset, $\Re$, of feasible anchors, i.e., $\Re \subseteq \mathcal{A}$, but also use $R.a$, $R.b$ and $R.c$ to represent the state of its configuration and its size, respectively. Let us assume that the set of anchors has been ordered so that their x-coordinates are in strictly increasing order. That is, $x_1 < x_2 < \ldots < x_n$. Let $\mathcal{A}_i$ denote the subset of anchors $\{A_1, A_2, \ldots, A_i\}$, for $i = 1, 2, \ldots, n$, and R^i denote a realization of $\mathcal{A}_i$ for some i. An optimal solution is a realization R^n such that $R^n.c$ is maximum among all possible realizations of $\mathcal{A}_n$.

We shall process the anchors, and their associated labels, in ascending order of their x-coordinates, i.e., in the order of $A_1, A_2, \ldots, A_n$. Given a realization R^{i-1} of $\mathcal{A}_{i-1}$, and the next anchor, A_i, $i > 1$, the placements of the label of A_i that do not overlap the last label both above and below the line in R^{i-1} are called *feasible label placements*. Before proceeding we define the notion of *equivalence* of two realizations:

Definition 1. *Given two realizations R_1^i and R_2^i of $\mathcal{A}_i$ such that $R_1^i.c$ and $R_2^i.c$ are equal, if $\{R_1^i.a, R_1^i.b\} = \{R_2^i.a, R_2^i.b\}$, we say that the two realizations are* **equivalent in size**, *or simply* **equivalent** *to each other.*

Based on the above definition, for a realization R^i with $R^i.a < R^i.b$, we always swap the upper and lower sides of the realization. That is, a realization will be represented in a *normal form* in which the coordinate *above* the line is no less than the coordinate *below* the line, i.e., $R^i.a \geq R^i.b$ without loss of generality. Here we define the *comparability* of two realizations.

Definition 2. *For any two realizations R_i^k and R_j^k, $1 \leq k \leq n$, if the following statements hold,*

(1). $R_i^k.c = R_j^k.c$, and

(2). $R_i^k.a \leq R_j^k.a$, and $R_i^k.b \leq R_j^k.b$ then we say that the two realizations are **comparable** *and R_i^k is* **better** *than R_j^k. Otherwise, they are* **incomparable**.

Lemma 1. *Let $\mathcal{R}$ be an optimal realization of $\mathcal{A}$. Suppose R_i^k and R_j^k are two comparable realizations for $\mathcal{A}_k$ and R_i^k is better than R_j^k for some $n \geq k \geq 1$. If $\mathcal{R}$ contains R_j^k as a subset, then there exists another optimal solution that contains R_i^k.*

By using a 2-tuple to represent the labeling state of a realization, we can transform it into a point in the two-dimensional plane. To be more precise, given a realization R represented by a 2-tuple $(R.a, R.b)$, we transform it into a point $P(x, y)$, where $x = R.a$ and $y = R.b$, in the plane. The two equivalent realizations will then be transformed into two points that are symmetric with respect to the line $x = y$. From now on, we use $P.x$ and $P.y$ to represent the x and y coordinates of a point P in the plane, and $P.c$ to represent its associated cardinality. We assume the point labeling on a line starts at the origin without loss of generality, which means the transformed points in the plane are all in the first quadrant. Using the normal form representation of a realization, all realizations will be mapped to points that are all located in the first quadrant below the line $x = y$.

Based on the *comparability* definition between two realizations, if one is better than the other, then the transformed points will carry the relationship of *domination*. That is, if realization R_s is *better* than realization R_t, then point P_t *dominates*[1] point P_s in the plane. On the other hand, if they are *incomparable* realizations, the transformed points in the plane do not dominate each other.

We shall also transform each anchor A_k with its position x_k and its given label size (or length) l_k, $1 \leq k \leq n$, into a point of another kind $P^{A_k}(x_k, y_k)$ located on line $x = y$ in the plane, that is, $x_k = y_k$. We define the operations in the plane as follows.

Definition 3. *Given a point $P(x, y)$ representing a realization and a point P^{A_k} (x_k, x_k) representing a new label $A_k.\ell$ of length l_k in the plane, $x_k - l_k \geq x$, we have the following operations depending on how we select the placement of label $A_k.\ell$ of the anchor A_k for the realization P.*

> *1. $A_k.\ell = 1$, then $P(x, y)$ generates $P'(x_k + l_k, y)$.*
> *2. $A_k.\ell = 2$, then $P(x, y)$ generates $P'(x_k, y)$.*
> *3. $A_k.\ell = 3$, then $P(x, y)$ generates $P'(x, x_k)$.*
> *4. $A_k.\ell = 4$, then $P(x, y)$ generates $P'(x, x_k + l_k)$.*

*The cardinality associated with point P' will be one more than that with point P. If the y-coordinate of P' is greater than the x-coordinate, we do the swapping operation to exchange the x- and y- coordinates. We call point P the **parent point** of P' and the generated point P' the **child point** of P.*

Property 1. Given a parent point $P(x, y)$ and a point $P^{A_k}(x_k, x_k)$ with label length l_k,

> 1. If $y > x_k$, the point cannot apply label at any position.
> 2. If $x_k \geq y > x_k - l_k$, the point can apply label at position 4.
> 3. If $x_k - l_k \geq y$, the point can apply label at positions 3 and 4.
> 4. If $x > x_k$, the point can apply label at neither position 1 nor 2.
> 5. If $x_k \geq x > x_k - l_k$, the point can apply label at positions 1 and 4.
> 6. If $x_k - l_k \geq x$, the point can apply label at positions 1, 2, 3 and 4.

---

[1] If $P_t.x \geq P_s.x$, and $P_t.y \geq P_s.y$, then P_t is said to *dominate* P_s.

3   Point Labeling on a Single Line

We adopt a greedy method to solve the model Max-1d4P, namely, we will process the anchors in sequential manner, and maintain a set of realizations that reflect the best possible labeling, ignoring those that are known to be no better than the present set of realizations, after each anchor is processed. The following lemma is obvious.

Lemma 2. *Given two realizations R_s^i and R_t^i, for $i = 1, 2, \ldots, n$, we will select R_s^i over R_t^i, either if $R_s^i.c > R_t^i.c$, or if $R_s^i.c = R_t^i.c$, and R_s^i is better than R_t^i.*

Due to space constraint, we shall skip the proofs. The details of the proofs can be found in Yu et al.[22].

Lemma 3. *Given a realization R^{i-1}, if both label placements of the next anchor A_i at positions 2 and 3 (respectively, positions 1 and 4) are feasible, the selection of label at position 2 (respectively, position 1), above the line will yield a better realization R^i.*

Agarwal et al.[1] provided a standard dynamic programming method to solve the fixed-height rectangle label placement model in $O(n \log n + n^{2k-1})$ time if all the rectangles in the plane are exactly stabbed by k horizontal lines. Poon et al.[19] used a similar approach to solving the fixed-height rectangle 4P model (we will specify an implicit difference between these two models later). They both associated a polygonal line consisting of $2k-1$ orthogonal segments to specify all the possible $(2k-1)$-dimensional realizations. For the Max-1d4P model, since all the rectangles are stabbed by two horizontal lines exactly, it only needs a two-dimensional table $R[x, y]$ which stores the cardinality of the realization R with $R.a = x$ and $R.b = y$ (as we introduced in Section 2), and their solution leads to an $O(n^2)$ time algorithm. Chan[3] presented a different form of dynamic programming and improved the time complexity to $O(n \log n + n\Delta^{k-1})$ for the same model. The form $R[i, S]$ stores the cardinality of the realization R associated with the vertical line $x = x_i$ and a set S of disjoint rectangles intersecting the line $x = x_i$, where x_i, $1 \leq i \leq n$, denotes the abscissas of the left boundaries of all the rectangles, and R is a maximum independent set of rectangles to the right of $x = x_i$ and intersecting none of S with $|S| \leq k - 1$. For the Max-1d4P model, since all the rectangles are exactly stabbed by two horizontal lines, there is at most Δ choices for $|S| = k - 1$ (as $k = 2$), and thus it takes $O(n\Delta)$ time. Although their dynamic programming methods are apparently different, the operations (inserting or discarding the next rectangle) are executed iteratively for every possible realization. We provide another form of dynamic programming strategy and tackle the operations for partial representative realizations instead. It solves the Max-1d4P model in $O(n \log \Delta)$ time based on a particular analysis and improves their quadratic time results ($O(n^2)$ and $O(n\Delta)$, respectively) in the worse case. We first observe some properties of the model Max-1d4P.

Proposition 1. *Given a realization R^{i-1}, when both placements at label positions 1 and 2 (respectively, positions 3 and 4) of next anchor A_i are feasible,* *the selection of label at position 2 (respectively, position 3) will yield a better realization R^i.*

Lemma 4. *Given a realization R^{i-1}, if the label placement of the next anchor A_i at position 3 is feasible, then label of A_i must be included in an optimal realization R^n that contains R^{i-1}, or $A_i.\ell \neq 0$.*

Corollary 1. *Given a realization R^{i-1}, when the label placement of next anchor A_i at position 2 is feasible, we have $A_i.\ell \neq 0$ in an optimal realization R^n.*

We introduce our main idea as follows. Let $S[i,j]$ denote a set of incomparable realizations R^j of cardinality i, for $1 \leq i \leq j \leq n$ ($i > 0$ since $A_1.\ell = 2$ without loss of generality). We shall apply a dynamic programming method to process the anchors and record the 'better' realizations of each possible cardinality. To find an optimal realization R^n, we may need to maintain intermediate realizations $S[i,j]$ for $1 \leq i \leq j \leq n$, that have the potential leading to an optimal realization. As we shall show later, for each $j \leq n$ we only need to maintain at most *five* subsets $S[k,j]$, $S[k+1,j]$, $S[k+2,j]$, $S[k+3,j]$, and $S[k+4,j]$ for some k, which is a *key* result of this paper. We shall process the table from $j = 1$ till n and fill each entry $S[i,j]$ with a set of incomparable realizations at each step.

The realizations in an incomparable set form a *"point chain"* in the plane without having any point in the chain dominate another. When we encounter a new anchor A_j, some of the points in this chain of cardinality k, for some k, will generate new child points, thus getting *upgraded* to a realization of cardinality $k+1$, some will remain as *non-upgraded* with cardinality k, and are kept as potential candidates without including A_j, leading possibly to an optimal solution, and some get eliminated due to some new child points upgraded from points of cardinality $k-1$. At the end after anchor A_n is processed, the realizations in the non-empty entry $S[i,n]$ with the largest i are optimal solutions.

To sum up, some points in the set $S[i,j]$ may simply move to $S[i,j+1]$ without increasing cardinality, following what we call a *non-upgrading process*. Other points in the set may generate points which are included in $S[i+1,j+1]$, whose cardinality is incremented, following what we call an *upgrading process*. When a point moves from one entry to another, it should be compared with other points in the target entry, and only *better* ones are kept. We repeat such operations until we have processed all anchors. The following is the algorithm for the model Max-1d4P.

Algorithm 1M4P. Find the maximum cardinality of map labeling for the model Max-1d4P.

Input. A set of anchors $\mathcal{A} = \{A_1, \ldots, A_n\}$ sorted by x-coordinates and associated set of labels.

Output. The maximum cardinality of an optimal realization R^n for Max-1d4P.

Method.

0. /*Use dynamic programming method on two parameters $S[i,j]$ with the anchor ordering in column and the cardinality of possible solutions in row.

Initialize the first entry $S[1,1]$ with the label placement of the first anchor at position 2, that is, $A_1.\ell = 2.*/$

1. For $j = 2$ **to** n

Let the largest cardinality of non-empty entries in column $j - 1$ be k;

For $i = k$ **down to** $\max\{k - 4, 0\}$

1-1. Classify the points in $S[i, j - 1]$ into upgrading and non-upgrading classes according to A_j;

1-2. Move the non-upgraded points into $S[i, j]$;

1-3. Move the upgraded points into $S[i + 1, j]$;

1-4. Compare the newly upgraded points with existing points in $S[i + 1, j]$ and keep the better ones;

2. Output the largest i of the nonempty entry $S[i, n]$;

In what follows we will prove a few results that help establish the correctness of our algorithm. Let P_s and P_t be two incomparable points, and $P^{A_k}(x_k, x_k)$ be a point associated with the next anchor A_k.

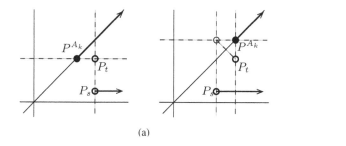 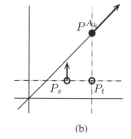

(a)   (b)

Fig. 1. Illustration of Lemma 5

Lemma 5. *For the following two cases, where P_s and P_t correspond to two realizations, and the next anchor is A_k associated with $P^{A_k}(x_k, x_k)$,*

(a) $P_s.x \geq P_t.x$, $P_s.y < P_t.y$, $P_s.c < P_t.c$ and $x_k \geq P_t.y$
(b) $P_s.y \geq P_t.y$, $P_s.x < P_t.x$, $P_s.c < P_t.c$ and $x_k \geq P_t.x$

P_t is better than P_s.

Lemma 6. *The points in an incomparable set going through an upgrading process collectively generate at most two incomparable child points.*

Theorem 1. *After processing an anchor A_u, if there is a point with cardinality k which is the ancestor of a point with cardinality $k + 5$, then no point with cardinality k will lead to an optimal solution. That is, the difference in cardinality of incomparable points is at most four.*

By Theorem 1, in computing $S[i, j]$ for $1 \leq i \leq j \leq n$, it is sufficient to maintain at most five consecutive sets of incomparable realizations $S[k, *]$, $S[k + 1, *]$, $S[k + 2, *]$, $S[k + 3, *]$ and $S[k + 4, *]$. We have the following based on the above two results.

Theorem 2. *The number of points in an incomparable set is bounded by $O(\Delta)$, where Δ is the maximum number of labels whose intersection is nonempty.*

Lemma 7. *The following operations each take $O(\log \Delta)$ time.*

(1) Classifying points into the upgrading and non-upgrading classes.
(2) Finding two incomparable points among all upgraded points.
(3) The comparison between upgraded points and $O(\Delta)$ incomparable points.

Theorem 3. *The time complexity of Algorithm 1M4P is $O(n \log \Delta)$.*

When the algorithm terminates, any point in nonempty $S[i, n]$ with the largest i is an optimal solution (of maximum cardinality i). The actual placement of labels can be obtained if we record the processing history when a point is upgraded.

4   Fixed-Height Rectangle Label Placement in the Plane

First, we point out an implicit difference between point labeling problem and label placement problem. As point labeling problem was considered, where a constant number of label positions is allowed for each anchor, all label positions of each anchor were regarded as pairwise intersecting and the reduction from point labeling problem to label placement problem seemed intuitive[1,19]. However, for common 4P model, if there are more than one anchor lying on some horizontal line (or vertical line) with nonempty label intersection, we have the following implicit difference. Figure 2 shows that for anchor A_i, the selection of label at position 1 will affect the selection of the label at position 3 for anchor A_j. They could be regarded as intersecting, but in fact they are not. This problem can be resolved by set manipulation instead. Let a selection set of each anchor consist of all its label positions. We then allow at most one label position of each selection set be included in the solution. This doesn't affect the asymptotic running time but increase the implementation complexity.

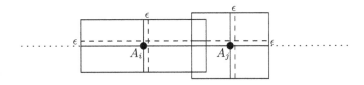

Fig. 2. An example shows an implicit difference between point labeling problem and label placement problem

For the fixed-height label placement model in the plane, Agarwal et al.[1] provided a $(1 + 1/k)$-factor PTAS algorithm running in $O(n \log n + n^{2k-1})$ time, and later Chan[3] improved it and presented a $(1 + 1/k)$-factor PTAS algorithm running in $O(n \log n + n\Delta^{k-1})$ time, for an integer $k \geq 1$. They both use the *line stabbing* technique and the shifting idea of Hochbaum and Maass[10]. Assume all the fixed-height rectangles are stabbed by m horizontal lines and each rectangle is stabbed by one line exactly, and let C^i, $0 \leq i \leq k$, be the sub-collection of all rectangles which are not stabbed by any horizontal line $y = y_l$ with $l \equiv i$ mod $(k+1)$. Then C^i is a union of groups of rectangles, where each group can be stabbed by k horizontal lines and no two rectangles from different groups intersect. If we can solve the fixed-height label placement model for each group of n' rectangles for a given integer $k \geq 1$ in $O(t(k, n'))$ time, then we take the union of all groups to obtain a solution of C^i, $\forall i$, and select one of the solutions, O', of maximum cardinality. The total time complexity is bounded by $O(t(k, n))$ time and it is trivial to show that $k|O^*| \leq (k+1)|O'|$, where O^* is an optimal solution[3]. Hence we consider the fixed-height label placement model for n rectangles stabbed by k horizontal lines, for a given integer $k \geq 1$, from now on. Given a set of n fixed-height rectangles $r_1, r_2, \ldots, r_n$ sorted by x-coordinates of the left boundaries of all rectangles, stabbed by k horizontal lines $L_1, \ldots, L_k$, the following property is immediate by the line stabbing technique.

Property 2. From left to right, as we process r_i stabbed by L_l, r_i intersects at most two rectangles among $r_1, \ldots, r_{i-1}$: r_f, r_g, or r_g, r_h, where r_f, r_g, and r_h, are last rectangles of $r_1, \ldots, r_{i-1}$ on lines L_{l-1}, L_l, and L_{l+1}, respectively.

We associate a polygonal line consisting of $2k - 1$ orthogonal segments to specify all the possible $(2k-1)$-dimensional realizations as used in Agarwal et al.[1]. However, based on the above property, we only need to consider the last rectangles on lines L_{l-1}, L_l, and L_{l+1} (i.e., three labels, and all the possible five-dimensional realizations), as we process each rectangle r_i, $1 \leq i \leq n$, stabbed by line L_l, for some l. In addition, we use the same dynamic programming method in Section 3 and extend our two-dimensional transformation for point labeling on a single line to $(2k-1)$-dimensional transformation for this model as follows. We transform a $(2k-1)$-tuple representation $(R.x_1, R.x_2, \ldots, R.x_{2k-1})$ of the labeling state of a realization R into a point $P(x_1, x_2, \ldots, x_{2k-1})$ in the $(2k-1)$-dimensional space. Therefore, as we process each rectangle r_i, $1 \leq i \leq n$, stabbed by line L_l, for some l, we only need to consider the five-dimensions $(x_{2l-2}, x_{2l-1}, x_{2l}, x_{2l+1}, x_{2l+2})$ of present incomparable points. The next lemma is an extension of Theorem 1.

Lemma 8. *After processing a rectangle r_i, if there is a point in $(2k - 1)$-dimensional space with cardinality c which is the ancestor of a point with cardinality $c + (4k + 1)$, then no point with cardinality c will lead to an optimal solution. That is, the difference in cardinality of incomparable points is at most $4k$. More precisely, after processing a rectangle r_i, if there is an ancestor point P with cardinality c upgraded by at least five rectangles stabbed by the same line, then no point with cardinality c will lead to an optimal solution.*

By Lemma 8, as we compute $S[i, j]$ for $1 \leq i \leq j \leq n$, it is sufficient to maintain at most $\min\{4k + 1, n\}$ consecutive sets of incomparable points $S[c, *], S[c + 1, *], \ldots, S[c + \min\{4k, n - 1\}, *]$. We further extend Theorem 2 to the following lemma.

Lemma 9. *The number of points in an incomparable set is bounded by $O(\Delta^{k-1})$, where Δ is the maximum number of rectangles whose intersection is nonempty.*

We construct two three-dimensional range trees T_1 and T_k, and $k - 2$ five-dimensional range trees $T_2, \ldots, T_{k-1}$ to determine whether the placement of next rectangle r_i stabbed by line L_j is feasible or not, by a range query of T_j in five-dimension $(x_{2j-2}, x_{2j-1}, x_{2j}, x_{2j+1}, x_{2j+2})$. According to the above lemmas, we extend Lemma 7 and Theorem 3 to obtain the result.

Theorem 4. *Given an integer $k \geq 1$, the label placement problem of n fixed-height rectangles stabbed by k horizontal lines can be solved in $O(n \log n + kn \log^4 \Delta + \Delta^{k-1})$ time and $O(k\Delta^3 \log^4 \Delta + k\Delta^{k-1})$ space.*

5 Concluding Remarks

We have extended the decision version of the map labeling problem on a horizontal line to an optimization version where the number of feasible labels is to be maximized. It is a variation of maximum independent set problem on interval graphs. Improving the previous related results, we have presented a faster $O(n \log \Delta)$ time algorithm for the Max-1d4P model by dynamic programming on two parameters: the anchor ordering and the cardinality of possible solutions. In addition, we have further extended our method to improve the previous results for the fixed-height rectangle label placement model in the plane and presented a $(1 + 1/k)$-factor PTAS algorithm that runs in $O(n \log n + kn \log^4 \Delta + \Delta^{k-1})$ time, using $O(k\Delta^3 \log^4 \Delta + k\Delta^{k-1})$ storage.

We conclude with two open questions concerning map label number maximization model. First, whether there exist solutions for Max-Slope4P fixed-height(or width) rectangle label model remains to be seen. Secondly, how to improve Algorithm 1M4P to obtain a linear time algorithm solving the Max-1d4P model given sorted anchors, and further extend it to reduce the time complexity of PTAS for label placement problem in the plane is worthwhile.

References

1. Agarwal, P.K., van Kreveld, M., Suri, S.: Label placement by maximum independent set in rectangles. Computational Geometry: Theory and Applications 11, 209–218 (1998)
2. Berman, P., DasGupta, B., Muthukrishnan, S., Ramaswami, S.: Efficient approximation algorithms for tiling and packing problems with rectangles. Journal of Algorithms 41, 443–470 (2001)
3. Chan, T.M.: A note on maximum independent sets in rectangle intersection graphs. Inform. Process. Letters 89, 19–23 (2004)
4. Chazelle, B., 36 co-authors: The computational geometry impact task force report. In: Chazelle, B., Goodman, J.E., Pollack, R. (eds.) Advances in Discrete and Computational Geometry, vol. 223, pp. 407–463. American Mathematical Society, Providence (1999)

5. Chen, Y.-S., Lee, D.T., Liao, C.-S.: Labeling points on a single line. International Journal of Computational Geometry and Applications (IJCGA) 15(3), 261–277 (2005)
6. Christensen, J., Marks, J., Shieber, S.: An empirical study of algorithms for point feature label placement. ACM Transactions on Graphics 14(3), 203–232 (1995)
7. Duncan, R., Qian, J., Vigneron, A., Zhu, B.: Polynomial time algorithms for three-label point labeling. Theoretical Computer Science 296(1), 75–87 (2003)
8. Formann, M., Wagner, F.: A packing problem with applications in lettering of maps. In: Proceedings of the 7th ACM Symposium on Computational Geometry, pp. 281–288. ACM Press, New York (1991)
9. Garrido, M.Á., Iturriaga, C., Márquez, A., Portillo, J.R., Reyes, P., Wolff, A.: Labeling subway lines. In: Eades, P., Takaoka, T. (eds.) ISAAC 2001. LNCS, vol. 2223, pp. 649–659. Springer, Heidelberg (2001)
10. Hochbaum, D.S., Maass, W.: Approximation schemes for covering and packing problems in image processing and VLSI. J. ACM 32(1), 130–136 (1985)
11. Iturriaga, C., Lubiw, A.: Elastic labels: The two-axis case. In: DiBattista, G. (ed.) GD 1997. LNCS, vol. 1353, pp. 181–192. Springer, Heidelberg (1997)
12. Iturriaga, C., Lubiw, A.: Elastic labels around the perimeter of a map. Journal of Algorithms 47(1), 14–39 (2003)
13. Jiang, M., Qian, J., Qin, Z., Zhu, B., Cimikowski, R.: A simple factor-3 approximation for labeling points with circles. Inform. Process. Letters 87(2), 101–105 (2003)
14. Kato, T., Imai, H.: The NP-completeness of the character placement problem of 2 or 3 degrees of freedom. In: Record of Joint Conference of Electrical and Electronic engineers in Kyushu, Japanese, p. 1138 (1988)
15. Knuth, D., Raghunathan, A.: The problem of compatible representatives. SIAM Disc. Math. 5(3), 422–427 (1992)
16. van Kreveld, M., Strijk, T., Wolff, A.: Point labeling with sliding labels. Computational Geometry: Theory and Applications 13, 21–47 (1999)
17. Kim, S.K., Shin, C.-S., Yang, T.-C.: Labeling a rectilinear map with sliding labels. International Journal of Computational Geometry and Applications 11(2), 167–179 (2001)
18. Marks, J., Shieber, S.: The computational complexity of cartographic label placement, Technical Report TR-05-91. Harvard University CS (1991)
19. Poon, S.-H., Shin, C.-S., Strijk, T., Uno, T., Wolff, A.: Labeling points with weights. Algorithmica 38(2), 341–362 (2003)
20. Strijk, T., Wolff, A.: Labeling points with circles. International Journal of Computational Geometry and Applications 11(2), 181–195 (2001)
21. Wagner, F., Wolff, A.: A practical map labeling algorithm. Computational Geometry: Theory and Applications. 7, 387–404 (1997)
22. Yu, K.-L., Liao, C.-S., Lee, D.T.: Maximizing the number of independent labels in the plane, manuscript, A preliminary version appeared in proceedings of the Frontiers of Algorithmics Workshop (April 2007), http://www.iis.sinica.edu.tw/~shou794/research/1DMIS_p15_0416.pdf

On the Fractional Chromatic Number
of Monotone Self-dual Boolean Functions

Daya Ram Gaur[1] and Kazuhisa Makino[2]

[1] University of Lethbridge, Lethbridge, AB, Canada, T1K 3M4
gaur@cs.uleth.ca
[2] Graduate School of Information Science and Technology, University of Tokyo,
Tokyo, 113-8656, Japan
makino@mist.i.u-tokyo.ac.jp

Abstract. We compute the exact fractional chromatic number for several classes of monotone self-dual Boolean functions. We characterize monotone self-dual Boolean functions in terms of the optimal value of a LP relaxation of a suitable strengthening of the standard IP formulation for the chromatic number. We also show that determining the self-duality of monotone Boolean function is equivalent to determining feasibility of a certain point in a polytope defined implicitly.

1 Introduction

A *Boolean function*, or a *function* in short, is a mapping $f : \{0,1\}^n \to \{0,1\}$, where $x = (x_1, \ldots, x_n) \in \{0,1\}^n$ is called a *Boolean vector* (a *vector* in short). A Boolean function is said to be *monotone* if $f(x) \leq f(y)$ for all vectors x and y with $x \leq y$, where $x \leq y$ denotes $x_i \leq y_i$ for all $i \in \{1, \ldots n\}$. It is known that a Boolean function is monotone if and only if it can be represented by a formula that contains no negative literal. Especially, any monotone function f has a unique prime disjunctive normal form (DNF) expression

$$f = \bigvee_{H \in \mathcal{H}} \left(\bigwedge_{j \in H} x_j \right), \tag{1}$$

where $\mathcal{H}$ is a *Sperner* (or *simple*) hypergraph on $V (= \{1, \ldots, n\})$, i.e., $\mathcal{H}$ is a subfamily of 2^V that satisfies $H \not\subseteq H'$ and $H \not\supseteq H'$ for all $H, H' \in \mathcal{H}$ with $H \neq H'$. It is well-known that $\mathcal{H}$ corresponds to the set of all prime implicants of f. Given a function f, we define its *dual* $f^d : \{0,1\}^n \to \{0,1\}$ by $f^d(x) = \overline{f(\overline{x})}$ for all vectors $x \in \{0,1\}^n$, where $\overline{x}$ is the componentwise complement of x, i.e., $\overline{x} = (\overline{x}_1, \ldots, \overline{x}_n)$. As is well-known, the formula defining f^d is obtained from that of f by exchanging $\vee$ and $\wedge$ as well as the constants 0 and 1. A function f is called *self-dual* if $f = f^d$ holds.

Monotone self-dual functions have been studied not only in Boolean algebra, but also in hypergraph theory [2,20], distributed systems [13,16], and game theory [29] under the names of strange hypergraphs, non-dominated coteries, and decisive games, respectively. For example, a Sperner hypergraph $\mathcal{H} \subseteq 2^V$ is called

F.P. Preparata and Q. Fang (Eds.): FAW 2007, LNCS 4613, pp. 148–159, 2007.

strange [20] (or *critical non-2-colorable* [1]) if it is intersecting (i.e., every pair in $\mathcal{H}$ has at least one element from V in common) and not 2-colorable (i.e., the chromatic number $\chi(\mathcal{H})$ of $\mathcal{H}$ satisfies $\chi(\mathcal{H}) > 2$). It is known (e.g., [2,8]) that a monotone function f is self-dual if and only if it can be represented by (1) for a strange hypergraph $\mathcal{H}$. Here we note that there exists a one-to-one correspondence between monotone self-dual functions and strange hypergraphs. Strange hypergraphs can also be characterized in terms of transversals. For a hypergraph $\mathcal{H} \subseteq 2^V$, $T \subseteq V$ is a *transversal* of $\mathcal{H}$ if $T \cap H \neq \emptyset$ for all $H \in \mathcal{H}$, and the family of minimal transversals of $\mathcal{H}$ is called the *transversal hypergraph* of $\mathcal{H}$, denoted by $Tr(\mathcal{H})$. Then $\mathcal{H}$ is strange if and only if $Tr(\mathcal{H}) = \mathcal{H}$ holds (e.g., [1,2,8]).

Another characterization of self-duality (i..e, strangeness) appears in the literature of distributed systems [13,16]. A *coterie* is an intersecting Sperner hypergraph. A coterie $\mathcal{H}$ *is dominated by* another coterie $\mathcal{H}'$ if for each $H \in \mathcal{H}$ there exists an $H' \in \mathcal{H}'$ such that $H' \subseteq H$, and is *non-dominated* if no such coterie $\mathcal{H}'$ exists. It is known (e.g., [13]) that $\mathcal{H}$ is strange if and only if it is a non-dominated coterie. In summery, the following equivalence is known.

Theorem 1 (E.g., [1,2,8,13,16]). *Let $\mathcal{H}$ be a Sperner hypergraph, and f be a monotone function defined by (1). Then the following statements are equivalent:*

1. $\mathcal{H}$ *is a non-dominated coterie.*
2. f *is self-dual.*
3. $\mathcal{H}$ *is strange (i.e., $\mathcal{H}$ is intersecting and $\chi(\mathcal{H}) > 2$).*
4. $Tr(\mathcal{H}) = \mathcal{H}$.

Given a monotone function f represented by (1), the *self-duality problem* is to determine whether $f^d = f$. By Theorem 1, the self-duality problem is to decide if a given hypergraph $\mathcal{H}$ satisfies $Tr(\mathcal{H}) = \mathcal{H}$, i.e., is strange (or a non-dominated coterie). Since it is known [3,8] that the self-duality problem is polynomially equivalent to the monotone duality problem, i.e., given two monotone DNFs φ and ψ, deciding if they are mutually dual (i.e., $\varphi^d \equiv \psi$), the self-duality problem has a host of applications in various areas such as database theory, machine learning, data mining, game theory, artificial intelligence, mathematical programming, and distributed systems (See surveys [11,9] for example).

While the self-duality problem is in co-NP, since for a non-self-dual function f, there exists a succinct certificate $x \in \{0,1\}^n$ such that $f(x) \neq f^d(x)$ (i.e., $f(x) = f(\overline{x})$), the exact complexity of the self-duality is still open for more than 25 years now (e.g., [11,19,17,26]). The best currently known upper timebound is quasi-polynomial time [12,14,31]. It is also known that the self-duality problem can be solved in polynomial time by using poly-logarithmically many nondeterministic steps [10,18]. These suggest that the self-duality problem is unlikely to be co-NP-complete, since it is widely believed that no co-NP-hard problem can be solved in quasi-polynomial time (without nondeterministic step) and in polynomial time with poly-logarithmically many nondeterministic steps. However the problem does not seem to lend itself to a polynomial time algorithm.

Much progress has been made in identifying special classes of monotone functions for which the self-duality problem can be solved in polynomial time (e.g., [4,6,7,8,10,15,19,21,22,23,27] and references therein). For example, Peled and Simeone [27] and Crama [6] presented polynomial time algorithms to dualize (and hence to determine the self-duality of) regular functions in polynomial time. Boros et al. [5] and Eiter and Gottlob [8] showed the self-duality for monotone k-DNFs (i.e, DNFs in which each term contains at most k variables) can be determined in polynomial time, and Gaur and Krishnamurti [15] improved upon it to have a polynomial-time algorithm for the self-duality for monotone $O(\sqrt{\log n})$-DNFs.

Our contributions. Motivated by Theorem 1 (that f is self-dual if and only if $\mathcal{H}$ is strange), we study the fractional chromatic number [30] of self-dual functions. We exactly characterize the fractional chromatic number of three classes of self-dual functions that arise in the context of distributed systems, namely, the functions associated with majority coteries [13], wheel coteries [24], and uniform Lovász coteries [25]. We also show that any threshold self-dual function has the fractional chromatic number greater than 2.

Since the fractional chromatic number of self-dual functions associated with uniform Lovász coteries is less than 2, it cannot be used to characterize self-dual functions, where we note that dual-minor (that corresponds to intersecting hypergraphs) and non-self-dual functions has chromatic number 2, and hence fractional chromatic number at most 2. Thus, by strengthening the standard integer programming formulation for chromatic number, we give another characterization of self-dual functions in terms of the optimal solution to an LP relaxation of the strengthening. This characterization also shows that the self-duality is equivalent to determining the feasibility of 'some' point in a suitably defined polytope.

2   Preliminaries

Let $\mathcal{H}$ be a hypergraph on vertex set V. A k-coloring of $\mathcal{H}$ is a partition $\{V_1, \ldots, V_k\}$ of V (i.e., $V = \bigcup_{i=1}^{k} V_i$ and $V_i \cap V_j = \emptyset$ for all i and j with $i \neq j$) such that every edge $H \in \mathcal{H}$ intersects at least two subsets V_i and V_j. Here the vertices that belong to V_i are assigned the color i. For a hypergraph, we denote by $\chi(\mathcal{H})$ the smallest integer k for which $\mathcal{H}$ admits a k-coloring. We define $\chi(\mathcal{H}) = +\infty$ if $\mathcal{H}$ contains a hyperedge H of size 1 (i.e., $|H| = 1$). A vertex subset $W \subseteq V$ is called *independent* if it does not contain any edge $H \in \mathcal{H}$; otherwise, *dependent*. Let $\mathcal{I}$ denote the family of all the (inclusionwise) maximal independent sets of $\mathcal{H}$. Then the following integer programming problem determines the chromatic number $\chi(\mathcal{H})$ of $\mathcal{H}$.

$$\text{IP: minimize} \quad \sum_{I \in \mathcal{I}} x_I$$

$$\text{subject to} \quad \sum_{I : I \ni v} x_I \geq 1 \quad \text{for all} \quad v \in V \tag{2}$$

$$x_I \in \{0, 1\} \quad \text{for all} \quad I \in \mathcal{I}, \tag{3}$$

where x_I takes 0/1 value (from constraint (3)) associated with maximal independent set $I \in \mathcal{I}$, constraint (2) ensures that each vertex is covered by some maximal independent set and the goal is to minimize the number of maximal independent sets needed. We note that $\chi(\mathcal{H})$ is the optimal value, since a $\chi(\mathcal{H})$-coloring can be constructed from a subfamily $\mathcal{I}^* = \{I \in \mathcal{I} \mid x_I = 1\}$.

Linear programming (LP) relaxation of the problem above is obtained by replacing (3) with non-negativity constraints:

$$x_I \geq 0 \quad \text{for all } I \in \mathcal{I}. \tag{4}$$

The optimal value of the LP relaxation, denoted $\chi_f(\mathcal{H})$, is the fractional chromatic number (see [30]). By definition, we have $\chi_f(\mathcal{H}) \leq \chi(\mathcal{H})$. Let us describe the dual (D) of the LP relaxation, where y_v denotes a variable associated with $v \in V$.

$$\text{D: maximize} \quad \sum_{v \in V} y_v$$
$$\text{subject to} \quad \sum_{v \in I} y_v \leq 1 \quad \text{for all } I \in \mathcal{I} \tag{5}$$
$$y_v \geq 0 \quad \text{for all } v \in V$$

The subsequent sections make use of the strong and the weak duality in linear programming extensively. Weak duality states that the value of a feasible dual solution is a lower bound on the optimal value of the primal, where the primal is a minimization problem. Strong duality states that the feasibility of the primal and the dual problems implies that two problems have the same optimal values. For details see Chapter 5 in [32], for example.

In general the number of maximal independent sets of a hypergraph $\mathcal{H}$ can be exponential in $|V|$ and $|\mathcal{H}|$, but by Theorem 1, we have the following nice characterization of maximal independent sets for strange hypergraphs, since I is a maximal independent set if and only if $\overline{I} (= V \setminus I)$ is a minimal transversal.

Lemma 1. *A hypergraph $\mathcal{H}$ is strange if and only if $\mathcal{I} = \{\overline{H} \mid H \in \mathcal{H}\}$ holds.*

This implies that the number of variables in the primal problem (the number of constraints (5) in the dual problem) is $|\mathcal{H}|$.

3   Fractional Chromatic Number of Strange Hypergraphs

In this section, we study the fractional chromatic number for well known classes of self-dual functions that have received considerable attention in the area of distributed systems.

3.1   Strange Hypergraphs $\mathcal{H}$ with $\chi_f(\mathcal{H}) > 2$

We show that the majority, wheel, and threshold hypergraph have fractional chromatic number greater than 2.

Let us first consider the majority hypergraphs. For a positive integer k, let $\mathcal{M}_{2k+1}$ be a majority hypergraph on V with $|V| = 2k + 1$ defined by

$$\mathcal{M}_{2k+1} = \{M \subseteq V \mid |M| = k + 1\}.$$

It is easy to see that $Tr(\mathcal{M}_{2k+1}) = \mathcal{M}_{2k+1}$ holds (i.e., $\mathcal{M}_{2k+1}$ is strange).

Theorem 2. *For a positive integer k, we have $\chi_f(\mathcal{M}_{2k+1}) = 2 + \frac{1}{k}$.*

Proof. It follows from Lemma 1 that we have $\binom{2k+1}{k+1}$ maximal independent sets I, each of which satisfies $|I| = k$. We can see that each vertex v in V belongs to $\binom{2k}{k-1}$ maximal independent sets. We construct feasible primal and dual solutions with value $2 + \frac{1}{k}$ to complete the proof.

For the primal problem, we assign $1/\binom{2k}{k-1}$ to each maximal independent set. Then we note that this is a feasible solution, and the value is $\binom{2k+1}{k+1}/\binom{2k}{k-1} = 2 + \frac{1}{k}$. On the other hand, for the dual problem, we assign $1/k$ to each vertex in V. This is again a feasible dual solution, and the value is $2 + 1/k$.   □

Let us next consider wheel hypergraphs. For a positive integer $n\,(> 3)$, $\mathcal{W}_n$ be a hypergraph on $V = \{1, \ldots, n\}$ defined by

$$\mathcal{W}_n = \{\{i, n\} \mid i = 1, \ldots, n-1\} \cup \{\{1, \ldots, n-1\}\}.$$

Clearly, $\mathcal{W}_n$ is strange, since $Tr(\mathcal{W}_n) = \mathcal{W}_n$ holds.

Theorem 3. *For a positive integer $n\,(> 3)$, we have $\chi_f(\mathcal{W}_n) = 2 + \frac{1}{n-2}$.*

Proof. We construct feasible primal and dual solutions with value $2 + 1/(n-2)$ to complete the proof.

For the primal problem, we assign 1 to maximal independent set $\{n\}$ and $1/(n-2)$ to all the other maximal independent sets. Then we can see that this is feasible whose value is $2 + 1/(n-2)$. On the other hand, for the dual problem, we assign $1/(n-2)$ to y_i, $i = 1, \ldots, n-1$, and 1 to y_n. Then this is a feasible dual solution with value $2 + 1/(n-2)$.   □

A function f is called *threshold* if it can be represented by

$$f(x) = \begin{cases} 1 & \text{if } \sum_i w_i x_i > 1 \\ 0 & \text{otherwise,} \end{cases} \tag{6}$$

for some nonnegative weights $w_1, \ldots, w_n$. We can see that functions $f_{\mathcal{M}_{2k+1}}$ and $f_{\mathcal{W}_n}$ associated with $\mathcal{M}_{2k+1}$ and $\mathcal{W}_n$ are threshold, since they can be represented by the following inequalities.

$$f_{\mathcal{M}_{2k+1}}(x) = \begin{cases} 1 & \text{if } \sum_{i=1}^{2k+1} \frac{1}{k} x_i > 1 \\ 0 & \text{otherwise} \end{cases} \quad \text{and} \quad f_{\mathcal{W}_n}(x) = \begin{cases} 1 & \text{if } \sum_{i=1}^{n-1} \frac{1}{n-2} x_i + x_n > 1 \\ 0 & \text{otherwise.} \end{cases}$$

As seen in Theorems 2 and 3, we have $\chi_f(\mathcal{M}_{2k+1}), \chi_f(\mathcal{W}_n) > 2$. The next theorem says that thresholdness ensures that the fractional chromatic number is greater than 2.

Theorem 4. *The fractional chromatic number of any threshold self-dual function is greater than* 2.

Proof. Let f be a threshold self-dual function defined by (6), and let $\mathcal{H}$ be a strange hypergraph corresponding to f. Let us consider the dual problem to have a lower bound on the fractional chromatic number. We assign the weights w_i in (6) to dual variables y_i. Then by (1), any independent set I satisfies $\sum_{i\in I} w_i \leq 1$, and hence it is feasible. We assume without loss of generality that there exists a vector $x \in \{0,1\}^n$ such that $\sum_i w_i x_i = 1$, i.e., a maximal independent set I^* of $\mathcal{H}$ such that $\sum_{i\in I^*} w_i = 1$. By Lemma 1, we have $\overline{I^*} \in \mathcal{H}$ and hence $\sum_{i\in \overline{I^*}} w_i > 1$. Thus the objective value of y_i is $\sum_{i=1}^{n} w_i = \sum_{i\in I^*} w_i + \sum_{i\in \overline{I^*}} w_i > 2$. $\square$

3.2 Strange Hypergraphs $\mathcal{H}$ with $\chi_f(\mathcal{H}) \leq 2$

This section shows that not every strange hypergraph has the fractional chromatic number greater than 2. Especially, we show that there exists an infinite family of strange hypergraphs $\mathcal{H}$ with $\chi_f(\mathcal{H}) < 2$.

Let us first see that the following strange hypergraph $\mathcal{H}$ has $\chi_f(\mathcal{H}) = 2$.

Example 1. Let $V = \{a,b\} \cup \{1,\ldots,7\}$, and $\mathcal{H}$ be a hypergraph on V given by

$$\mathcal{H} = \{\{a,b\}\} \cup \{\{1,2,3,c\},\{3,4,5,c\},\{1,5,6,c\},$$
$$\{1,4,7,c\},\{2,5,7,c\},\{3,6,7,c\},\{2,4,6,c\} \mid c \in \{a,b\}\}.$$

Note that this $\mathcal{H}$ satisfies $Tr(\mathcal{H}) = \mathcal{H}$ (i.e., $\mathcal{H}$ is strange), and hence we have $\chi(\mathcal{H}) = 3$. For its fractional chromatic number, we have a feasible primal solution with value 2 obtained by assigning $\frac{1}{2}$ to four maximal independent sets $\{1,2,6,7,b\}, \{2,3,5,6,b\}, \{4,5,6,7,a\}$, and $\{1,3,4,6,a\}$, and 0 to all the others. A dual solution of value 2 is obtained by assigning 1 to x_a and x_b, and 0 to all the others. Thus we have $\chi_f(\mathcal{H}) = 2$.

We next show that there exist a strange hypergraph $\mathcal{H}$ with $\chi_f(\mathcal{H}) < 2$.

A finite projective plane of order n is defined as a set of $n^2 + n + 1$ points with the properties that:

1. Any two points determine a line,
2. Any two lines determine a point,
3. Every point has $n + 1$ lines on it, and
4. Every line contains $n + 1$ points.

When n is a power of a prime, finite projective planes can be constructed as follows, where the existence of finite projective planes when n is not a power of a prime is an important open problem in combinatorics.

There are three types of points:

1. a single point p,
2. n points $p(0), p(1)$, and $p(n-1)$,
3. n^2 points $p(i,j)$ for all $i,j \in \{0,\ldots,n-1\}$.

The lines are of the following types:

1. one line $\{p, p(0), p(1), \ldots, p(n-1)\}$,
2. n lines of the type $\{p, p(0, c), p(1, c), \ldots, p(n-1, c)\}$ for all c's,
3. n^2 lines of the type $\{p(c), p(0 \times c + r \pmod{n}), 0), p(1 \times c + r \pmod{n}, 1), \ldots,$
   $p((n-1) \times c + r \pmod{n}, n-1)\}$ for all c's and r's.

For example, the finite projective plane of order 2, called *Fano plane*, has 7 points

$$p, p(0), p(1), p(0, 0), p(0, 1), p(1, 0), p(1, 1),$$

and 7 lines

$$\{p, p(0), p(1)\}, \{p, p(0, 0), p(1, 0))\}, \{(p, p(0, 1), p(1, 1))\}, \{p(0), p(0, 0), p(0, 1)\},$$
$$\{(p(0), p(1, 0), p(1, 1)\}, \{p(1), p(0, 0), p(1, 1)\}, \{p(1), p(1, 0), p(0, 1)\}.$$

It is known [2] that Fano plane is a strange hypergraph, if we regard points and lines as vertices and hyperedges, respectively, but no finite projective plane of order $n \, (> 2)$ is strange.

Theorem 5. *Let $\mathcal{F}_n$ be a finite projective plane of order n. Then we have*
$\chi_f(\mathcal{F}_n) \leq 1 + \frac{n+1}{n^2}$ *if $n \geq 3$, and* $\chi_f(\mathcal{F}_2) = \frac{7}{4}$.

Proof. Let us first show that $\chi_f(\mathcal{F}_n) \leq 1 + (n+1)/n^2$ for $n \, (\geq 2)$ by constructing a prime feasible solution with value $1 + (n+1)/n^2$. By the definition of finite projective plane, $F \in \mathcal{F}_n$ is a minimal transversal of $\mathcal{F}_n$, and hence $\overline{F}$ is a maximal independent set of $\mathcal{F}_n$. By assigning $1/n^2$ to each maximal independent set $\overline{F}$ with $F \in \mathcal{F}_n$, and 0 to each maximal independent set I with $\overline{I} \notin \mathcal{F}_n$, we have a feasible primal solution with value $1 + (n+1)/n^2$.

We next prove $\chi_f(\mathcal{F}_2) = \frac{7}{4}$ by constructing a dual feasible solutions with value $7/4$. Since $\mathcal{F}_2$ is strange, Lemma 1 implies that $\mathcal{F}_2$ has 7 maximal independent sets, each of which has size 4. Thus by assigning $1/4$ to each vertex, we have a feasible solution with value $7/4$, which completes the proof. ∎

We now describe an infinite family of strange hypergraphs (obtained from Crumbling Walls coteries) whose fractional chromatic number goes to 1 as $n \, (= |V|)$ does.

Crumbling walls due to Peleg and Wool [28] are coteries that generalize the triangular coteries, grids, hollow grids, and wheel coteries. Let $V = \{1, 2, \ldots, n\}$, and let $U_0, U_1, \ldots, U_d$ be a partition of V, where we denote $|U_i|$ by n_i. Then crumbling wall $\mathcal{H}$ is defined by $\mathcal{H} = \bigcup_{i=0}^{d} \mathcal{H}_i$ such that

$$\mathcal{H}_i = \{U_i \cup \{u_{i+1}, \ldots, u_d\} \mid u_j \in U_j \text{ for all } j = i+1, \ldots, n\}. \tag{7}$$

Note that $H \cap H' \neq \emptyset$ holds for all $H, H' \in \mathcal{H}$ (i.e., $\mathcal{H}$ is a coterie). It is known that a crumbling wall is strange if and only if $n_0 = 1$ and $n_i \geq 2$ for all $i \geq 1$ [28]. Crumbling walls with $n_0 = 1$ are also known as *Lovász* hypergraphs (or coteries), as the construction was first proposed by Lovász [20].

We consider the class of *uniform Lovász* hypergraphs, denoted by $\mathcal{L}_{k,d}$, which are crumbling walls with $n_0 = 1$ and $n_i = k (\geq 2)$ for all $i = 1, \ldots, d$. For example, if $k, d = 2$, then we have

$$\mathcal{L}_{2,2} = \{\{1, 2, 4\}, \{1, 2, 5\}, \{1, 3, 4\}, \{1, 3, 5\}, \{2, 3, 4\}, \{2, 3, 5\}, \{4, 5\}\},$$

where $U_0 = \{1\}$, $U_1 = \{2, 3\}$, and $U_2 = \{4, 5\}$.

Theorem 6. *Let k and d be positive integers with $k \geq 2$. Then we have*

$$\chi_f(\mathcal{L}_{k,d}) = 1 + \frac{k^d}{\displaystyle\sum_{i=1}^{d}(k-1)^i k^{d-i}}, \tag{8}$$

which satisfies $\chi_f(\mathcal{L}_{k,d}) \rightarrow +1$ as $k, d \rightarrow +\infty$. □

On the positive note, we show that at least half of strange hypergraphs have fractional chromatic number at least 2.

Lemma 2. *Let $\mathcal{H}$ be an intersecting hypergraph having an hyperedge with exactly k elements. Then we have $\chi_f(\mathcal{H}) \geq \frac{k}{k-1}$.*

Proof. Since the value of a feasible dual solution is a lower bound on $\chi_f(\mathcal{H})$, we construct a dual feasible solution with value $\frac{k}{k-1}$. Let H be a hyperedge of size k. Assign $\frac{1}{k-1}$ to variables y_j with $j \in H$, and 0 to y_j with $j \notin H$. Suppose that the solution is not feasible. Then for some maximal independent set I, we have $\sum_{j \in I} y_j > 1$, which implies that H is contained in the independent set I. This contradicts that I is independent. □

It follows from the lemma that all the intersecting hypergraph with a hyperedge of size 2 has fractional chromatic number at least 2.

Let $\mathcal{H}$ be a hypergraph on V, and let a and b are new vertices, i.e., $a, b \notin V$. Define $\mathcal{H}_{a,b}$ by

$$\mathcal{H}_{a,b} = \{\{a, b\}\} \cup \{H \cup \{c\} \mid H \in \mathcal{H}, c \in \{a, b\}\}.$$

It is easy to see that the strangeness of $\mathcal{H}$ implies that of $\mathcal{H}_{a,b}$. We say that two hypergraphs are different, if they are not identical up to isomorphism (i.e., renaming of the vertices).

Theorem 7. *At least half of strange hypergraphs have fractional chromatic number at least 2.*

Proof. Let $\mathbb{S}$ be the family of all strange hypergraphs (unique up to isomorphism). Let $\mathbb{S}_3 \subseteq \mathbb{S}$ be the family of strange hypergraphs such that all the hyperedges contain at least 3 vertices, and let $\mathbb{S}_2 = \mathbb{S} \setminus \mathbb{S}_3$. Then for each $\mathcal{H} \in \mathbb{S}_3$, we have $\mathcal{H}_{a,b} \in \mathbb{S}_2$, and $\mathcal{H}_{a,b}$ is different from $\mathcal{H}'_{a,b}$ if $\mathcal{H}$ and $\mathcal{H}'$ are different.

Therefore, $|\mathbb{S}_2| \geq |\mathbb{S}_3|$ holds. By Lemma 2, every $\mathcal{H} \in \mathbb{S}_2$ has $\chi_f(\mathcal{H}) \geq 2$. This completes the proof. □

Remark. Let $\mathcal{H}$ be an intersecting Sperner hypergraph with a hyperedge of size 2. If $\mathcal{H}$ is not strange, then $\chi(\mathcal{H}) = 2$ holds by Theorem 1, and hence $\chi_f(\mathcal{H}) \leq 2$. By this, together with Lemma 2, we have $\chi_f(\mathcal{H}) = 2$, which implies that there exists an infinite family of non-strange hypergraphs $\mathcal{H}$ with $\chi_f(\mathcal{H}) = 2$. Moreover, by Theorem 6, we know the existence of an infinite family of strange hypergraphs $\mathcal{H}$ with $\chi_f(\mathcal{H}) < 2$. These imply that the fractional chromatic number cannot be used to separate strange hypergraphs from non-strange hypergraph. Hence, we need to strengthen the integer program by adding additional inequalities, which is discussed in the next section.

4   A LP Characterization of Strange Hypergraphs

In this section, we show how to strengthen the LP-relaxation using derived constraints. The strengthened relaxation (which contains the derived constraints) has optimal value $\chi_f^*(\mathcal{H}) > 2$, provided that $\mathcal{H}$ is strange. Let us consider the following integer program SIP for the chromatic number. The linear programming relaxation to SIP is denoted by SLP.

$$\text{SIP:} \quad \text{minimize} \quad \sum_{I \in \mathcal{I}} x_I$$

$$\text{subject to} \quad \sum_{I:I \ni v} x_I \geq 1 \quad \text{for all} \ \ v \in V \tag{9}$$

$$\sum_{I:I \cap H \neq \emptyset} x_I \geq 2 \quad \text{for all} \ \ H \in \mathcal{H} \tag{10}$$

$$x_I \in \{0,1\} \quad \text{for all} \ \ I \in \mathcal{I}, \tag{11}$$

Lemma 3. *x is a feasible solution of SIP if and only it is a feasible solution of IP.*

Proof. We show the lemma by proving that constraint (10) is implied by constraints (9) and (11).

Let H be a hyperedge of $\mathcal{H}$. Then from (9), the following inequality holds:

$$\sum_{v \in H} \sum_{I:I \ni v} x_I \geq \sum_{v \in H} 1 \ (= |H|). \tag{12}$$

We note that the coefficient α_I of variable x_I in inequality (12) satisfies the following properties

1. $\alpha_I = 0$, if $I \cap H = \emptyset$,
2. $\alpha_I < |H| - 1$, if $I \cap H \neq \emptyset$.

Since each variable x_i takes only $0/1$, we can replace the previous constraint by (10). □

Here we prove analogue of Theorem 1.

Theorem 8. *Let $\mathcal{H}$ be an intersecting Sperner hypergraph. Then $\mathcal{H}$ is strange if and only if $\chi_f^*(\mathcal{H}) > 2$.*

Proof . ($\Rightarrow$) We construct a dual solution to have a lower bound on the optimal value of the primal problem. Let us assign $1/(|\mathcal{H}|-1)$ to dual variables associated with (10), and 0 to dual variables associated with (9). By lemma 1, $I \in \mathcal{I}$ if and only if $\overline{I} \in \mathcal{H}$. This implies that for each $I \in \mathcal{I}$, there exists a hyperedge $H \in \mathcal{H}$ such that $I \cap H = \emptyset$, where such a hyperedge is $\overline{I}$. Thus the assignment above constructs a feasible solution with value $2(1 + 1/(|E| - 1))$, which implies $\chi_f^*(\mathcal{H}) > 2$.

($\Leftarrow$) If $\chi_f^*(\mathcal{H}) > 2$, then we have $2 < \chi_f^*(\mathcal{H}) \leq \chi^*(\mathcal{H}) = \chi(\mathcal{H})$ by Lemma 3, where $\chi^*(\mathcal{H})$ denotes the optimal value of SIP. It follows from Theorem 1 that $\mathcal{H}$ is strange. □

Remark. The implicitly defined LP relaxation of SIP can be solved in polynomial time, provided that there exists a separation oracle for the LP relaxation. This would imply a polynomial time algorithm for the self-duality problem. However, the arguments used in the proof of Theorem 8 can be used to show that determining the feasibility of a point in the polytope associated with SLP is equivalent to the self-duality problem.

Let $\mathcal{P}$ be a polytope defined by the dual of the LP relaxation of SIP. Let y be dual variables associated with constraint (9), and let z be dual variables associated with constraint (10). Let (y^*, z^*) be a vector obtained by assigning 0 to each y_i, and $\frac{1}{|\mathcal{H}| - 1}$ to each z_i.

Theorem 9. *Let $\mathcal{H}$ be an intersecting Sperner hypergraph. Then $(y^*, z^*) \in \mathcal{P}$ if and only if $\mathcal{H}$ is strange.*

Proof. ($\Leftarrow$) Suppose that $\mathcal{H}$ is strange. Then by the proof (of forward direction) of Theorem 8, (y^*, z^*) is a feasible solution to the dual of SLP and hence it is contained in $\mathcal{P}$.

($\Rightarrow$) Suppose that $\mathcal{H}$ is not strange. Then by Lemma 1, there exists a maximal independent set I such that $I \neq \overline{H}$ for all $H \in \mathcal{H}$. Since $\mathcal{H}$ is intersecting, each $\overline{H}$ with $H \in \mathcal{H}$ is independent. Thus such an I satisfies $I \not\subseteq \overline{H}$ (i.e., $I \cap H \neq \emptyset$) for all $H \in \mathcal{H}$. This implies that, if we consider assignment (y^*, z^*) to the dual variables, the dual constraint associated with I is not satisfied, since $\frac{|\mathcal{H}|}{|\mathcal{H}| - 1} > 1$. which implies $(y^*, z^*) \notin \mathcal{P}$. □

5   Conclusion

In this paper, we have characterized self-dual functions in terms of optimal value of a certain linear programming problem. The linear programming problem is a relaxation of a strengthened version of the standard IP formulation for the chromatic number and its dual is defined implicitly with exponentially many constraints. The linear programming problem could in principle be solved in polynomial time, if there exists a separation oracle. However, we have also shown that the problem for determining feasibility of a given point in the associated polytope is equivalent to the self-duality problem. We have computed the exact fractional chromatic number for well-known classes of self-dual functions arising from majority coteries, wheel coteries, and uniform Lovász coteries. The existence of a polynomial time algorithm for determining self-duality of monotone Boolean functions remains open.

References

1. Benzaken, C.: Critical hypergraphs for the weak chromatic number. Journal of Combinatorial Theory B 29, 328–338 (1980)
2. Berge, C.: Hypergraphs: Combinatorics of Finite Sets. North-Holland, Amsterdam (1989)
3. Bioch, C., Ibaraki, T.: Complexity of identification and dualization of positive Boolean functions. Information and Computation 123, 50–63 (1995)
4. Boros, E., Elbassioni, K.M., Gurvich, V., Khachiyan, L.: Generating maximal independent sets for hypergraphs with bounded edge-intersections. In: Farach-Colton, M. (ed.) LATIN 2004. LNCS, vol. 2976, pp. 488–498. Springer, Heidelberg (2004)
5. Boros, E., Gurvich, V., Hammer, P.L.: Dual subimplicants of positive Boolean functions. Optimization Methods and Software 10, 147–156 (1998)
6. Crama, Y.: Dualization of regular boolean functions. Discrete Appl. Math. 16, 79–85 (1987)
7. Domingo, C., Mishra, N., Pitt, L.: Efficient read-restricted monotone CNF/DNF dualization by learning with membership queries. Machine Learning 37, 89–110 (1999)
8. Eiter, T., Gottlob, G.: Identifying the minimal transversals of a hypergraph and related problems. SIAM J. Comput. 24, 1278–1304 (1995)
9. Eiter, T., Gottlob, G.: Hypergraph transversal computation and related problems in logic and AI. In: Flesca, S., Greco, S., Leone, N., Ianni, G. (eds.) JELIA 2002. LNCS (LNAI), vol. 2424, pp. 549–564. Springer, Heidelberg (2002)
10. Eiter, T., Gottlob, G., Makino, K.: New results on monotone dualization and generating hypergraph transversals. SIAM J. Comput. 32, 514–537 (Extended abstract appeared in STOC-02) (2003)
11. Eiter, T., Makino, K., Gottlob, G.: Computational aspects of monotone dualization: A brief survey. KBS Research Report INFSYS RR-1843-06-01, Institute of Information Systems, Vienna University of Technology, Austria (2006)
12. Fredman, M.L., Khachiyan, L.: On the complexity of dualization of monotone disjunctive normal forms. J. Algorithms 21, 618–628 (1996)
13. Garcia-Molina, H., Barbara, D.: How to assign votes in a distributed system. J. Assoc. Comput. Mach. 32, 841–860 (1985)
14. Gaur, D.: Satisfiability and self-duality of monotone Boolean functions. Dissertation, School of Computing Science, Simon Fraser University (January 1999)
15. Gaur, D., Krishnamurti, R.: Self-duality of bounded monotone Boolean functions and related problems. In: Arimura, H., Sharma, A.K., Jain, S. (eds.) ALT 2000. LNCS (LNAI), vol. 1968, pp. 209–223. Springer, Heidelberg (2000)
16. Ibaraki, T., Kameda, T.: A theory of coteries: Mutual exclusion in distributed systems. IEEE Trans. Parallel Distrib. Syst. 4, 779–794 (1993)

17. Johnson, D.S.: Open and closed problems in NP-completeness. Lecture given at the International School of Mathematics "G. Stampacchia": Summer School "NP-Completeness: The First 20 Years", Erice (Sicily), Italy (June 20 - 27, 1991)
18. Kavvadias, D.J., Stavropoulos, E.C.: Monotone Boolean dualization is in co-NP$[\log^2 n]$. Information Processing Letters 85, 1–6 (2003)
19. Lawler, E., Lenstra, J., Kan, A.R.: Generating all maximal independent sets: NP-hardness and polynomial-time algorithms. SIAM Journal on Computing 9, 558–565 (1980)
20. Lovász, L.: Coverings and coloring of hypergraphs. In: Proceedings of the 4th Southeastern Conference on Combinatorics, Graph Theory and Computing, pp. 3–12 (1973)
21. Makino, K.: Efficient dualization of $O(\log n)$-term monotone disjunctive normal forms. Discrete Appl. Math. 126, 305–312 (2003)
22. Makino, K., Ibaraki, T.: The maximum latency and identification of positive boolean functions. SIAM J. Comput. 26, 1363–1383 (1997)
23. Makino, K., Uno, T.: New algorithms for enumerating all maximal cliques. In: Hagerup, T., Katajainen, J. (eds.) SWAT 2004. LNCS, vol. 3111, pp. 260–272. Springer, Heidelberg (2004)
24. Marcus, Y., Peleg, D.: Construction methods for quorum systems. Technical Report CS 92-33, The Weizmann Institute of Science, Rehovot, Israel (1992)
25. Neilsen, M.L.: Quorum structures in distributed systems. PhD thesis, Manhattan, KS, USA (1992)
26. Papadimitriou, C.: NP-completeness: A retrospective. In: Degano, P., Gorrieri, R., Marchetti-Spaccamela, A. (eds.) ICALP 1997. LNCS, vol. 1256, pp. 2–6. Springer, Heidelberg (1997)
27. Peled, U.N., Simeone, B.: An $O(nm)$-time algorithm for computing the dual of a regular boolean function. Discrete Appl. Math. 49, 309–323 (1994)
28. Peleg, D., Wool, A.: The availability of crumbling wall quorum systems. Discrete Appl. Math. 74, 69–83 (1997)
29. Ramamurthy, K.G.: Coherent Structures and Simple Games. Kluwer Academic Publishers, Dordrecht (1990)
30. Schneiderman, E.R., Ullman, D.H.: Fractional Graph Theory. Wiley Interscience, Chichester (1997)
31. Tamaki, H.: Space-efficient enumeration of minimal transversals of a hypergraph. IPSJ-AL 75, 29–36 (2000)
32. Vanderbei, R.J.: Linear programming: Foundations and extensions. In: International Series in Operations Research & Management Science, 2nd edn., vol. 37, Kluwer Academic Publishers, Boston, MA (2001)

On the Complexity of Approximation Streaming Algorithms for the k-Center Problem[*]

Mahdi Abdelguerfi[1], Zhixiang Chen[2], and Bin Fu[2]

[1] Dept. of Computer Science, University of New Orleans, LA 70148, USA
mahdi@cs.uno.edu
[2] Dept. of Computer Science, University of Texas - Pan American
TX 78539, USA
{chen,binfu}@cs.panam.edu

Abstract. We study approximation streaming algorithms for the k-center problem in the fixed dimensional Euclidean space. Given an integer $k \geq 1$ and a set S of n points in the d-dimensional Euclidean space, the k-center problem is to cover those points in S with k congruent balls with the smallest possible radius. For any $\epsilon > 0$, we devise an $O(\frac{k}{\epsilon^d})$-space $(1 + \epsilon)$-approximation streaming algorithm for the k-center problem, and prove that the updating time of the algorithm is $O(\frac{k}{\epsilon^d} \log k) + 2^{O(\frac{k^{1-1/d}}{\epsilon^d})}$. On the other hand, we prove that any $(1 + \epsilon)$-approximation streaming algorithm for the k-center problem must use $\Omega(\frac{k}{\epsilon^{(d-1)/2}})$-bits memory. Our approximation streaming algorithm is obtained by first designing an off-line $(1+\epsilon)$-approximation algorithm with $O(n \log k) + 2^{O(\frac{k^{1-1/d}}{\epsilon^d})}$ time complexity, and then applying this off-line algorithm repeatedly to a sketch of the input data stream. If ϵ is fixed, our off-line algorithm improves the best-known off-line approximation algorithm for the k-center problem by Agarwal and Procopiuc [1] that has $O(n \log k) + (\frac{k}{\epsilon})^{O(k^{1-1/d})}$ time complexity. Our approximate streaming algorithm for the k-center problem is different from another streaming algorithm by Har-Peled [16], which maintains a core set of size $O(\frac{k}{\epsilon^d})$, but does not provide approximate solution for small $\epsilon > 0$.

1 Introduction

In recent years, a new class of data-intensive applications that require managing data streams (e.g. [6]) has become widely recognized. Streaming algorithms have applications in data management, network monitoring, stock market, sensor networks, astronomy, telecommunications and others. Processing data streams is typically performed via a single pass over the stream. In the streaming model of computation, the input data items can only be accessed in the order they arrive. A streaming algorithm cannot save the input data stream, but only has a rough sketch (or a summary) of the input data stream. So, a streaming algorithm

---

[*] This research is supported by Louisiana Board of Regents fund under contract number LEQSF(2004-07)-RD-A-35, and in part by NSF Grant CNS-0521585.

F.P. Preparata and Q. Fang (Eds.): FAW 2007, LNCS 4613, pp. 160–171, 2007.
© Springer-Verlag Berlin Heidelberg 2007

performs sub-linear space computation. Due to such sub-linear space limited computation, the design of streaming algorithms poses unique challenges.

In this paper, we study approximation streaming algorithms for the k-center problem, a well-studied clustering problem in data management, in the fixed dimensional Euclidean space. Given an integer $k \geq 1$ and a set S of n points in the d-dimensional Euclidean space $\Re^d$, the k-center problem is to cover these n points in S with k congruent balls of the smallest possible radius. For any $\epsilon > 0$, we develop an $O(\frac{k}{\epsilon^d})$-space $(1+\epsilon)$-approximation streaming algorithm for the k-center problem in $\Re^d$. The algorithm has updating time $O(\frac{k}{\epsilon^d} \log k) + 2^{O(\frac{k^{1-1/d}}{\epsilon^d})}$. On the other hand, we prove that any $(1 + \epsilon)$-approximation streaming algorithm for the k-center problem must use $\Omega(\frac{k}{\epsilon^{(d-1)/2}})$-bits of memory. Our approximation streaming algorithm is obtained by first designing an off-line $(1 + \epsilon)$-approximation algorithm with $O(n \log k) + 2^{O(\frac{k^{1-1/d}}{\epsilon^d})}$ time complexity, and then applying this off-line algorithm repeatedly to a sketch of the input data stream. Interestingly, if ϵ is fixed, our off-line algorithm improves the best-known off-line approximation algorithm for the k-center problem by Agarwal and Procopiuc [1] that has $O(n \log k) + (\frac{k}{\epsilon})^{O(k^{1-1/d})}$ time complexity.

1.1   Related Works

Early research on streaming algorithms dealt with simple statistics of the input data streams, such as the median [18], the number of distinct elements [10], or frequency moments [2]. Streaming algorithms with constant-factor approximation for the k-mean clustering problem are reported in [19,15]. A constant-factor approximation streaming algorithm for the k-median problem was obtained in [4], which requires space $O(k(\log n)^{O(1)})$. Constant-factor approximation streaming algorithms for the k-center problem in the Euclidean space or metric space were reported in [5,4]. A $(1+\epsilon)$-approximation algorithm was presented in [3], which is more suitable for higher dimensional space with a smaller k because its computational time is $2^{O((k \log k)/\epsilon^2)} \cdot dn$. The streaming algorithm and its complexity for computing the diameter of 2D geometric points are studied in [9]. The diameter problem is similar to the 1-center problem.

The streaming algorithm for the k-center problem was also studied by Har-Peled [16]. He showed that a set of core points of size $O(\frac{k}{\epsilon^d})$ can be maintained with updating time $O(nk/\epsilon^d)$. The approximate solution for the core set is also an approximate solution for the original problem, but he did not provide an approximation scheme for the k-center problem with ratio $(1+\epsilon)$ for small $\epsilon > 0$. This paper gives the approximation streaming algorithm for any fixed $\epsilon > 0$.

2   Notations

A data stream is an ordered sequence of data items (or points) $p_1, p_2, \cdots, p_n$. Here, n denotes the number of data points in the stream. A *streaming algorithm* is an algorithm that computes some function over a data stream and has the following properties: 1. The input data are accessed in the sequential order of the data stream. 2. The order of the data points in the stream is not controlled by the algorithm.

Given an integer $k \geq 1$ and a set S of n points in the d-dimensional Euclidean space $\Re^d$, the k-center problem is to find the minimum radius r, denoted by $opt(S)$, so that those n points in S can be covered by k congruent balls of radius r. When approximation to the k-center problem is concerned with, any approximation to $opt(S)$, denoted by $app(S)$, should satisfy $app(S) \geq opt(S)$. The approximation ratio is defined to be $\max \frac{app(S)}{opt(S)}$.

For a data stream $S = p_1, p_2, \cdots, p_n$, when no confusing arises, we also use S to refer the set $\{p_1, p_2, \ldots, p_n\}$. When a new data point p_{n+1} arrives at the stream, we let $S \circ p_{n+1}$ denote the stream $p_1, p_2, \cdots, p_n, p_{n+1}$.

Let $o = (a_1, \cdots, a_d)$ be a point in $\Re^d$ and $r > 0$. Define $S_d(r, o)$ to be the sphere centered at o with radius r in $\Re^d$. Namely, $S_d(r, o) = \{(x_1, \cdots, x_d) | (x_1 - a_1)^2 + \cdots + (x_d - a_d)^2 = r^2\}$. Define $B_d(o, r)$ to be the ball centered at o with radius r in $\Re^d$. Similarly, $B_d(o, r) = \{(x_1, \cdots, x_d) | (x_1 - a_1)^2 + \cdots + (x_d - a_d)^2 \leq r^2\}$. For a real number x, $\lfloor x \rfloor$ is the largest integer $y \leq x$ and $\lceil x \rceil$ is the least integer $z \geq x$. Throughout this paper, the dimension number d is fixed.

3   A Space Upper Bound for the k-Center Problem

In this section, we will devise an approximation streaming algorithm for the k-center problem and analyze its space complexity and updating time complexity.

We will use geometric separators to design a divide and conquer algorithm for the off-line k-center problem. We find that two types of geometric separators can be applied to this problem.

Miller et al.'s Separator. A d-dimensional neighborhood system $\Phi = \{B_1, \cdots, B_n\}$ is a finite collection of balls in $\Re^d$. For each point $p \in \Re^d$, let ply of p, denoted by $ply_\Phi(p)$, be the number of balls from Φ that contains p. Φ is a k-ply neighborhood system if for all $p, ply_\Phi(p) \leq k$. Each $(d-1)$-dimensional sphere S in $\Re^d$ partitions Φ into three subsets: $\Phi_I(S), \Phi_E(S)$ and $\Phi_O(S)$, those are the balls that are in the interior of S, in the exterior of S, or intersect S, respectively.

Given a function $f(n) \geq 0$ and a constant $\delta > 0$, a $(d-1)$-dimensional sphere S in $\Re^d$ is an $f(n)$-sphere separator that δ-splits a neighborhood system Φ with n balls in $\Re^d$, if $|\Phi_O(S)| \leq f(n)$ and $|\Phi_I(S)|, |\Phi_E(S)| \leq \delta n$.

Theorem 1 ([17,7]). *Suppose $\Phi = \{B_1, \cdots, B_n\}$ is a k-ply neighborhood system in $\Re^d$. Then there is an $O(n)$ time algorithm that finds an $O(k^{1/d}n^{1-1/d})$-sphere separator S that δ-splits the system for any $(d+1)/(d+2) < \delta < 1$.*

Fu's Width-Bounded Separator. For two points p_1, p_2 in $\Re^d$, $dist(p_1, p_2)$ denotes the Euclidean distance between p_1 and p_2. For a set $A \subseteq \Re^d$, let $dist(p_1, A) = \min_{q \in A} dist(p_1, q)$.

A *hyper-plane* in $\Re^d$ through a fixed point $p_0 \in \Re^d$ is defined by the equation $(p - p_0) \cdot v = 0$, where v is a normal vector of the plane and "." is the usual vector inner product $(u \cdot v = \sum_{i=1}^{d} u_i v_i$ for $u = (u_1, \cdots, u_d)$ and $v = (v_1, \cdots, v_d))$. Given any $Q \subseteq \Re^d$ and any constant $a > 0$, an *a-wide-separator* is determined by a hyper-plane L. This separator has two measurements to determine its quality of separation: (1) balance$(L, Q) = \frac{\max(|Q_1|, |Q_2|)}{|Q|}$, where $Q_1 = \{q \in Q | (q - p_0) \cdot v < 0\}$ and $Q_2 = \{q \in Q | (q - p_0) \cdot v > 0\}$, and we assume that L is through p_0 with a normal vector v; and (2) size$(L, P, \frac{a}{2}, w)$, which is the number of points in Q with distance at most $\frac{a}{2}$ to the hyper-plane L.

It is easy to see from the definition that the first measurement determines the balance quality of a separator. A well-balanced separator can reduce the problem size efficiently to facilitate the application of a divide and conquer algorithm. The other measurement determines the number of points that are either on or near the separator hyper-plane. Those points form a boundary between two sides of the hyper-plane, and shall be considered when combining solutions to the sub-problems on two sides of the hyper-plane into a solution to the original problem. Therefore, the fewer the number of those points, the more efficient it is to combine solutions of sub-problems.

A point in $\Re^d$ is a grid point if all of its coordinates are integers. For a d-dimensional point $p = (i_1, i_2, \cdots, i_d)$ and $a > 0$, define $\text{grid}_a(p)$ to be the set $\{(x_1, x_2, \cdots, x_d) | i_j - \frac{a}{2} \leq x_j < i_j + \frac{a}{2}, j = 1, 2, \cdots, d\}$, which is a half open and half close d-dimensional axis-parallel box with a volume of a^d. In particular, when $a = 1$, we let $\text{grid}(p)$ to stand for $\text{grid}_1(p)$. For $a_1, \cdots, a_d > 0$, a $(a_1, \cdots, a_d)$-grid point is a point $(i_1 a_1, \cdots, i_d a_d)$ for some integers $i_1, \cdots, i_d$. For a set of points Q in $\Re^d$, $GD_a(Q) = \{p | p$ is a $(a, \cdots, a)$-grid point with $q \in \text{grid}_a(p)$ for some $q \in Q\}$.

Theorem 2 ([12]). *Let $w = O(1)$ be a positive real parameter and $\delta > 0$ be a small constant. Let P be a set of n grid points in $\Re^d$. Then there is an $O(n^{d+1})$-time and $O(n)$-space algorithm that finds a separator hyper-plane L such that each side of L has $\leq \frac{d}{d+1} n$ points from P, and the number of points of P with distance $\leq w$ to L is $\leq (c_d + \delta) w \cdot n^{\frac{d-1}{d}}$ for all large n, where c_d is a constant for a fixed dimension d.*

A sub-linear time randomized algorithm was presented in [13] for finding such a separator as outlined in the above theorem. The deterministic $O(n^{d+1})$-time algorithm of Theorem 2 is also sufficient for deriving the time bound in our approximation streaming algorithm for the k-center problem.

A Separator for Our Algorithm. Using Theorem 1 or Theorem 2, we derive a separator that will form the basis of the algorithm in section 3.1.

Lemma 1. *Let $r = O(1)$ be a positive real parameter and P be a set of n grid points in $\Re^d$. There is a surface L in $\Re^d$ such that, 1) L is either a hyper-plane or a sphere; 2) L partitions $\Re^d$ into regions E and O; 3) $|E \cap P|, |O \cap P| \leq \alpha n$ and the number of points of P within distance at most r to L is at most $\beta r n^{1-1/d}$; and 4) L can be found in time $O(n^\gamma)$ and space $O(n)$, where α, β, and γ are positive constants for a fixed dimension d. Furthermore, $\alpha < 1$.*

Proof. We prove this lemma by Theorem 1. Let $P = \{p_1, \cdots, p_n\}$ be the set of points in $\Re^d$. Let $\Phi = \{B_1, \cdots, B_n\}$ be the set of balls such that B_i has center at p_i and radius r for $i = 1, \cdots, n$. For each point q in $\Re^d$, there are at most $k = c_d r^d$ grid points with distance at most r to q, where c_d is a constant for a fixed d. Therefore, Φ is a $c_d r^d$-ply neighborhood system. By Theorem 1, we can find a sphere L in $O(n)$ time that satisfies $|E \cap P|, |O \cap P| \leq \alpha n$ for some $\alpha < \frac{d+1}{d+2} + \epsilon$ for any constant ϵ, and the number of points with distance at most r to L is $O(k^{1/d} n^{1-1/d}) = O(rn^{1-1/d}) \leq \beta r n^{1-1/d}$, where β can be chosen from the definition of the Big-O notation. Since the computational time is $O(n)$, the space cost is $O(n)$ and we also have $\gamma = 1$.   □

Lemma 3 can also be proved by a direct application of Theorem 2. In this proof, we have $\alpha < \frac{d}{d+1} + \delta$, $\beta \leq c_d + \delta$, and $\gamma = d + 1$, where δ is a small positive constant.

The two different proofs of Lemma 3 show that we will get different constants α, β and γ for the lemma. We expect all of them to be as small as possible. The constant γ can be as small as $\frac{2}{d}$ if the sub-linear time randomized algorithm of [13] is used. These three constants affect the analysis of the algorithms for the k-center problem in the next two subsections. In the reminder of this paper, we will refer to L as a $2r$-wide separator for the grid point set P.

3.1   An Off-Line Approximation Algorithm

Motivated by Agarwal and Procopiuc [1], we use the algorithm by Feder and Greene [8] to find a 2-approximation radius r to the optimal radius to cover a given set of points with k balls. We try all radii $r_i = (1 + \epsilon)^i \frac{r}{2}$ to r. One of these radii is $(1 + \epsilon)$-close to the optimal radius. We use a small number of grid points to characterize the distribution of the input points. Covering those grid points roughly means covering the input points. The grid size is dynamically increased so that the approach does not lose much accuracy. We need a linear space constant factor approximation algorithm for the k-center problem. For the completeness, we follow the proof of [14,8] to prove Theorem 3 (see section 6 in the Appendix). Using Theorem 3, we can achieve the same space upper bound for the streaming algorithm. We can speed up the updating time via the algorithm of Theorem 4, which has a more involved proof [8].

Theorem 3 ([14,8]). *There is an $O(n \cdot k)$ time and $O(n)$ space algorithm such that, given a set of n points in $\Re^d$ and integer $k > 0$, it outputs a radius r with $r^* \leq r \leq 2r^*$, where r^* is the optimal radius for covering those points with k balls.*

Theorem 4 ([8]). *There is an $O(n \log k)$ time and $O(n)$ space algorithm such that, given a set of n points in $\Re^d$ and integer $k > 0$, it outputs a radius r with $r^* \leq r \leq 2r^*$, where r^* is the optimal radius for covering those points with k balls.*

Lemma 2. *Assume $r > 0$ and $s > 0$. There exists a $2^{O(r^d m^{1-\frac{1}{d}})}$-time and $O(r^d m^{1-\frac{1}{d}} + m)$-space algorithm such that, given a set of m $(s, s, \cdots, s)$-grid points Q in $\Re^d$, it outputs a minimal number of balls of radius $r \cdot s$ with centers all at $(s, s, \cdots, s)$-grid points to cover all the points in Q.*

Proof. Since the radius is propositional to the grid size, we assume $s = 1$ without loss of generality. Our algorithm is a divide and conquer approach. Using a width-bounded separator, the problem is decomposed into 3 parts. The separator region has a width larger than the diameter of the covering balls, and it has a small number of points in Q. The other two parts are on the two sides of the separator region, and are separated by a distance larger than the diameter. Therefore, points of Q in these two parts can be covered independently.

    Algorithm $Cover(Q, r)$
    Input: Q is a set of grid points in $\Re^d$ and r is the radius of balls to cover the points in Q.
    Output: the minimal number of balls of radius r that center at grid points and cover all the points in Q.
        If $|Q|$ is small, use a brute force method to obtain and return the answer;
        Let $w = 2r$ and $least = \infty$.
        Find a w-wide separator L with the algorithm of Lemma 1.
        Let Q_0 be the set of all the grid points in Q within distance r to L.
        Let H be the set of all the grid points such that each point $p \in H$
            has another point $q \in Q_0$ with $\text{dist}(p, q) \leq r$.
        For each $t \leq |H|$, select t grid points from H for the centers of t balls.
            If those t balls cover all the points in Q_0 then
                Let $Q_1, Q_2 \subseteq Q$ be the set of all the uncovered points on one
                    side of L with distance $> r$ to L, respectively.
                Let $u = Cover(Q_1, r) + Cover(Q_2, r) + t$.
                If $least > u$ then $least = u$.
        Return $least$.
    End of Algorithm

    By Lemma 3, we can find a w-wide separator L such that each side of L has $\leq \alpha m$ points from Q and the number of points of Q with distance $\leq \frac{w}{2} = r$ to L is bounded by $\beta w m^{1-\frac{1}{d}}$. The separator can be found in $O(m^\gamma)$ time. This follows, given the definitions of Q_0, H, Q_1 and Q_2 in the algorithm, that $|H| = O(r^d|Q_0|) = O(r^d m^{1-\frac{1}{d}})$, $|Q_1| \leq \alpha m$, and $|Q_2| \leq \alpha m$. For each $t \leq |H|$, the algorithm selects t grid points from H as centers for t balls of radius r to cover all the points in Q_0. If all those points in Q_0 are covered, the algorithm then recursively calls $Cover(Q_1, r)$ and $Cover(Q_2, r)$ to cover the rest of the points in $Q_1 \cup Q_2$. Any set of grid point centered, congruent balls covering Q with radius r shall be composed of three subsets of balls for covering Q_0, Q_1 and Q_2, respectively. Those balls covering Q_0 shall have grid point centers from H, and any ball covering Q_1 does not overlap with any ball covering Q_2, because the separator has a width of $2r$. Hence, the correctness of the algorithm is easy to see.

Let $T(m)$ be the computational time. Since there are at most $2^{|H|} = 2^{O(r^d m^{1-\frac{1}{d}})}$ ways to select $t \leq |H|$ balls of radius r, we have $T(m) \leq 2^{O(r^d m^{1-\frac{1}{d}})} T(\alpha m) + O(m^\gamma)$. This implies that $T(m) = 2^{O(r^d m^{1-\frac{1}{d}})}$ for a fixed d. Let $S(m)$ be the space. We have that $S(m) = S(\alpha m) + O(r^d m^{1-1/d}) + O(m)$, which implies that $S(m) = O(r^d m^{1-\frac{1}{d}} + m)$. $\qquad\square$

Theorem 5. *For parameter $1 > \epsilon > 0$ and integer $k > 0$, there exists an $O(n \log k + 2^{O(\frac{k^{1-\frac{1}{d}}}{\epsilon^d})})$ time and $O(n)$ space $(1 + \epsilon)$-approximation algorithm for the d-dimensional k-center problem for a set of n points P in $\Re^d$.*

Proof. We first use the algorithm of Theorem 4 to find the approximate radius r_0 such that $r^* \leq r_0 \leq 2r^*$, where $r^* = opt(P)$. Let $u = \frac{\epsilon r_0}{10\sqrt{d}}$ be the unit length. We have $r_0 = \frac{10\sqrt{d}}{\epsilon} u$. All the grid points mentioned in this proof are $(u, u, \cdots, u)$-grid points. We find a set of $(u, u, \cdots, u)$-grid points Q that each q in Q has $\mathrm{grid}(q) \cap P \neq \emptyset$. Initially, $Q = \emptyset$. For each point $p \in P$, add all the $(u, u, \cdots, u)$-grid points q with $p \in \mathrm{grid}(q)$ to the set Q. At this time, we complete the construction of Q and its size is $O(\frac{k}{\epsilon^d})$, because each ball of radius r_0 only has $O(\frac{1}{\epsilon^d})$ $(u, u, \cdots, u)$-grid points.

Next, we will try the radii $r = \frac{r_0}{2}, (1+\frac{\epsilon}{3})\frac{r_0}{2}, \cdots, (1+\frac{\epsilon}{3})^s \frac{r_0}{2}$, where s is the least integer with $(1+\frac{\epsilon}{3})^s > 4$. Clearly, $s = O(\frac{1}{\epsilon})$. Since $r^* \leq r_0 \leq 2r^*$, there exists an integer $i \leq s$ such that $(1+\frac{\epsilon}{3})^{i-1}\frac{r_0}{2} \leq r^* \leq (1+\frac{\epsilon}{3})^i \frac{r_0}{2}$. Let $r_i = (1+\frac{\epsilon}{3})^{i+1} \frac{r_0}{2}$. Then $r^*(1+\frac{\epsilon}{3}) \leq r_i \leq (1+\frac{\epsilon}{3})^2 r^* \leq (1+\epsilon)r^*$ (if the constant ϵ is selected small enough). Let $D_1, \cdots, D_k$ be the balls of the optimal radius r^* to cover all the points in P. The center of D_i is at p_i $(i = 1, 2, \cdots, k)$. Let q_j be the $(u, u, \cdots, u)$-grid point with $p_j \in \mathrm{grid}(q_j)$ $(j = 1, \cdots, k)$. $\mathrm{dist}(q_j, p_j) \leq \sqrt{d}u \leq \frac{\epsilon}{6}r_0 \leq \frac{\epsilon}{3}r^*$. Assume that p is a point with $\mathrm{dist}(p_j, p) \leq r^*$. Then $\mathrm{dist}(q_j, p) \leq \mathrm{dist}(q_j, p_j) + \mathrm{dist}(p_j, p) \leq \sqrt{d}u + r^* \leq \frac{\epsilon}{3}r^* + r^* \leq (1 + \frac{\epsilon}{3})r^* \leq r_i \leq (1 + \epsilon)r^*$. This means that all the points covered by $D_1, \cdots, D_k$ can be covered by the other k balls of radius r_i with centers at $(u, u, \cdots, u)$-grid points.

We use the algorithm of Lemma 2 to check if all the points in P can be covered by k balls of radius $r_i(i = 1, \cdots, s)$. We select the least r_i such that all the points in P can be covered by k balls of radius r_i. The computational time of this part is at most $s \cdot 2^{O((\frac{r_s}{u})^d k^{1-\frac{1}{d}})} = 2^{O(\frac{k^{1-\frac{1}{d}}}{\epsilon^d})}$. This gives, with the additional $O(n \log k)$ for finding r_0 by Theorem 4, the desired time complexity of the whole process. The algorithm of Theorem 4 has $O(n)$ space cost, so is the algorithm of Lemma 2. It is easy to see that we have $O(n)$ space complexity. $\qquad\square$

3.2   Converting the Off Line Algorithm into Streaming Algorithm

Using Theorem 5, we derive a streaming approximate algorithm for the k-center problem.

Theorem 6. *Let d be a fixed dimension. For any integer $k > 0$ and any real parameter $\epsilon > 0$, there exists a streaming algorithm such that given a stream* *of data points in $\Re^d$, the algorithm returns $(1 + \epsilon)$-approximation to the minimum radius of k congruent balls to cover all of the data points. Moreover, the algorithm's updating time is $O(\frac{k}{\epsilon^d} \log k) + 2^{O(\frac{k^{1-\frac{1}{d}}}{\epsilon^d})}$ and its space complexity is $O(\frac{k}{\epsilon^d})$.*

4   Space Lower Bound for the k-Center Problem

In this section, we prove $\Omega\left(\frac{k}{\epsilon^{(d-1)/2}}\right)$ space lower bound for the streaming algorithm for the k-center problem. The proof of our lower bound is self-contained.

Theorem 7. *Every $(1 + \epsilon)$-approximation streaming algorithm for the k-center problem requires $\Omega\left(\frac{k}{\epsilon^{(d-1)/2}}\right)$ space.*

Before presenting the proof, we briefly explain the method for deriving the lower bound. In the d-dimensional space $\Re^d$, we select k centers $o_1, o_2, \cdots, o_k$ of k congruent balls with radius r such that the distance between every two of these ball is much larger than r. For each center point o_i, we can arrange $\Omega(\frac{1}{\epsilon^{(d-1)/2}})$ points on the sphere $S_d(o_i, r)$ such that the distance between every two of these points is at least ϵr. For k balls of radius r, there are all together $\Omega(\frac{k}{\epsilon^{(d-1)/2}})$ such points on the spheres, and we let H be the set of all these points. For each subset $H_1 = \{p_1, \cdots, p_m\} \subseteq H$, a stream of input points $o_1, o_2, \cdots, o_k, p_1, \cdots, p_m$ is derived. Thus we get $2^{|H|} = 2^{\Omega(\frac{k}{\epsilon^{(d-1)/2}})}$ different streams of input points. If A is a $(1 + \epsilon)$-approximation stream algorithm that only uses $o(\frac{k}{\epsilon^{(d-1)/2}})$ bits of memory, there will be two different streams derived respectively from two different subsets H_1, H_2 of H such that A has the same space configuration after running on the two input streams. Let $p \in H_1$ and $p \notin H_2$. Assume p is on the sphere of the ball $S_d(o_i, r)$. We can select a point p^* on the line through p and o_i such that, adding p^* towards the end of the streams derived from H_1 and H_2 will cause the algorithm to produce two different approximation solutions for the two new input streams. Precisely, the two approximation solutions differ by a factor larger than $(1 + \epsilon)$. On the other hand, the algorithm A shall produce the same approximation solution for the two new input streams, because it has the same memory configuration before p^* arrives. The above contradiction implies that any $(1 + \epsilon)$-approximation streaming algorithm for the k-center problem needs $\Omega\left(\frac{k}{\epsilon^{(d-1)/2}}\right)$ bits of memory.

Definition 1. *Let P be a set of points in $\Re^d$. Define miniBall(P) as the least radius of a ball that can cover all the points in P.*

Lemma 3. *Let P be a set of points on the $S_d(o, r)$. Assume that for every two points $p_1, p_2 \in P$, the angle $\angle p_1 o p_2$ is at least θ. Let P_1 and P_2 be different subsets of P. For every $R > r$, there exists a point $p^* \in \Re^d$ such that miniBall$(P_1 \cup \{o, p^*\}) = \frac{R+r}{2}$ and miniBall$(P_1 \cup \{o, p^*\}) -$ miniBall$(P_2 \cup \{o, p^*\}) \geq \frac{(1 - \cos\theta)r(R-r)}{2(r\cos\theta + R)}$. In particular, if $R = 2r$, then miniBall$(P_1 \cup \{o, p^*\}) -$ miniBall $(P_2 \cup \{o, p^*\}) \geq \frac{2\sin^2\frac{\theta}{2}}{9} \cdot$ miniBall$(P_1 \cup \{o, p^*\})$.*

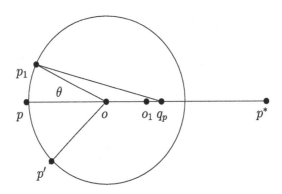

Fig. 1. $r = dist(p, o) = dist(p_1, o) = dist(p', o), R = dist(o, p^*), x = dist(o, q_p)$

Proof. Since $P_1 \neq P_2$, we assume that $p \in P_1$ and $p \notin P_2$. Let p_1 be a point on $S_d(o, r)$ such that the $\angle pop_1 = \theta$ (see Figure 1). Let p^* be a point on the line through p and o such that $dist(p^*, o) = R$ and $dist(p^*, p) = R + r$. We look for a point q_p on the line po such that o is between p and q_p, and $dist(p_1, q_p) = dist(p^*, q_p)$. Let $z = dist(q_p, p_1)$ and $x = dist(o, q_p)$. We have $z^2 = x^2 + r^2 + 2xr \cos \theta$. On the other hand, we also let $z^2 = (R - x)^2$. From the two equations, we have $x = \frac{R^2 - r^2}{2(r \cos \theta + R)}$.

Let o_1 be the median between p and p^*. For every point p_2 in $S_d(o, r)$, $dist(p_2, o_1) \leq dist(p_2, o) + dist(o, o_1) = r + dist(o, o_1) = \frac{R+r}{2}$. Recall that $dist(p, o_1) = dist(o_1, p^*) = \frac{R+r}{2}$. Therefore, $miniBall(P_1 \cup \{o, p^*\}) = \frac{R+r}{2}$. For every point $p' \in P_2$, consider the triangle $\triangle p'oq_p$. We have $dist(p', q_p)^2 = r^2 + x^2 + 2rx \cos \theta'$, where θ' is angle $\angle p'op$ ($\angle p'oq_p + \theta' = \pi$). Since θ' is at least θ, $dist(p', q_p)^2 \leq r^2 + x^2 + 2rx \cos \theta = dist(p_1, q_p)^2 = (R - x)^2$. Hence, $dist(p', q_p) \leq R - x$, this implies that $miniBall(P_2 \cup \{o, p^*\}) \leq R - x$.

Now we have that $miniBall(P_1 \cup \{o, p^*\}) - miniBall(P_2 \cup \{o, p^*\}) \geq \frac{(r+R)}{2} - (R - x) = \frac{r-R}{2} + x = \frac{r-R}{2} + \frac{R^2 - r^2}{2(r \cos \theta + R)} = \frac{(r-R)(r \cos \theta + R) + (R^2 - r^2)}{2(r \cos \theta + R)} = \frac{r^2 \cos \theta + Rr - Rr \cos \theta - r^2}{2(r \cos \theta + R)} = \frac{(1 - \cos \theta) Rr - (1 - \cos \theta) r^2}{2(r \cos \theta + R)} = \frac{(1 - \cos \theta) r (R - r)}{2(r \cos \theta + R)}$. If $R = 2r$, then $miniBall(P_1 \cup \{o, p^*\}) = \frac{R+r}{2} = \frac{3r}{2}$. Furthermore, $miniBall(P_1 \cup \{o, p^*\}) - miniBall(P_2 \cup \{o, p^*\}) \geq \frac{(1 - \cos \theta) r (R - r)}{2(r \cos \theta + R)} = \frac{2r^2 \sin^2 \frac{\theta}{2}}{2(r \cos \theta + 2r)} \geq \frac{2r^2 \sin^2 \frac{\theta}{2}}{6r} = \frac{2 \sin^2 \frac{\theta}{2}}{9}$. $\square$

Lemma 4. *Let $d \geq 2$. Let r and δ be positive real numbers. There is a set of points A_d on the sphere $S_d(o, r)$ such that A_d contains $c_d(\frac{r^{d-1}}{\delta^{d-1}})$ points and the distance between every two points in A_d is at least δ, where c_d is a positive constant for fixed d.*

Proof. Without loss of generality, we assume that o is the origin point. The lemma is true for the case $d = 2$. Assume that for each $d' < d$, there exists $A_{d'}$ on $S_{d'}(o', r')$ with at least $c_{d'}(\frac{r'^{d'-1}}{\delta^{d'-1}})$ points and every two points in $A_{d'}$ have a distance at least δ, where $c_{d'}$ is a positive constant for fixed d'. The sphere $S_d(o, r)$

has equation $x_1^2 + x_2^2 + \cdots + x_d = r^2$. Let $h_i = \frac{r}{10} + i\delta$, where $i = 0, 1, \cdots, \lfloor \frac{8r}{10\delta} \rfloor$. Let L_i be the sphere $\{(x_1, \cdots, x_{d-1}, h_i) | x_1^2 + \cdots + x_{d-1}^2 = r^2 - h_i^2\}$, which is the intersection of the hyper-plane $x_d = h_i$ and $S_d(o, r)$. The radius of L_i is $r_i = \sqrt{r^2 - h_i^2} \geq \frac{r}{10}$. By the induction hypothesis, one can arrange $\geq c_{d-1} \cdot \frac{r_i^{d-2}}{\delta^{d-2}}$ points on L_i with a pair-wise distance at least δ.

Since the distance between the hyper-planes $x_d = h_i$ and $x_d = h_j$ $(i \neq j)$ is at least δ, every two points with one from L_i and the other from L_j have a distance at least δ. The set $\{0, 1, \cdots, \lfloor \frac{8r}{10\delta} \rfloor\}$, the range of i for h_i, has at least $\lceil \frac{8r}{10\delta} \rceil \geq \frac{8r}{10\delta}$ elements. Therefore, the total number of points on $S_d(o, r)$ is

$\geq \sum_{i \in \{0, 1, \cdots, \lfloor \frac{8r}{10\delta} \rfloor\}} c_{d-1} \frac{r_i^{d-2}}{\delta^{d-2}} \geq \frac{8r}{10\delta} \frac{c_{d-1}(\frac{r}{10})^{d-2}}{\delta^{d-2}} = \frac{8c_{d-1}}{10^{d-1}} \frac{r^{d-1}}{\delta^{d-1}}$. The part $\frac{8c_{d-1}}{10^{d-1}}$ is a constant for fixed dimension d. $\qquad\square$

Proof (for Theorem 7). Let r be a positive constant, which will be used as the radius for k spheres in the input set construction. Let $\delta = \sqrt{18\epsilon}r$. We select k centers $o_1, \cdots, o_k$ in $\Re^d$ such that $o_i = (20 \cdot i \cdot r, 0, \cdots, 0)$. The distance between different o_i and o_j is at least $20r$. For each center o_i we select a set of points on the sphere $S_d(o_i, r)$ such that the pair-wise distance of these points is at least δ.

Let o be the origin point. We consider the sphere $S_d(o, r)$. Select two points p_1 and p_2 on the sphere such that $dist(p_1, p_2) = \delta$. Let θ be the angle $\angle p_1 o p_2$. It is easy to see that

$$\sin\frac{\theta}{2} = \frac{\delta}{2r} = \frac{\sqrt{18\epsilon}}{2}. \qquad (1)$$

By Lemma 4, we can select a set of points F of size $c_d(\frac{r^{d-1}}{\delta^{d-1}})$ on the sphere $S_d(o, r)$ such that every two points of F have a distance at least δ.

We shift k copies of $S_d(o, r)$ to $S_d(o_i, r)$ for $i = 1, \cdots, k$. For each $S_d(o_i, r)$, let F_i be the set of points shifted from F on $S_d(o, r)$. Clearly, miniBall$(F_i \cup \{o_i\}) \leq r$. Let $H = F_1 \cup \cdots F_k$, which is the set of all the chosen points on the k spheres. By Lemma 4 and (1), the set H has totally at least $b_d\left(\frac{k}{\epsilon^{(d-1)/2}}\right)$ points, where b_d is a positive constant for fixed d. For each subset $H_1 = \{p_1, \cdots, p_m\} \subseteq H$, construct a stream of input points $S(H_1) = o_1, \cdots, o_k, p_1, \cdots, p_m$. There are totally $2^{|H|}$ such streams of input points, each of which is uniquely derived from a subset of H and contains the k centers in the first k points.

Assume there is a $(1 + \epsilon)$-approximation streaming algorithm that uses $< b_d\left(\frac{k}{\epsilon^{(d-1)/2}}\right)$ bits of memory. Those bits of memory has less than $2^{b_d\left(\frac{k}{\epsilon^{(d-1)/2}}\right)}$ different configurations. As the total number of subsets of H is more than $2^{b_d\left(\frac{k}{\epsilon^{(d-1)/2}}\right)}$, there are two different subsets H_1 and H_2 of H such that the algorithm has the same space configuration after running on both streams derived from H_1 and H_2. We assume that H_1 and H_2 are different at $S_d(o_i, r)$. Let $p \in S_d(o_i, r)$ be a point in H_1 but not in H_2. Let $H_{1,i}$ be the set of all points of H_1 on $S_d(o_i, r)$ and $H_{2,i}$ be the set of all the points of H_2 on $S_d(o_i, r)$.

Let $R = 2r$. By Lemma 3 and (1), there is a point p^* such that miniBall$(H_{1,i} \cup \{o_i, p^*\}) = \frac{R+r}{2} = \frac{3r}{2}$, and miniBall$(H_{1,i} \cup \{o, p^*\}) - $ miniBall$(H_{2,i} \cup \{o, p^*\}) \geq \frac{2\sin^2\frac{\theta}{2}}{9} \cdot$ miniBall$(H_{1,i} \cup \{o, p^*\}) = \epsilon \cdot$ miniBall$(H_{1,i} \cup \{o, p^*\})$.

We append the point p^* to both data streams $S(H_1)$ and $S(H_2)$ to obtain data streams $S(H_1) \circ p^*$ and $S(H_2) \circ p^*$, respectively. On the one hand, since the approximation algorithm has the same memory configuration for both $S(H_1)$ and $S(H_2)$, it will produce the same approximation for $S(H_1) \circ p^*$ and $S(H_2) \circ p^*$. On the other hand, the optimal solution for $S(H_1) \circ p^*$ is miniBall($H_{1,i} \cup \{o, p^*\}$) and the optimal solution for $S(H_2) \circ p^*$ is miniBall($H_{2,i} \cup \{o, p^*\}$). We have that opt($S(H_1) \circ p^*$) − opt($S(H_2) \circ p^*$) $\geq \epsilon \cdot$ miniBall($H_{1,i}$) $\cup \{o, p^*\}$) = $\epsilon \cdot$ opt($S(H_1) \circ p^*$). So, opt($S(H_2) \circ p^*$) $\leq (1-\epsilon) \cdot$ opt($S(H_1) \circ p^*$). A $(1+\epsilon)$-approximation app($S(H_2) \circ p^*$) for $S(H_2) \circ p^*$ has opt($S(H_2) \circ p^*$) $\leq$ app($S(H_2) \circ p^*$) $\leq (1+\epsilon) \cdot$ opt($S(H_2) \circ p^*$) $\leq (1+\epsilon)(1-\epsilon) \cdot$ opt($S(H_1) \circ p^*$) $<$ opt($S(H_1) \circ p^*$). So, app($S(H_2) \circ p^*$) is not a $(1+\epsilon)$-approximation for opt($S(H_1) \circ p^*$), a contradiction to the early statement that the algorithm produces the approximation for both streams. □

5 Conclusions

Data streams present novel and exciting challenges for algorithm design. Clustering is a well establish tool used in classification, data mining and other fields. For the k-center clustering problem, we show an $O(\frac{k}{\epsilon^d})$ space $(1 + \epsilon)$-approximation streaming algorithm. We also prove that every $(1 + \epsilon)$ approximation streaming algorithm requires $\Omega(\frac{k}{\epsilon^{(d-1)/2}})$ bits space. An interesting open problem is to close the gap between the lower bound and upper bound for the space for the approximation streaming algorithm for the k-center problem.

References

1. Agarwal, P., Procopiuc, C.M.: Exact and approximation algorithms for clustering. Algorithmica 33, 201–226 (2002)
2. Alon, N., Matias, Y., Szegedy, M.: The space complexity of approximating the frequency moments. In: STOC 1996, pp. 20–29 (1996)
3. Badoiu, M., Har-Peled, S.: Approximate clustering via core-sets. In: STOC 2002, pp. 250–257 (2002)
4. Charikar, M., O'Callaghan, L., Panigrahy, R.: Better streaming algorithms for clustering problems. In: STOC 2003, pp. 30–39 (2003)
5. Charikar, M., Chekuri, C., Feder, T., Motwani, R.: Incremental clustering and dynamic infomration retrieval. In: STOC 1997, pp. 626–635 (1997)
6. Chaudhry, N., Shaw, K., Abdelguerfi, M. (eds.): Data stream management. Springer, Heidelberg (2005)
7. Eppstein, D., Miller, G.L., Teng, S.-H.: A deterministic linear time algorithm for geometric separator and its applications. In: STOC 1993, pp. 99–108 (1993)
8. Feder, T., Greene, D.: Optimal algorithms for approximate clustering. In: STOC 1988, pp. 434–444 (1988)
9. Feigenbaum, J., Kannan, S., Zhang, J.: Computing Diameter in the Streaming and Sliding-Window Models. Algorithmica 41(1), 25–41 (2004)
10. Flajolet, P., Martin, G.: Probabilistic counting algorithms for data base application. Journal of computer and system sciences 31, 182–209 (1985)

11. Fu, B., Wang, W.: A $2^{O(n^{1-1/d}\log n)}$-time algorithm for d-dimensional protein folding in the HP-model. In: Díaz, J., Karhumäki, J., Lepistö, A., Sannella, D. (eds.) ICALP 2004. LNCS, vol. 3142, pp. 630–644. Springer, Heidelberg (2004)
12. Fu, B.: Theory and Application of Width Bounded Geometric Separator, The draft is available at Electronic Colloquium on Computational Complexity 2005, TR05-13. In: Fleischer, R., Trippen, G. (eds.) ISAAC 2004. LNCS, vol. 3341, pp. 427–441. Springer, Heidelberg (2004)
13. Fu, B., Chen, Z.: Sublinear-time algorithms for width-bounded geometric separator and their application to protein side-chain packing problem. In: Cheng, S.-W., Poon, C.K. (eds.) AAIM 2006. LNCS, vol. 4041, pp. 149–160. Springer, Heidelberg (2006)
14. González, T.: Clustering to minimize the maximum intercluster distance. Theoretical Computer Science 38, 293–306 (1985)
15. Guha, S., Mishra, N., Motwani, R., O'Callaghan, L.: Custering Data Streams. IEEE Trans. Knowl. Data Eng. 15(3), 515–528 (2003)
16. Har-Peled, S.: Clustering Motion. Discrete and Computational Geometry 4(31), 545–565 (2004)
17. Miller, G.L., Teng, S.-H., Thurston, W.P., Vavasis, S.A.: Separators for sphere-packings and nearest neighbor graphs. J. ACM 44(1), 1–29 (1997)
18. Munro, J.I., Paterson, M.S.: Selection and sorting with limited storage. Theoretical computer science 12, 315–323 (1980)
19. O'Callaghan, L., Mishra, N., Meyerson, A., Guha, S., Motawani, R.: Streaming-Data Algorithms for High-Quality Clustering. In: Proceedings of the 18th International Conference on Data Engineering (ICDE'02), pp. 685–696 (2002)

6   Appendix: Proof of Theorem 3

Proof. Let S be a set of points in $\Re^d$. We construct a subset $T \subseteq S$. For each point p not in T, define $neighbor(p)$ to be the nearest point in T to p, and $\text{dist}(p)$ to be the distance from p to $neighbor(p)$.

$T = \emptyset$;
$\text{dist}(p) = \infty$ for all $p \in S$;
while $|T| \le k$ do {
    $D = \max\{\text{dist}(p)|p \in S - T\}$;
    let $p' \in S - T$ have $\text{dist}(p') = D$;
    $T = T \cup \{p'\}$;
    update $neighbor(p)$ and $\text{dist}(p)$ for each $p \in S - T$
}

By the end of the algorithm, there are $k+1$ points in T and every two points in T have distance at least D. By pigeon hole principle, at least two of them are in the same ball. Therefore, the optimal radius is at least $D/2$. Let each point of T before the last assignment to D be the center of one ball. Since every point p in $S - T$ has $\text{dist}(p) \le D$, we have that each ball has radius at most D. The cycle in the algorithm repeats at most $k+1$ times. Each time it costs $O(n)$ steps to update $neighbor(p)$ and $\text{dist}(p)$ to check the new element just added to the set T. The total number of steps is $O(n \cdot k)$. The space is used for saving S, T, $neighbor(p)$ and $\text{dist}(p)$ for each point $p \in S$. Therefore, the space is $O(n)$.   □

Scheduling an Unbounded Batch Machine to Minimize Maximum Lateness

Shuyan Bai[1], Fuzeng Zhang[2], Shuguang Li[3,*], and Qiming Liu[2]

[1] Division of Basic Courses, Yantai Vocational College,
Yantai 264000, China
[2] School of Computer Science and Technology, Ludong University,
Yantai 264025, China
[3] School of Mathematics and Information Science, Yantai University,
Yantai 264005, China
sgliytu@hotmail.com

Abstract. We consider the problem of scheduling n jobs with release times on an unbounded batch machine to minimize the maximum lateness. An unbounded batch machine is a machine that can process up to b ($b \geq n$) jobs simultaneously. The jobs that are processed together form a batch, and all jobs in a batch start and complete at the same time. The processing time of a batch is the time required for processing the longest job in the batch. We present a linear time approximation scheme for this problem.

1 Introduction

A batch machine is a machine that can process up to b jobs simultaneously as a batch. The processing time of a batch is the time required for processing the longest job in the batch. All jobs contained in the same batch start and complete at the same time, since the completion time of a job is equal to the completion time of the batch to which it belongs. The scheduling problems involve assigning all n jobs to batches and determining the batch sequence in such a way that certain objective function is optimized [2]. See [8] for definitions of scheduling objective functions.

There are two distinct models for batch scheduling problems. In the *bounded* model, the bound b for each batch size is effective, i.e., $b < n$. Problems of this model arise in the manufacture of integrated circuits [9]. In the *unbounded* model, there is effectively no limit on the sizes of batches, i.e., $b \geq n$. Scheduling problems of this model arise, for instance, in situations where compositions need to be hardened in kilns, and a kiln is sufficiently large that it does not restrict batch sizes [2]. In this paper, we are interested in the unbounded model.

Much research has been done on scheduling an unbounded batch machine. Brucker et al. [2] provided a thorough discussion on minimizing various regular cost functions when all jobs have the same release time. For the problem of

---

* Corresponding author.

F.P. Preparata and Q. Fang (Eds.): FAW 2007, LNCS 4613, pp. 172–177, 2007.
© Springer-Verlag Berlin Heidelberg 2007

minimizing total weighted completion time with release times, Deng and Zhang [5] established its NP-hardness and presented polynomial time algorithms for several special cases. Deng et al. [4] presented the first polynomial time approximation scheme (PTAS) for the problem of minimizing total completion time with release times. Cheng et al. [3] investigated the problem of scheduling jobs with release times and deadlines and established its NP-hardness. They also showed the polynomial solvability of several special cases. Z. Liu et al. [10] proved that minimizing total tardiness is NP-hard. In addition, they established the pseudopolynomial solvability of the problem with release times and any regular objective.

Cheng et al. [3] pointed out that the problem of minimizing the maximum lateness on an unbounded batch machine is NP-hard, but left the question of algorithms open. In this paper we present a polynomial time approximation scheme for this problem that runs in $O(n) + C$ time, where the additive constant C depends only on the accuracy ϵ. We apply the techniques in [1,2,6] to get this result.

The remainder of this paper is organized as follows. In Section 2, we formulate the problem and simplify it by applying the rounding method. In Section 3, we present the linear time approximation scheme.

2   Preliminaries

The problem of scheduling an unbounded batch machine to minimize the maximum lateness can be formulated as follows. There are n independent jobs to be processed on an unbounded batch machine that can process up to b ($b \geq n$) jobs simultaneously. Each job j is associated with a processing time p_j, a release time r_j, and a due date d_j. Each job is constrained to start after its release time, and should be ideally completed before its due date. We assume that preemption is not allowed. The processing time of a batch is the time required for processing the longest job in the batch. All jobs contained in the same batch start and complete at the same time, since the completion time of a job is equal to the completion time of the batch to which it belongs. The goal is to schedule the jobs on the unbounded batch machine so as to minimize the maximum lateness $\max\{C_j - d_j\}$, where C_j denotes the completion time of job j in the schedule.

By setting a delivery time $q_j = d_{\max} - d_j$ for all jobs j (where $d_{\max} = \max_j d_j$), we get an equivalent model of the problem. The objective is now to find a schedule that minimizes $\max\{C_j + q_j\}$. When considering the performance of approximation algorithms, the delivery-time model is preferable [7]. Hence in the sequel, we use instead the delivery-time formulation.

An algorithm A is a ρ-approximation algorithm for a minimization problem if it produces a solution which is at most ρ times the optimal one, in time that is polynomial in the input size. We also say that ρ is the worst-case ratio of algorithm A. The worst-case ratio is the usual measure for the quality of an approximation algorithm for a minimization problem: the smaller the ratio is, the better the approximation algorithm is. A family of algorithms $\{A_\epsilon\}$ is called a *polynomial time approximation scheme* (PTAS) if, for each $\epsilon > 0$, the algorithm A_ϵ is a $(1 + \epsilon)$-approximation algorithm running in time that is polynomial in the input size. Note that ϵ is a fixed number, and thus the running time may depend upon $1/\epsilon$ in any manner.

In the following, we will apply the rounding method to form an approximate version of the problem that has a simpler structure. We say, as in [1], that a transformation produces $1 + O(\epsilon)$ *loss* if it potentially increases the objective function value by a $1 + O(\epsilon)$ factor. We assume without loss of generality that $1/\epsilon$ is integral. Let $r_{\max} = \max_j r_j$, $p_{\max} = \max_j p_j$, $q_{\max} = \max_j q_j$.

We start by presenting the lower and upper bounds on the original optimal value *opt*.

Lemma 1. $\max\{r_{\max}, p_{\max}, q_{\max}\} \leq opt \leq r_{\max} + p_{\max} + q_{\max}.$

Proof. It is obvious that $opt \geq \max\{r_{\max}, p_{\max}, q_{\max}\}$. On the other hand, we can get a feasible schedule with objective value no more than $r_{\max} + p_{\max} + q_{\max}$, by starting a batch at time $r_{\max}$ to process all the jobs. Hence we get $opt \leq r_{\max} + p_{\max} + q_{\max}$. □

Lemma 1 shows that any job will be delivered by time $r_{\max} + p_{\max} + q_{\max}$ in an optimal schedule. Let $M = \epsilon \cdot \max\{r_{\max}, p_{\max}, q_{\max}\}$. We partition the time interval $[0, r_{\max} + p_{\max} + q_{\max})$ into $1/\epsilon + 1$ disjoint intervals $[0, M), [M, 2M), \ldots, [(1/\epsilon - 1)M, (1/\epsilon)M), [(1/\epsilon)M, r_{\max} + p_{\max} + q_{\max})$. We then round up each release time r_j to $\lceil r_j/M \rceil \cdot M$. Since $M \leq \epsilon \cdot opt$, the objective value may increase by at most $\epsilon \cdot opt$.

Next, we discuss small and large jobs (batches). We say that a job (or a batch) is *small* if its processing time is less than ϵM, and *large* otherwise.

Lemma 2. *Let k be the number of distinct processing times of large jobs. With $1 + \epsilon$ loss, we assume that $k \leq 1/\epsilon^3 - 1/\varepsilon + 1$.*

Proof. Round each processing time of large jobs up to the nearest integer multiple of $\epsilon^2 M$. This increases the objective value by at most $\epsilon \cdot opt$ due to the definition of large jobs. For each large job j, we know that $\epsilon M \leq p_j \leq (1/\epsilon)M$. Hence we get $k \leq [(1/\epsilon)M]/(\epsilon^2 M) - [\epsilon M/(\epsilon^2 M) - 1] = 1/\epsilon^3 - 1/\epsilon + 1$. □

A job is called *available* if it has been released but not yet been scheduled. Based on the special property of an unbounded batch machine, we have the following observation.

Lemma 3. *There is an optimal schedule for the rounded problem (with rounded release times and processing times) that has the following properties:*

1) Any two processing times of the batches started in the same interval are distinct; and

2) the batches started in the same interval are processed successively in the order of increasing processing times; and

3) from time zero onwards, the batches are filled one by one such that each of them contains all the currently available jobs with processing times no more than the processing time of the batch.

3   A Linear Time Approximation Scheme

In this section we will present the linear time approximation scheme for the problem of scheduling an unbounded batch machine to minimize the maximum lateness.

Consider a feasible schedule for the rounded problem. Remove from this schedule all the batches that are started after time $(1/\epsilon)M$. The remaining part is called a *partial schedule*.

Given a partial schedule, we remove from it all the jobs and the small batches, but retain all the empty large batches. Thereby we get an *outline*, which specifies the processing times of the empty large batches started (but not necessarily finished) in each of the first $1/\epsilon$ intervals. The concept is motivated from [6].

Note that there are at most $1/\epsilon$ large batches started in each of the first $1/\epsilon$ intervals. By Lemma 2, we know that the processing time of each large batch is chosen from $k \leq 1/\epsilon^3 - 1/\epsilon + 1$ values. Combining Lemma 3, we can bound the number of different outlines from above by

$$\Gamma = \left[\binom{k}{0} + \binom{k}{1} + \cdots + \binom{k}{1/\epsilon} \right]^{1/\epsilon} < k^{1/\epsilon^2}.$$

Next, we give a description of the algorithm in [2] that solves the problem of scheduling an unbounded batch machine to minimize the maximum lateness exactly if all the jobs are released at time zero. Assume that jobs are indexed according to the Shortest Processing Time (SPT) rule, so that $p_1 \leq p_2 \leq \cdots \leq p_n$. An *SPT-batch schedule*, is one in which adjacent jobs in the sequence $(1, 2, \ldots, n)$ may be grouped to form batches. Brucker et al. [2] proved that there exists an optimal schedule that is an SPT-batch schedule. They proposed a backward dynamic programming algorithm that searches for such an optimal schedule. Let F_j be the minimum value of the maximum lateness for SPT-batch schedules containing jobs $j, j+1, \ldots, n$, where processing starts at time zero. The initialization is $F_{n+1} = -\infty$, and the recursion for $j = n, n-1, \ldots, 1$ is $F_j = \min_{j < k \leq n+1} \{ \max\{ F_k + p_{k-1}, \max_{j \leq i \leq k-1} \{ p_{k-1} + q_i \} \} \}$. The optimal solution value is then equal to F_1, and the corresponding optimal schedule is found by backtracking. The algorithm requires $O(n^2)$ time. We call this algorithm *BDP Algorithm*.

Roughly speaking, our algorithm consists of two phases. In the first phase, we generate partial schedules by enumerating all possible outlines. In the second phase, we apply BDP Algorithm to schedule the remaining jobs at the end of each partial schedule. From among the feasible schedules generated, we choose the one with the minimum objective value.

Algorithm MML (Minimizing Maximum Lateness)

Step 1. Round the release times and processing times of the jobs as described in the previous section. Group the small jobs with the same release time together. Group the large jobs with the same release time and processing time together.

Step 2. Find the maximum delivery time among the jobs of each group. (The information is useful for computing the objective values of the generated schedules.)

Step 3. Get all possible outlines.

Step 4. For each of the outlines, do the following:

(a) Start the specified empty large batches as early as possible in the order of increasing processing times. If some batch has to be delayed to start in one of the next intervals, then delete the outline.

(b) From time zero onwards, fill the empty large batches one by one such that each of them contains all the currently available large jobs with processing times no more than the processing time of the batch.

(c) Start a batch with processing time at most ϵM, right before the first large batch started in each of the first $1/\epsilon$ intervals, to schedule all the currently available small jobs. Stretch each of the first $1/\epsilon$ intervals to make an extra space with length ϵM to accommodate the scheduled small jobs. Certainly, if an interval is covered entirely by a large batch, then we need not to stretch it for scheduling small jobs.

Step 5. For each of the partial schedules generated above, do the following:

(a) View the remaining large jobs with the same processing time as an aggregate large job whose delivery time is defined to be the maximum delivery time among those jobs.

(b) At the end of the partial schedule (after time $(1/\epsilon)M$), first start a batch with processing time at most ϵM to schedule all the remaining small jobs, and then run BDP Algorithm to schedule the aggregate large jobs.

(c) Compute the objective value of the obtained feasible schedule.

Step 6. From among the generated feasible schedules, choose the one with the minimum objective value.

Theorem 1. *Algorithm MML is a polynomial time approximation scheme that runs in linear time for the problem of scheduling an unbounded batch machine to minimize the maximum lateness.*

Proof. Any feasible schedule for the rounded problem is associated with an outline. Given an outline that is associated with an optimal schedule for the rounded problem, the way of Algorithm MML to schedule large jobs in the first $1/\epsilon$ intervals is optimal (by Lemma 3), the way to schedule aggregate large jobs in the last interval is also optimal (by the optimality of BDP Algorithm), and the way to schedule the small jobs increases the objective value by at most $\epsilon M \cdot (1/\epsilon + 1) \le (\epsilon + \epsilon^2) \cdot opt$. Rounding the release times and processing times of the jobs increases the objective value by at most $2\epsilon \cdot opt$. Hence Algorithm MML outputs a $(1 + 3\epsilon + \epsilon^2)$-approximate schedule for the original problem.

We now discuss the time complexity of Algorithm MML. Steps 1 and 2 requires $O(n)$ time. Each outline can be generated in $O(1/\epsilon^4)$ time, since there are at most $k \leq 1/\epsilon^3 - 1/\varepsilon + 1$ different processing times of large batches (by Lemma 2). Recall that there are at most $\Gamma < k^{1/\epsilon^2}$ different outlines. Step 3 can thus be executed in $O(1/\epsilon^{3/\epsilon^2+4})$ time. Generating a partial schedule from an outline needs $O(1/\epsilon^4)$ time. Generating a feasible schedule from a partial schedule needs $O(1/\epsilon^6)$ time, since there are at most $k \leq 1/\epsilon^3 - 1/\varepsilon + 1$ aggregate large jobs. Steps 4 and 5 can thus be executed in $O(1/\epsilon^{3/\epsilon^2+6})$ time. Hence, Algorithm MML runs in $O(n + 1/\epsilon^{3/\epsilon^2+6})$ time. □

References

1. Afrati, F., Bampis, E., Chekuri, C., Karger, D., Kenyon, C., Khanna, S., Milis, I., Queyranne, M., Skutella, M., Stein, C., Sviridenko, M.: Approximation schemes for minimizing average weighted completion time with release dates. In: The proceeding of the 40th Annual Symposium on Foundations of Computer Science, New York, pp. 32–43 (October 1999)
2. Brucker, P., Gladky, A., Hoogeveen, H., Kovalyvov, M.Y., Potts, C.N., Tautenhahn, T., van de Velde, S.L.: Scheduling a batching machine. Journal of Scheduling 1, 31–54 (1998)
3. Cheng, T.C.E., Liu, Z., Yu, W.: Scheduling jobs with release dates and deadlines on a batch processing machine. IIE Transactions 33, 685–690 (2001)
4. Deng, X., Feng, H., Zhang, P., Zhao, H.: A polynomial time approximation scheme for minimizing total completion time of unbounded batch scheduling. In: Eades, P., Takaoka, T. (eds.) ISAAC 2001. LNCS, vol. 2223, pp. 26–35. Springer, Heidelberg (2001)
5. Deng, X., Zhang, Y.: Minimizing mean response time in batch processing system. In: Asano, T., Imai, H., Lee, D.T., Nakano, S.-i., Tokuyama, T. (eds.) COCOON 1999. LNCS, vol. 1627, pp. 231–240. Springer, Heidelberg (1999)
6. Hall, L.A., Shmoys, D.B.: Approximation schemes for constrained scheduling problems. In: Proceedings of the 30th annual IEEE Symposium on Foundations of Computer Science, pp. 134–139. IEEE Computer Society Press, Los Alamitos (1989)
7. Kise, H., Ibaraki, T., Mine, H.: Performance analysis of six approximation algorithms for the one-machine maximum lateness scheduling problem with ready times. Journal of the Operations Research Society of Japan 22, 205–224 (1979)
8. Lawler, E.L., Lenstra, J.K., Kan, A.H.G.R., Shmoys, D.B.: Sequencing and scheduling: algorithms and complexity. In: Graves, S.C., Kan, A.H.G.R., Zipkin, P.H. (eds.) Handbooks in Operations Research and Management Science. Logistics of Production and Inventory, vol. 4, pp. 445–522. North-Holland, Amsterdam (1993)
9. Lee, C Y., Uzsoy, R., Martin, L.A.: Vega, Efficient algorithms for scheduling semiconductor burn-in operations. Operations Research 40, 764–775 (1992)
10. Liu, Z., Yuan, J., Cheng, T.C.E.: On scheduling an unbounded batch machine. Operations Research Letters 31, 42–48 (2003)

A Non-interleaving Denotational Semantics of Value Passing CCS with Action Refinement*

Guang Zheng[1], Shaorong Li[2], Jinzhao Wu[2], and Lian Li[1]

[1] School of Information Science and Engineering, Lanzhou University,
Lanzhou 730000, China
[2] College of opto-electronic Information, University of Electronic Science and
Technology of China, Chengdu 610054, China

Abstract. Process algebra provides essential tools for studying concurrent systems. An important branch of process algebra is value passing CCS. However, value passing CCS lacks not only action refinement, which is an essential operation in the design of concurrent systems, but also non-interleaving semantics, which is appropriate to specify the partial order and equivalence relations. In this paper, we will define action refinement and non-interleaving semantics by valued stable event structures and valued labeled configuration structures in which special valued actions will be executed. The refinement operation and semantics are useful for the hierarchical design methodology in value passing CCS, e.g. top-down system design.

Keywords: value passing CCS, valued stable event structures, valued labeled configuration structures, action refinement, semantics.

1 Introduction

Process algebra is a widely accepted language of specifying concurrent systems. The fundamental work is done by Milner in CCS[25]. He defined the language of value passing processes, called *value passing CCS* or *full CCS*. By far, the simple pure CCS (without value passing) is well studied, and some researchers began to study the bisimulations and equivalences of value passing CCS [1,11,26,17], and the behavioral semantics for abstract data types in [4]. However, there has no work done on the non-interleaving semantics of value passing CCS with action refinement.

Action refinement is an essential operation in the top-down design of concurrent system. Actions are considered as the basic blocks of the process algebras [14,15,16,8,2,24,19,20,22,7,5,6,18]. Through action, consider the system behavior as an entity in certain level of abstraction. Action refinement theory [13,14,15,16,8,2,24,19,20,22] gives the way of top-down design in concurrent systems. This makes it possible to design a system from an abstract level to a level more concrete, and can be done level by level until no refinement is needed.

---

* This work was supported by the National Natural Science Foundation of China (No. 90612016 and 60473095).

F.P. Preparata and Q. Fang (Eds.): FAW 2007, LNCS 4613, pp. 178–190, 2007.
© Springer-Verlag Berlin Heidelberg 2007

Action refinement of pure CCS has been well studied [13,15,16,20,22,23,24], and it has also been applied to process algebra extended with probability, time [4,5,6,18,19]. However, as we know, there have no work done on value passing processes with action refinement so far, so we first propose it as a function on value passing CCS.

Non-interleaving semantics is appropriate for true concurrency semantics. We select valued stable event structures and valued labeled configuration structures for the models of true concurrency in value passing CCS. The reasons are as following: (1), the concurrency models can roughly be distinguished in two kinds: interleaving and non-interleaving. The interleaving model specifies the independent execution of processes, and the non-interleaving one can represent the casual relations between actions more explicitly; (2), some equivalence relations are not preserved under the refinement of interleaving approach [23]. These equivalence relations are often used to proof correctness of the implementation of concurrent systems. This problem can be well solved in non-interleaving models, (3), non-interleaving is useful in model checking for partial order reduction [9].

We give a non-interleaving denotational semantics of value passing CCS with action refinement in this paper, which makes it possible to design valued concurrent systems in the hierarchical methodology known as top-down system design in software engineering.

This paper is organized as follows. Section 2 introduces the language of a value passing CCS. Section 3 introduces valued event structures and operators on them. Section 4 introduces action refinement both in syntax and semantics. Section 5 shows the non-interleaving semantics. Section 6 concludes the paper.

2   Value Passing CCS

In this section we introduce the syntax of the language we consider in this paper. The language is essentially the same as the one originally proposed by Milner in [25].

Our language is given by the BNF-definition as follows:

$$t ::= 0 \mid \sqrt{} \mid \alpha.t \mid \text{if } b \text{ then } t \mid t+t \mid t;t \mid t|t \mid t \setminus L \mid t[\lambda]$$
$$\alpha ::= \tau \mid c?x \mid c!e$$

0 is the inactive process capable of doing nothing; $\sqrt{}$ is the termination predicate; $\alpha.t$ is action prefixing where we have three kinds of actions: the invisible action τ, representing internal communication, and the input and output actions $c?x$ and $c!e$ respectively; if b then t is the (one armed) conditional construction; $t+t$ is nondeterministic choice (or summation); $t;t$ is sequential composition; $t|t$ is parallel composition; $t \setminus L$, where L is a finite subset of *Chan*, is channel restriction; $t[\lambda]$, where λ is a partial function from *Chan* to *Chan* with finite domain, is channel renaming.

Example 2.1 Here we give out the language of the system of Attendance Register with the input device of *FingerprintReader* (short for *ARF*), the ARF

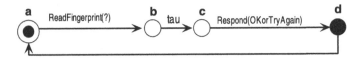

Fig. 1. The event structure of ARF

system can read *fingerprint*, after some inner action τ(**tau**), then respond with result of *'OK'* or *'TryAgain'*, after that, system returns to the state of initialization. $ARF := ReadFingerprint(?).\tau.Respond(!v).ARF$.

3   Valued Event Structures and Operators on Them

In this section, we introduce valued stable event structures and valued labeled configuration structures, we define a compositional operator for general action refinement on event structures and show that the subclass of stable event structures is closed under these operators [23,21,12].

3.1   Valued Stable Event Structures

Definition 3.1. (*Valued stable event structures*) *A labeled event structure (over an alphabet Act) with value is a 5-tuple* $\mathcal{E}_{V\restriction\mathcal{E}} = (E_{V\restriction E}, Con_{V\restriction Con}, \mapsto_{V\restriction\mapsto}, \sqrt{}, l_v)$ *where*

- $E_{V\restriction E}$, a set of *valued events*, the value of certain step can carry any possible values of actions;
- $Con_{V\restriction Con}$, is a non-empty set of finite subsets of $E_{V\restriction E}$(the *valued consistency predicate*) satisfying $Y_{V\restriction Y} \subseteq X_{V\restriction X} \in Con_{V\restriction Con} \Rightarrow Y_{V\restriction Y} \in Con_{V\restriction V}$, here the time of $Y_{V\restriction Y}$ is earlier than that of $X_{V\restriction X}$ and the $_{V\restriction X}$ maybe different from $_{V\restriction Y}$ the same as that in [10];
- $\mapsto_{V\restriction\mapsto} \subseteq Con_{V\restriction Con} \times E_{V\restriction E}$, the *enabling relation with value*, satisfying $X_{V\restriction X} \vdash e_{v'-v} \wedge X_{V\restriction X} \subseteq Y_{V\restriction Y} \in Con_{V\restriction Con} \Rightarrow Y_{V\restriction Y} \vdash e'_v$, here $e_{v'-v}$ refers to the value changed between the state of v' and v;
- $\sqrt{} \subseteq Con_{V\restriction Con}$, the *termination predicate*, satisfying $e_v \notin Con_{V\restriction Con} \in \sqrt{} \Rightarrow X_{V\restriction X} \cup e_v \notin Con_{V\restriction Con}$, the symbol $\sqrt{}$ can do nothing but predicating the termination, it neither does anything further in the system, nor carries value;
- $l_v : E_{V\restriction E} \rightarrow Act_{V\restriction E+v}$, the *labeling function with value*.

Example 3.1 Here, we depict the event structure of example 2.1 bye Fig.1, from Fig.1, we can see clearly the data flow of ARF.

3.2   Operators on Valued Stable Event Structures

Similar with the operators on stable event structures, for detail in [1,10] we describe the semantics of the operators in valued stable event structures as following:

Action prefix. Let $\alpha_v \mathcal{E}_{1_{V|\mathcal{E}_1}} = (\mathcal{E}_{1_{V|\mathcal{E}_1}} \cup \{\alpha_v\}, Con_{1_{V|Con_1}} \cup \{\alpha_v\}, \mapsto_V, \sqrt{_1} l_{1_{v|l_1}} \cup \{\alpha_v\})$ (up to isomorphism), where $\alpha_v \in Obs_{V|Obs} \cup \{\tau\}$, and $\alpha_v \in EVENT_{V|EVENT} \setminus E_1$; $\mapsto_{V|\mapsto} = \mapsto_{1_{V|\mapsto_1}} \cup (\{\{\alpha_v\}\} \times E_{1_{V|E_1}})$.

Sequential composition. Let $\mathcal{E}_{1_{V|\mathcal{E}_1}}; \mathcal{E}_{2_{V|\mathcal{E}_2}} = (E_{1_{V|E_1}} \cup E_{2_{V|E_2}}, Con_{V|Con}, \mapsto_{V|\mapsto}, \sqrt{_2}, l_{1_{v|l_1}} \cup l_{2_{v|l_2}})$, and $Con_{V|Con} = Con_{1_{V|Con_1}} \cup Con_{2_{V|Con_2}} \cup (\sqrt{_1} \times E_{2_{V|E_2}})$; $\mapsto_{V|\mapsto} = \mapsto_{1_{V|\mapsto_1}} \cup \mapsto_{2_{V|\mapsto_2}} \cup \{\sqrt{_1} \times \mapsto_{2_{V|\mapsto_2}}\}$.

Choice. Let $\mathcal{E}_{1_{V|\mathcal{E}_1}} + \mathcal{E}_{2_{V|\mathcal{E}_2}} = (E_{1_{V|E_1}} \cup E_{2_{V|E_2}}, Con_{1_{V|Con_1}} \cup Con_{2_{V|Con_2}}, \mapsto_{1_{V|\mapsto_1}} \cup \mapsto_{2_{V|\mapsto_2}}, \sqrt{_1} \cup \sqrt{_2}, l_{1|l_1} \cup l_{2_{v|l_2}})$

Parallel composition. Let $\mathcal{E}_{1_{V|\mathcal{E}_1}} | \mathcal{E}_{2_{V|\mathcal{E}_2}} = (E_{V|E}, Con_{1_{V|Con_1}} \cup Con_{2_{V|Con_2}}, \mapsto_{1_{V|\mapsto_1}} \times \mapsto_{2_{V|\mapsto_2}}, \sqrt{_1} \cup \sqrt{_2}, l_{1_{v|l_1}} \times l_{2_{v|l_2}})$, and $E_{V|E} = (E_{1_{V|E_1}} \times \{*\}) \cup (\{*\} \times E_{2_{V|E_2}}) \cup (E_{1_{V|E_1}} \times E_{2_{V|E_2}})$

Relabeling. For relabeling function $r : Act_V \to Act_{V'}$ with $r(0) = 0$, $r(\tau) = \tau$ and $r(\sqrt{}) = \sqrt{}$, let $\alpha_v \in Obs_{V|Obs}$, let $\mathcal{E}_{1_{V|\mathcal{E}_1}}[r] = (E_{1_{V|E_1}}, Con_{V|Con}, \mapsto_{V|\mapsto}, \sqrt{}, l_v)$, where l_v is the function composition of r and $l(e_v) = r(l_1(e_{v'}))$.

3.3   Valued Labeled Configuration Structures

Definition 3.2. A *valued labeled configuration structure* is a triple $\mathcal{C}_{V|C} = (C_{V|C}, \sqrt{}, l_v)$ where $C_{V|C}$ is a family of finite sets (the configurations), $\sqrt{} \subseteq C_{V|C}$ is a *termination predicate*, satisfying $X \in \sqrt{} \cap X \subseteq Y \in C_{\mathcal{C}_{V|c}} \Rightarrow X = Y$ (i.e. terminating configurations must be maximal), and $l_v : \bigcup_{X_{V|X} \in C_{V|C}} \to Act_V$ is a *labeling function*.

Let $\mathbb{C}$ denote the domain of configuration structures labeled over Act, and $\mathbb{C}_{V|C}$ denote the domain of valued configuration structures labeled over Act_V. The set $E_{\mathcal{C}}$ of *events* of $\mathcal{C} \in \mathbb{C}$ is defined by $E_{\mathcal{C}} = \bigcup_{X \in C_{\mathcal{C}}} X$, and the set $E_{\mathcal{C}_{V|c}}$ of *event* with *value* of $\mathcal{C}_{V|c} \in \mathbb{C}_{V|C}$ is defined by $E_{\mathcal{C}_{V|c}} = \bigcup_{X_{V|X} \in C_{\mathcal{C}_{V|c}}} X_{V|X}$.

3.4   Operators on Valued Configuration Structures

Compared with the operators on stable event structures, the operators on valued configuration structures can only be described stepwise, action(s) can only be taken by configurations step by step, we describe them as following:

Action prefix. Let $\alpha_v.\mathcal{E}_{V|\mathcal{E}}$ be the valued action prefix, and $C_{\mathcal{E}''_{V|\mathcal{E}}} - C_{\mathcal{E}'_{V|\mathcal{E}}} = e_v$ with $\alpha_v = l(e_v)$

Sequential composition. Let $\mathcal{E}_{1_{V|\mathcal{E}_1}}; \mathcal{E}_{2_{V|\mathcal{E}_2}}$ be the valued sequential composition, and $\Sigma(C_{\mathcal{E}''_{V|\mathcal{E}}} - C_{\mathcal{E}'_{V|\mathcal{E}}}) = e_v^*$, with $e_v \in \{e_{1_{v_1}} \vee e_{2_{v_2}} | e_{1_{v_1}} \in E_{1_{V|E_1}} \wedge e_{2_{v_2}} \in E_{2_{V|E_2}} \wedge e_{1_{v_1}} < e_{2_{v_2}}\}$

Choice. Let $\mathcal{E}_{1_{V|\mathcal{E}_1}} + \mathcal{E}_{2_{V|\mathcal{E}_2}}$ be valued choice, and $\Sigma(C_{\mathcal{E}''_{V|\mathcal{E}}} - C_{\mathcal{E}'_{V|\mathcal{E}}}) = e_v^*$, with $e_v \in \{e_{1_{v_1}} \vee e_{2_{v_2}} | e_{1_{v_1}} \in E_{1_{V|E_1}} \wedge e_{2_{v_2}} \in E_{2_{V|E_2}}\}$ and $e_v^* \in e_{1_{v|e_1}}^*$ or $e_v^* \in e_{2_{v|e_2}}^*$

Parallel composition. Let $\mathcal{E}_{1_{V|\mathcal{E}_1}} | \mathcal{E}_{2_{V|\mathcal{E}_2}}$ be the valued parallel composition, and $\Sigma(C_{\mathcal{E}''_{V|\mathcal{E}}} - C_{\mathcal{E}'_{V|\mathcal{E}}}) = e_v^*$, with $e_v \in \{(e_{1_{v_1}}, *) \vee (*, e_{2_{v_2}}) \vee (e_{1_{v_1}}, e_{2_{v_2}}) | e_{1_{v_1}} \in E_{1_{V|E_1}} \wedge e_{2_{v_2}} \in E_{2_{V|E_2}}\}$

4   Action Refinement

In this section, we will introduce action refinement into value passing CCS both in syntax and semantics.

4.1   Value Passing CCS with Action Refinement

We will introduce a new calculator into value passing CCS: f value passing CCS with *action refinement*. This syntactic action refinement in the value passing CCS. And valued action could not affect the behavior of the action.

Definition 4.1. In order to make sure that there is no confusion of communication levels, like in [25,3], we have to sort out some "bound" actions. Formally, the *sort* of a given expression P and $P_{V \upharpoonright P}$ (valued P) is defined as follows:

$$Sort(0) = Sort(\sqrt{}) = Sort(\alpha_v) = Sort(\alpha) = \emptyset;$$
$$Sort(\alpha_v.P_{V \upharpoonright P}) = Sort(\alpha.P) = Sort(P);$$
$$Sort(P1_{V \upharpoonright P1} \circ P2_{V \upharpoonright P2}) = Sort(P1 \circ P2) = Sort(P1) \cup Sort(P2),$$
$$for \circ \in \{; , +, |\};$$
$$Sort(P_{V \upharpoonright P} \backslash A) = Sort(P \backslash A) = Sort(P) \cup A;$$
$$Sort(P_{V \upharpoonright P}[\lambda]) = Sort(P) \cup dom(\lambda)$$

By Act_0 we denote a subset of Act with $Sort(P) \cup \{\tau, \sqrt{}\} \subseteq Act_0$ that cannot be refined. Let $f : Act \backslash Act_0 \rightarrow VPEpr$ be a function. We are interested in the situation where the semantic model of $f(\alpha)$ is a refinement of action α. We call $f(\alpha)$ *a refinement of action* α and $f(\alpha_v)$ *a refinement of action* α_v (valued α).

Example 4.1 Revisit Example 2.1, if we want to design the system ARF, perhaps we would like the ARF to act in such a way in detail: when someone put his/her finger on the system, the system would try to recognize the people by the fingerprint, and at the same time, record the fingerprint no matter the fingerprint can be recognized or not. If the system can recognized the fingerprint, it will respond OK, otherwise, respond $TryAgain$ instead, then, the system returns to its initialized state.

Then, the refined ARF of Example 1.1 would be:

$$ARF := ReadFingerprint(fingerprint).ARF'$$
$$ARF' := (CheckFingerprint(fingerprint)|RecordLog(fingerprint)).ARF''$$
$$ARF'' := ((CheckFingerprint(fingerprint) = True.Respond(OK)) +$$
$$(CheckFingerprint(fingerprint) <> True.Respond(TryAgain))|$$
$$RecordLog(fingerprint)).ARF'''$$
$$ARF''' := ((Respond(OK) + Respond(TryAgain))|RecordLog(fingerprint))$$
$$.ARF''''$$
$$ARF'''' := Done.ARF$$

4.2   Action Refinement of Valued Event Structures

A refinement operator on valued event structures will be given by a function ref specifying for each action α_v an event structure $ref(\alpha_v)$ which is to be substituted for α_v. We only consider non-forgetful refinements in this paper, where as forgetful refinements refine some actions into an event structure with $\emptyset \in \sqrt{}$, or equivalently $Con = \sqrt{} = \{\emptyset\}$.

Definition 4.2. (*Action refinement of stable event structures*)

(i) A function $ref' : Act \to \mathbb{E}$ is called a refinement function *(for stable event structures)* iff $\forall a \in Act : \emptyset \notin \sqrt{}_{ref(a)}$

(ii) Let $\mathcal{E} \in \mathbb{E}$ and let ref' be a refinement function.

Then the *refinement* of $\mathcal{E}$ by ref', $ref'(\mathbb{E})$, is the event structures defined by the following:

- $E_{ref'(\mathcal{E})} := \{(e,e') | e \in E_{\mathcal{E}}, e' \in E_{ref'(l_{\mathcal{E}}(e))}\}$,
- $\widetilde{X} \in Con_{ref'(\mathcal{E})}$ iff $\pi_1(\widetilde{X}) \in Con_{\mathcal{E}}$ and $\forall e \in \pi_1(\widetilde{X}) : \pi_2^e(\widetilde{X}) \in Con_{ref'(l_{\mathcal{E}}(e))}$; $Conf(\mathcal{E})$ by ref;
- $\widetilde{X} \vdash_{ref'(\mathcal{E})} (e,e')$ iff $ready(\widetilde{X}) \vdash_{\mathcal{E}} e$ and $\pi_2^e(\widetilde{X}) \vdash_{ref'(l_{\mathcal{E}}(e))} e'$;
- $\widetilde{X} \in \sqrt{}_{ref'(\mathcal{E})}$ iff $\pi_1(\widetilde{X}) \in \sqrt{}_{\mathcal{E}}$ and $\forall e \in \pi_1(\widetilde{X}) : \pi_2^e(\widetilde{X}) \in \sqrt{}_{ref'(l_{\mathcal{E}}(e))}$;
- $l_{ref'(\mathcal{E})}(e,e') := l_{ref'(l_{\mathcal{E}}(e))}(e')$;

with $\pi_1(\widetilde{X}) := \{e | \exists f : (e,f) \in \widetilde{X}\}, \pi_2^e(\widetilde{X}) := \{f | (e,f) \in \widetilde{X}\}$
and $ready(\widetilde{X}) := e \in \pi_1(\widetilde{X}) | \pi_2^e(\widetilde{X}) \in \sqrt{}_{ref'(l_{\mathcal{E}}(e))}$ for $\widetilde{X} \subseteq E_{ref'(\mathcal{E})}$
from the Definition 4.2, we define *action refinement of valued event structures* with the symbol of v in the form like $v_{\upharpoonright C}$ to represent the *value* bounded to action ranging over scope $_{\upharpoonright C}$.

Definition 4.3. *Action refinement of valued stable event structures*

(i) A function $ref : Act_V \to \mathbb{E}_{V \upharpoonright \mathbb{E}}$ is called a refinement function *(for valued event structures)* iff $\forall a_v \in Act_V : \emptyset \notin \sqrt{}_{ref(a)}$

(ii)Let $\mathcal{E}_{V \upharpoonright \mathcal{E}} \in \mathbb{E}_{V \upharpoonright \mathbb{E}}$ and let ref be a refinement function with value.

Then the *refinement* of $\mathcal{E}_{V \upharpoonright \mathcal{E}}$ by ref, $ref(\mathcal{E}_{V \upharpoonright \mathcal{E}})$, is the event structure defined by the following:

- $E_{ref(\mathcal{E}_{V \upharpoonright \mathcal{E}})} := \{(e_v, e'_{v'}) | e_v \in E_{\mathcal{E}_{V \upharpoonright \mathcal{E}}}, e'_{v'} \in E_{ref(l_{\mathcal{E}}(e_v))}\}$, a set of *valued events*, the value is in the form of matrix;
- $\widetilde{X}_{\widetilde{V} \upharpoonright \widetilde{X}} \in Con_{ref(\mathcal{E}_{V \upharpoonright \mathcal{E}})}$ iff $\pi_1(\widetilde{X}_{\widetilde{V} \upharpoonright \widetilde{X}}) \in Con_{\mathcal{E}_{V \upharpoonright \mathcal{E}}}$ and $\forall e_{v1} \in \pi_1(\widetilde{X}_{\widetilde{V} \upharpoonright \widetilde{X}}) :$ $\pi_2^{e_{v1}}(\widetilde{X}_{\widetilde{V} \upharpoonright \widetilde{X}}) \in Con_{ref(l_{\mathcal{E}_{V \upharpoonright \mathcal{E}}}(e_v))}$;
- $\widetilde{X}_{\widetilde{V} \upharpoonright \widetilde{X}} \vdash_{ref(\mathcal{E}_{V \upharpoonright \mathcal{E}})} (e_{v1}, e'_{v2})$ iff $ready(\widetilde{X}_{\widetilde{V} \upharpoonright \widetilde{X}}) \vdash_{\mathcal{E}_{V \upharpoonright \mathcal{E}}} e_{v1}$ and $\pi_2^{e_{v1}}(\widetilde{X}_{\widetilde{V} \upharpoonright \widetilde{X}}) \vdash_{ref(l_{\mathcal{E}_{V \upharpoonright \mathcal{E}}} e_{v1})} e'_{v2}$;
- $\widetilde{X} \in \sqrt{}_{ref(\mathcal{E})}$ iff $\pi_1(\widetilde{X}) \in \sqrt{}_{\mathcal{E}}$ and $\forall e \in \pi_1(\widetilde{X}) : \pi_2^e(\widetilde{X}) \in \sqrt{}_{ref(l_{\mathcal{E}_{V \upharpoonright \mathcal{E}}}(e))}$;
- $l_{ref(\mathcal{E}_{V \upharpoonright \mathcal{E}})}(e_{v1}, e'_{v2}) := l_{ref(l_{\mathcal{E}_{V \upharpoonright \mathcal{E}}}(e_{v1}))}(e'_{v2})$

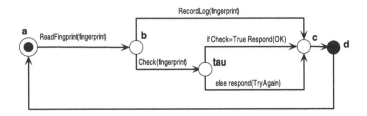

Fig. 2. The refined valued event structure of ARF

with $\mathcal{E}_{V\upharpoonright\mathcal{E}} \in \mathbb{E}_V$, let ref be a refinement function, we call $\widetilde{X}_{\widetilde{V}\upharpoonright\widetilde{X}}$ a refinement of configuration $X_{V\upharpoonright X} \in Conf(\mathcal{E}_{V\upharpoonright\mathcal{E}})$ by ref iff

- $\widetilde{X}_{\widetilde{V}\upharpoonright\widetilde{X}} = \bigcup_{e_v \in X_{V\upharpoonright X}} \{e_v\} \times X_{e_v}$ where
  $\forall e_v \in X_{V\upharpoonright X} : X_{e_v} \in Conf(ref(l_{\mathcal{E}_{V\upharpoonright\mathcal{E}}}(e_v))) - \{\emptyset\}$
- $e_v \in busy(\widetilde{X}_{\widetilde{V}\upharpoonright\widetilde{X}}) \Rightarrow e_v$ maximal in $X_{V\upharpoonright X}$ with respect to $<_{\mathcal{E}_{V\upharpoonright\mathcal{E}}}$ where
  $busy(\widetilde{X}_{\widetilde{V}\upharpoonright\widetilde{X}}) := \{e_v \in X_{V\upharpoonright X}|X_{e_v}$ not terminated $\}$

then $Conf(ref(\mathcal{E}_{V\upharpoonright\mathcal{E}})) = \{\widetilde{X}_{\widetilde{V}\upharpoonright\widetilde{X}}|\widetilde{X}_{\widetilde{V}\upharpoonright\widetilde{X}}$ is a refinement of a configuration $X_{V\upharpoonright X} \in Conf(\mathcal{E}_{V\upharpoonright\mathcal{E}})\}$

and $\pi_1(\widetilde{X}_{\widetilde{V}\upharpoonright\widetilde{X}}) := \{e_{v1}|\exists f_{v2} : (e_{v1}, f_{v2}) \in \widetilde{X}_{\widetilde{V}\upharpoonright\widetilde{X}}\}, \pi_2^{e_{v1}}(\widetilde{X}_{\widetilde{V}\upharpoonright\widetilde{X}}) := \{f_{v2}|(e_{v1}, f_{v2}) \in \widetilde{X}_{\widetilde{V}\upharpoonright\widetilde{X}}\}$

and $ready(\widetilde{X}_{\widetilde{V}\upharpoonright\widetilde{X}}) := e_{v1} \in \pi_1(\widetilde{X}_{\widetilde{V}\upharpoonright\widetilde{X}})|\pi_2^{e_{v1}}(\widetilde{X}_{\widetilde{V}\upharpoonright\widetilde{X}}) \in \sqrt{}_{ref(l_{\mathcal{E}_{V\upharpoonright\mathcal{E}}}(e))}$ for

$\widetilde{X}_{\widetilde{V}\upharpoonright\widetilde{X}} \subseteq E_{ref(\mathcal{E}_{V\upharpoonright\mathcal{E}})}$

As usual, we verify the that $ref(\mathcal{E}_{V\upharpoonright\mathcal{E}})$ is well-defined, even when isomorphic event structures are identified. In addition we check that stability of $\mathcal{E}_{V\upharpoonright\mathcal{E}}$ and ref is preserved.

Example 4.2, Here, we present Fig.2 as the refined system of ARF(Fig.1), compared with Fig.1, Fig.2 has a state of **tau**, which is refined from the transition of tau in Fig.1, from which we can see intuitively the refinement of valued ARF.

Proposition 4.1

- If $\mathcal{E}_{V\upharpoonright\mathcal{E}} \in \mathbb{E}_{V\upharpoonright\mathbb{E}}$ and ref is a refinement function then $ref(\mathcal{E}_{V\upharpoonright\mathcal{E}})$ is a valued event structure indeed;
- If $\mathcal{E}_{V\upharpoonright\mathcal{E}} \in \mathbb{E}_{V\upharpoonright\mathbb{E}}$ and ref, ref' are refinement functions with $ref(\alpha_v) \cong ref'(\alpha_v)$ for all $\alpha_v \in Act_V$ then $ref(\mathcal{E}_{V\upharpoonright\mathcal{E}}) \cong ref'(\mathcal{E}_{V\upharpoonright\mathcal{E}})$;
- If $\mathcal{E}_{V\upharpoonright\mathcal{E}}, \mathcal{F}_{V\upharpoonright\mathcal{F}} \in \mathbb{E}_{V\upharpoonright\mathbb{E}}, \mathcal{E}_{V\upharpoonright\mathcal{E}} \cong \mathcal{F}_{V\upharpoonright\mathcal{F}}$, and ref is a refinement function then $ref(\mathcal{E}_{V\upharpoonright\mathcal{E}}) \cong ref(\mathcal{F}_{V\upharpoonright\mathcal{F}})$;
- If $\mathcal{E}_{V\upharpoonright\mathcal{E}}$ is stable and $ref(a_v)$ is stable for all $a_v \in Act_V$ then $ref(\mathcal{E}_{V\upharpoonright\mathcal{E}})$ is stable

Proof. Straightforward.   □

Definition 4.4

- A function $ref : Act_V \rightarrow \mathbb{C}_{\mathbb{V} \upharpoonright \mathbb{C}} - \{\varepsilon\}$ is called a *refinement function (for labeled configuration structures with value)*.
- Let $\mathcal{C}_{\mathcal{V} \upharpoonright \mathcal{C}}$ be a valued configuration structure and let ref be a refinement function.

  We call $\widetilde{X}_{\widetilde{V} \upharpoonright \widetilde{X}}$ a *refinement of a valued configuration* $X_{V \upharpoonright X} \in C_{\mathcal{C}_{\mathcal{V} \upharpoonright \mathcal{C}}}$ by ref iff

  - $\widetilde{X}_{\widetilde{V} \upharpoonright \widetilde{X}} = \bigcup_{e_v \in X_{V \upharpoonright X}} \{e_v\} \times X_{e_v}$ where $\forall e_v \in X_{V \upharpoonright X} : X_{e_v} \in C_{ref(l_c(e_v)) - \{\emptyset\}}$;

  - $\forall Y_{V \upharpoonright Y} \subseteq busy(\widetilde{X}_{\widetilde{V} \upharpoonright \widetilde{X}}) : X_{V' \upharpoonright X'} - Y_{V \upharpoonright Y} \in C_{\mathcal{C}_{\mathcal{V} \upharpoonright \mathcal{C}}}$ where $busy\,(\widetilde{X}_{\widetilde{V} \upharpoonright \widetilde{X}}) := \{e_v \in X_{V \upharpoonright X} | X_{e_v} \notin \sqrt{}_{ref(l_{\mathcal{C}_{\mathcal{V} \upharpoonright \mathcal{C}}}(e))}\}$; Such a refinement is *terminated* iff

    $\forall e \in X : X_e \in \sqrt{}_{ref(l_c(e))}\}$, i.e. iff $busy\,(\widetilde{X}) = \emptyset$

    The *refinement of* $\mathcal{C}_{\mathcal{V} \upharpoonright \mathcal{C}}$ by ref is defined as $ref(\mathcal{C}_{\mathcal{V} \upharpoonright \mathcal{C}}) = (C_{ref(\mathcal{C}_{\mathcal{V} \upharpoonright \mathcal{C}})}, \sqrt{}_{ref(\mathcal{C})}, l_{ref(\mathcal{C}_{\mathcal{V} \upharpoonright \mathcal{C}})})$ with

    * $C_{ref(\mathcal{C}_{\mathcal{V} \upharpoonright \mathcal{C}})} := \{\widetilde{X}_{\widetilde{V} \upharpoonright \widetilde{X}} | \widetilde{X}_{\widetilde{V} \upharpoonright \widetilde{X}}$ is a refinement of some $X_{V \upharpoonright X} \in C_{\mathcal{C}_{\mathcal{V} \upharpoonright \mathcal{C}}}$ by $ref\}$;

    * $\sqrt{}_{ref(\mathcal{C})} := \{\widetilde{X} | \widetilde{X}$ is a terminated refinement of some $X \in \sqrt{}_{\mathcal{C}}$ by $ref\}$;

    * and $l_{ref(\mathcal{C}_{\mathcal{V} \upharpoonright \mathcal{C}})}(e_v, e'_{v'}) := l_{ref(\mathcal{C}_{\mathcal{V} \upharpoonright \mathcal{C}}(e_v))}(e'_{v'})$ for all $(e_v, e'_{v'}) \in E_{ref(\mathcal{C}_{\mathcal{V} \upharpoonright \mathcal{C}})}$.

All the actions except those predicating termination carry no value, actions satisfy the condition of $\forall e \in \{X : X_e \in \sqrt{}_{ref(l_{\mathcal{C} \upharpoonright \mathcal{C}}(e))}\}$ just evolve without value ($\xrightarrow{\sqrt{}} 0$), or they carry the value $\{\emptyset\}$, we omit it for short here.

Example 4.3 Here we depict the refined system of *ARF* on labeled configuration structure by Fig.3.

Now, it is time for us to show that refinement is a well-defined operation on labeled configuration structures, even when isomorphic configuration structures are identified.

Proposition 4.2

- If $\mathcal{C}_{\mathcal{V} \upharpoonright \mathcal{C}} \in \mathbb{C}_{\mathbb{V} \upharpoonright \mathbb{C}}$ and ref is a refinement function then also $ref(\mathcal{C}_{\mathcal{V} \upharpoonright \mathcal{C}})$ is a valued configuration structure;
- If $\mathcal{C}_{\mathcal{V} \upharpoonright \mathcal{C}} \in \mathbb{C}_{\mathbb{V} \upharpoonright \mathbb{C}}$ and ref, ref' are refinement functions with $ref(a_v) \cong ref'(a_v)$ for all $a_v \in Act_V$ then $ref(\mathcal{C}_{\mathcal{V} \upharpoonright \mathcal{C}}) \cong ref'(\mathcal{C}_{\mathcal{V} \upharpoonright \mathcal{C}})$;
- If $\mathcal{C}_{\mathcal{V} \upharpoonright \mathcal{C}}, \mathcal{D}_{\mathcal{V} \upharpoonright \mathcal{D}} \in \mathbb{C}_{\mathbb{V} \upharpoonright \mathbb{C}}$ and ref is a refinement function and $\mathcal{C}_{\mathcal{V} \upharpoonright \mathcal{C}} \cong \mathcal{D}_{\mathcal{V} \upharpoonright \mathcal{D}}$ then $ref(\mathcal{C}_{\mathcal{V} \upharpoonright \mathcal{C}}) \cong ref'(\mathcal{D}_{\mathcal{V} \upharpoonright \mathcal{D}})$.

Proof. Straightforward.   □

Lemma 4.1. *Let* $\mathcal{C}_{\mathcal{V} \upharpoonright \mathcal{C}}$ *be a valued configuration structure and let ref be a refinements function. Then* $\widetilde{X}_{\widetilde{V} \upharpoonright \widetilde{X}} \in C_{ref(\mathcal{C}_{\mathcal{V} \upharpoonright \mathcal{C}})}$ *iff there is a configuration* $X_{C \upharpoonright X} \in \mathcal{C}_{\mathcal{V} \upharpoonright \mathcal{C}}$ *such that*

- $\widetilde{X}_{\widetilde{V} \upharpoonright \widetilde{X}} = \bigcup_{e_v \in X_{V \upharpoonright X}} \{e_v\} \times X_{e_v}$ where $\forall e_v \in X_{V \upharpoonright X} : X_{e_v} \in C_{ref(l_{\mathcal{C}_{\mathcal{V} \upharpoonright \mathcal{C}}}(e_v))}$;

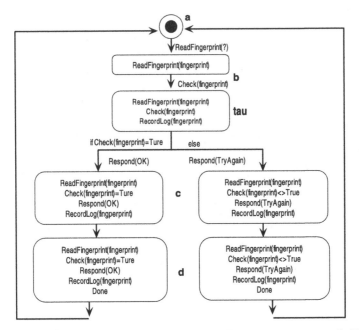

Fig. 3. The refinement on valued labeled configuration structure of ARF

- $\forall Y_{V\restriction Y} \subseteq busy_{X_{V\restriction X}}(\widetilde{X}_{\widetilde{V}\restriction\widetilde{X}}) : X_{V\restriction X} - Y_{V\restriction Y} \in C_{\mathcal{C}_{V\restriction c}}$ where

  $busy_{X_{V\restriction X}}(\widetilde{X}_{\widetilde{V}\restriction\widetilde{C}}) := \{e_v \in X_{V\restriction X} | X_{e_v} \notin \sqrt{}_{ref(l_{c\restriction c})(e)}\}$.

Proof. "Only if" is given by Definition, but "if" is not, because we do not require here that $X_{e_v} \neq \emptyset$. So it is easy to get $\widetilde{X}_{\widetilde{V}\restriction\widetilde{X}} \in C_{ref(\mathcal{C}_{V\restriction c})}$.   □

5 A Non-interleaving Semantics

The refinement function f introduced in Sect 4.1 is defined on a syntactic level, and we call it syntactic. The refinement function ref presented in Sect 4.2 uses semantic object, i.e. valued stable event structures and valued labeled configuration structures, as the definition domain, and we call it semantic.

Let $f : Act \backslash Act_0 \to VPEpr$ be a syntactic refinement function for expression P, and $ref : Act_V \to \mathbb{E}_{V\restriction\mathbb{E}}$ be a semantic refinement function. Since we are interested in the refinement of an already given expression P, we suppose in the following that all the expressions and operators mentioned occur in P.

Lemma 5.1

$$\mathcal{C}(f(\alpha_v.P1_{V\restriction P1})) = \begin{cases} \mathcal{C}(\alpha_v.f(P1_{V\restriction P1})), & \text{if } \alpha \in Act_0; \\ \mathcal{C}(f(\alpha_v.P1_{V\restriction P1})) = \mathcal{C}(f(\alpha_v)); f(P1_{V\restriction P1})), & \text{otherwise} \end{cases} \quad (1)$$

$$\mathcal{C}(f(\circ(P1_{V\restriction P1}))) = \mathcal{C}(\circ f(P1_{V\restriction P1})), \quad \circ \in \{\backslash A, [\lambda]\}; \quad (2)$$

$$\mathcal{C}(f(P1_{V\restriction P1} \circ P2_{V\restriction P2})) = \mathcal{C}(f(P1_{V\restriction P1}) \circ f(P2_{V\restriction P2})), \circ \in \{;, +, |\}. \quad (3)$$

Proof. straight forward   □

In the sequel, we show that the claim also holds for the case of parallel composition and a similar result holds for recursion.

Lemma 5.2. $\mathcal{C}(f(P1_{V \restriction P1} | P2_{V \restriction P2})) \approx \mathcal{C}(f(P1_{V \restriction P1}) | f(P2_{V \restriction P2}))$.

Proof. straight forward   □

Lemma 5.3. $\mathcal{C}(f(P1(\bar{e}_{\bar{v}}))) \approx \mathcal{C}(ref(P1(\bar{e}_{\bar{v}})))$

Proof. For convenience, we denote $f(P1(\bar{e}_{\bar{v}}))$ by $\mathcal{E}^f$ and $ref(P1(\bar{e}_{\bar{v}}))$ by $\mathcal{E}^{ref}_{V \restriction \mathcal{E}}$. We need to prove $\mathcal{C}(\mathcal{E}^f_{V \restriction \mathcal{E}}) \approx \mathcal{C}(\mathcal{E}^{ref}_{V \restriction \mathcal{E}})$. Let $\mathcal{E}_{V \restriction \mathcal{E}} = P1(\bar{e}_{\bar{v}})$, $\mathcal{E}i_{V_i \restriction \mathcal{E}i} = vi_{P1_{V \restriction P1}}(\bot)$, $\mathcal{E}_{fi} = f(\mathcal{E}i)$ and $\mathcal{E}_{refi} \cong vi_{ref(\mathcal{E}i)}(\bot)$.

We can write any $\sigma^f_v \in \mathcal{C}(\mathcal{E}^f_{V \restriction \mathcal{E}})$ as $f(\sigma_v, \cup_{e_v} \sigma_{e_v})$, where $\sigma_v \in \mathcal{C}(\mathcal{E}_{V \restriction \mathcal{E}})$ and $\sigma_{e_v} \in \mathcal{C}(f(l(e_v)))$ with $e_v \in E(\sigma_v)$ and let $l(e_v) \notin Act_{0_V}$, and σ_{e_v} successfully terminates if e_v is not maximal in σ_v. Clearly, there exists $i \geq 0$ such that $\sigma_v \in \mathcal{C}(\mathcal{E}i_{V_i \restriction \mathcal{E}i})$, and $\sigma_{e_v} \in \mathcal{C}(f(li(e_v)))$, and $\sigma^f_v \in \mathcal{C}(\mathcal{E}^{fi}_{V \restriction \mathcal{E}})$. However, there exists $\sigma^{ref}_v \in \mathcal{C}(\mathcal{E}^{refi}_{V \restriction \mathcal{E}}) \subseteq \mathcal{C}(\mathcal{E}^f_{V \restriction \mathcal{E}})$, such that $\sigma^f_v \approx \sigma^{ref}_v$.

Symmetrically, for an arbitrary $\sigma^{ref}_v \in \mathcal{C}(\mathcal{E}^{ref}_{V \restriction \mathcal{E}})$, there exists $\sigma^f_v \in \mathcal{C}(\mathcal{E}^f_{V \restriction \mathcal{E}})$, such that $\sigma^{ref}_v \approx \sigma^f_v$.   □

If we denote by $\mathcal{E}^f_{V \restriction \mathcal{E}}$ and $\mathcal{E}^{ref}_{V \restriction \mathcal{E}}$ the event structure defined in Lemmas 5.2 and 5.3, there the set $H = \{(\sigma^f_v, \sigma^{ref}_v, h) | \sigma^f_v \in \mathcal{C}(\mathcal{E}^f_{V \restriction \mathcal{E}})\}$, where σ^{ref}_v and h are constructed from σ^f_v as in Lemmas 6.2 and 6.3, is a history preserving bisimulation between $\mathcal{E}^f_{V \restriction \mathcal{E}}$ and $\mathcal{E}^{ref}_{V \restriction \mathcal{E}}$.

Theorem 5.1. Let $P \in VPEpr$, $f : Act \setminus Act_0 \mapsto VPEpr$ a syntactic refinement function for P, and $ref : Act \setminus Act_0 \mapsto VES$ (*VES* short for valued event structures), $f(\alpha_v) = ref(\alpha_v)$ for for all α_v, $\alpha_v \in Act_V \setminus Act_0$. Then ref is a semantic refinement function, and $f(P_{V \restriction P}) \cong ref(P_{V \restriction P})$.

Proof. We have already demonstrated that f is a semantic refinement function. Thereby we prove the second statement. It is obviously true for 0 and $\sqrt{}$. The conclusion for recursion has already been proved. Now suppose that it is true for $P1$ and $P2$. We look at the other operators.

If $\alpha_v \notin Act_{0_V}$,

$$f(\alpha_v.P1_{V \restriction P1}) = \alpha_v.f(P1_{V \restriction P1});$$
$$f(P1_{V \restriction P1}) \cong ref(P1_{V \restriction P1});$$
$$ref(\alpha_v.P1_{V \restriction P1}) = \alpha_v.ref(P1_{V \restriction P1}).$$

The conclusion for the case where $\alpha_v \in Act_{0_V}$ follows analogously.

For $\circ \in \{\setminus A, [\lambda]\}$,
$$f(\circ P1_{V \restriction P1}) = \circ f(P1_{V \restriction P1}) \cong \circ(ref(P1_{V \restriction P1})) = ref(\circ P1_{V \restriction P1})$$
For $\circ \in \{;, +, |\}$,

$$f(P1_{V \restriction P1} \circ P2_{V \restriction P2}) = f(P1_{V \restriction P1}) \circ f(P2_{V \restriction P2})$$
$$\cong ref(P1_{V \restriction P1}) \circ ref(P2_{V \restriction P2})$$
$$= ref(P1_{V \restriction P1} \circ P2_{V \restriction P2})$$ □

Table 1. The coincident states between syntax and non-interleaving semantics of refined *ARF*

State Sequence	Example 4.1	Example 4.2	Example 4.3
State 1	ARF	a	a
State 2	ARF'	b	b
State 3	ARF''	tau	tau
State 4	ARF'''	c	c
State 5	ARF''''	d	d

Example 5.1 Revisit examples of 4.1, 4.2 and 4.3, we can intuitively get the non-interleaving semantics of example 2.1 by example 4.2 and 4.3, and they can be show clearly by table 1:

6   Conclusions

In this paper, we have introduced the language of value passing CCS. In order to show the non-interleaving denotational semantics of value passing CCS, we defined valued stable event structures and valued labeled configuration structures. Value passing CCS has several operators, and we gave denotational definition of them with event structures mentioned above. In order to improve the expressiveness of value passing CCS to meet the need of hierarchical methodology in the system design, we added action refinement operation into value passing CCS without distinguishing semantics between the early and the late as [1]. We also showed the syntactic and non-interleaving semantic refinement of the value passing CCS.

In this paper, we treated the value as binding to action. Some actions can be executed without value i.e. 0 and $\sqrt{}$, others must run with value of certain form and type. As value plays an center role in the development of the computational theory, this work is much more significant in practice.

References

1. Ingólfsdóttir, A., Lin, H.: A symbolic Approach to value passing Processes. In: Bergstra, J.A., Ponse, A., Smolka, S.A. (eds.) Handbook of Process Algebra, pp. 427–478. Elsevier, Amsterdam (2001)
2. Bossi, A., Piazza, C., Rossi, S.: Preserving (Security) Properties under Action Refinement. In: CILC 2004, Parma (June 2004)
3. Hoare, C.A.R.: Communicating Sequential Processes. Prentice Hall International, Englewood Cliffs (1985)
4. Bolton, C., Davies, J., Woodcock, J.: On the Refinement and Simulation of Data Types and Processes. In: Proceedings of the 1st International Conference on Integrated Formal Methods (IFM '99), pp. 273–292. Springer, Heidelberg (1999)

5. Fecher, H., Majster-Cederbaum, M., Wu, J.: Action refinement for probabilistic processes with true concurrency models. In: Hermanns, H., Segala, R. (eds.) PROB-MIV 2002, PAPM-PROBMIV 2002, and PAPM 2002. LNCS, vol. 2399, pp. 77–94. Springer, Heidelberg (2002)
6. Fecher, H., Majster-Cederbaum, M., Wu, J.: Refinement of actions in a real-time process algebra with a true concurrency model. Electronic Notes in Theoretical computer Science 70(3) (2002)
7. Goltz, U., Gorrieri, R., Rensink, A.: Techinical Report UBLCS-99-09, University of Bologna, 1999. In: Handbook of Process Algebra, ch. XVI, pp. 1047–1147. Elsevier Science, Amsterdam (2001)
8. Bergstra, J.A., Klop, J.W.: Algebra of Communitating Processes with Abstraction. TCS 37(1), 77–121 (1985)
9. Katoen, J.P.: Concepts, Algorithms and Tools for Model Checking. Friedrich-Alexander Universität Erlangen-Nürnberg (1999)
10. Katoen, J.P.: Quantitative and qualitative extensions of event structures, PhD thesis, University of Twente (1996)
11. Rathke, J., Hennessy, M.: Local Model Checking for value passing Processes (Extended Abstract). In: Ito, T., Abadi, M. (eds.) TACS 1997. LNCS, vol. 1281, Springer, Heidelberg (1997)
12. Fecher, H.: Event Structures for Interrupt Process Algebras. In: EXPRESS'03. Electronic Notes in Theoretical Computer Science, vol. 96, pp. 113–127. Elsevier Science Publishers, Amsterdam (2004)
13. Aceto, L.: Action-refinement in Process Algebra. Cambridge Univ. Press, Cambridge (1992)
14. Aceto, L., Hennessy, M.: Adding action refinement to a finite process algebra. In: Leach Albert, J., Monien, B., Rodríguez-Artalejo, M. (eds.) Automata, Languages and Programming. LNCS, vol. 510, pp. 506–519. Springer, Heidelberg (1991)
15. Aceto, L., Hennessy, M.: Towards action refinement in process algebras. Information and Computation 103(2), 204–269 (1993)
16. Aceto, L., Hennessy, M.: action refinement to a finite process algebra. Information and Computation 115(2), L179–247 (1994)
17. Hennessy, M., Lin, H.: Proof systems for message-passing process algebras. In: Formal Aspects of Computing, vol. 8, pp. 379–407. Springer, London (1996)
18. Majster-Cederbaum, M., Wu, J.: Adding action refinement to stochastic true concurrency models. In: Dong, J.S., Woodcock, J. (eds.) ICFEM 2003. LNCS, vol. 2885, pp. 226–245. Springer, Heidelberg (2003)
19. Majster-Cederbaum, M., Wu, J.: Action refinement for true concurrent real time. In: Proc. ICECCS 2001, pp. 58–68. IEEE Computer Society Press, Los Alamitos (2001)
20. Majster-Cederbaum, M., Wu, J.: Towards action refinement for true concurrent real time. Acta Informatica 39, 531–577 (2003)
21. van Glabbeek, R., Vaandrager, F.: Bundle Event Structures and CCSP. In: Amadio, R.M., Lugiez, D. (eds.) CONCUR 2003. LNCS, vol. 2761, pp. 57–71. Springer, Heidelberg (2003)
22. van Glabbeek, R., Goltz, U.: Refinement of Actions in Causality Based models. In: de Bakker, J.W., de Roever, W.-P., Rozenberg, G. (eds.) Stepwise Refinement of Distributed Systems. LNCS, vol. 430, pp. 267–300. Springer, Heidelberg (1990)
23. van Glabbeek, R., Goltz, U.: Refinement of actions and equivalence notions for concurrent systems. Acta Informatica 37, 29–327 (2001)
24. Gorrieri, R., Rensink, A.: Action Refinement. In: Handbook of Process Algebra (2001)

25. Milner, R.: Communication and Concurrency. Prentice-Hall, Englewood Cliffs (1989)
26. Deng, W., Lin, H.: Extended Symbolic Transition Graphs with Assignment. In: Proceedings of the 29th Annual International Computer Software and Applications Conference (COMPSAC'05), vol. 1 (July 2005)
27. Wirth, N.: Program Development by Stepwise Refinement. Communications of ACM 14(4), 221–227 (1971)

Constraints Solution for Time Sensitive Security Protocols*

Zhoujun Li[1,2], Ti Zhou[1], Mengjun Li[1], and Huowang Chen[1]

[1] School of Computer Science, National University of Defence Technology, Changsha,
410073, China
tzhou@nudt.edu.cn
[2] School of Computer, BeiHang University, Beijing, 100083, China

Abstract. With the development of network and distributed systems,
more and more security protocols rely heavily on time stamps, which are
taken into account by a few formal methods. Generally, these methods
use constraints to describe the characteristic of time variables. However,
few of them give a feasible solution to the corresponding constraints solving problem. An effective framework to model and verify time sensitive
security protocols is introduced in [1], which doesn't give an automatic
algorithm for constraints solution. In this paper, an effective method is
presented to determine whether the constraints system has a solution,
and then implemented in our verifying tool SPVT. Finally, Denning-
Sacco protocol is taken as an example to show that security protocols
with time constraints can be modeled naturally and verified automatically and efficiently in our models.

Keywords: time sensitive; security protocol; constraint; algorithm.

1 Introduction

With the development of network and distributed systems, more and more security protocols rely heavily upon time stamps or counters instead of nonces to
prevent from replaying and redirecting messages. However, the problem is that
while the value of a nonce is not predictable, the value of a counter or a time
stamp is. Hence, new attacks could be produced by counters or time stamps
instead of nonces. Therefore, the verification method has to take into account
this predictable feature. The complexity of security protocols and the size of distributed systems have often been too great to analyze without formal methods.
However, introducing time into a protocol analysis framework brings complica-
tions, because the verification method should deal with some time-dependent
behavior and concern the required property with time. So it is very urgent to
find an appropriate analysis for sensitivity to time.

The use of time stamps can be described as follows: the sender issues a fresh
message with the time stamp that marks its time of issue; then, the receiver

---

* Supported by the National Natural Science Foundation of China under Grant No.
60473057, 90604007.

F.P. Preparata and Q. Fang (Eds.): FAW 2007, LNCS 4613, pp. 191–203, 2007.

checks whether the time stamp has not expired to establish the validity of the message. For the relation between times is the powerful one to determine whether messages are valid or not, the order of these time stamps must be researched. Using constraints of time variables and clocks is the best way to build this relation. In fact, time sensitive security protocols can be naturally specified by constraints programming and verified by symbolic operation. The constraints solution in verification of protocols is too important but few work on them. If they want to verify time sensitive protocols with constraints in current formal methods, they need to use the tools which solve general constraints. So the efficiencies of those methods will drop down. To improve the efficiency of verification, this paper researches the constraints system in the verification of security protocols and presents a simple and natural method to determine the solution of constraints in Horn logic model.

Related work. Many approaches focus on the study of protocols that use time stamps [2,3,4,5,6,7,8,9]. [2] also analyzes protocols with timing information, and his approach is based on a discrete time model. [3] uses a model with discrete time and an upper bound on the time window, and finds a timed-authentication attack on Wide Mouthed Frog protocol. [4] makes a semi-automated analysis on a Timed CSP model of Wide Mouth Frog protocol, and uses PVS to discharge proof obligations so as to find an invariant property. [5] presents a real-time process algebra for the analysis of time-dependent properties. It focuses on compositional results and show how to model timeouts. The theory of timewise refinement [6] allows results to be translated between the untimed and timed models, enabling verifications to be carried out at their most appropriate level of abstraction and then combined if necessary from different models. [7] proposes a method for design and analysis of security protocols that are aware of timing issues. It models security protocols using timed automata, and uses UPPAAL to simulate, debug and verify protocols. The verification method in [8] is based on the combination of constraint solving technology and first order term manipulation, and it uses Sicstus Prolog to solve constraints. He assumes a global clock and verify the Wide Mouthed Frog protocol. [9] presents a symbolic decision procedure for time-sensitive cryptographic protocols with time stamps. It uses logic formulae to describe symbolic constraints. But it doesn't give an automatic method to determine how to pick time variables for obtaining an attack. So this is the first paper concerning with the constraints solution in verification of security protocols.

Bruno Blanchet and Martin Abadi have proposed a security protocol model based on Horn logic in [10,11,12]. Based on this model, a very efficient verification method is also presented which fits for verifying interleaving runs of the security protocol's infinite sessions and terminates for many security protocols. In [13,14], an extended Horn logic model for security protocols is proposed, and the modified-version verification method to construct counter-examples automatically is presented. The counter-examples are represented in standard notation, which is more elegant and intelligible than the representation form based on traces. Based on these principles and methods, an effective verifier SPVT

(Security Protocol Verifying Tool) is developed, and many classic protocols are verified by it. To verify time sensitive security protocols, we extends this model to verify protocols with time stamps effectively in [1]. [1] also verifies Wide Mouthed Frog protocol. However, [1] doesn't give an automated method to find whether or not the constraints have a solution. This paper will discuss how to determine this problem automatically and effectively. For discussing the constraints system in detail, the principle of verifying time sensitive security protocols on Horn logic will be omitted. If the reader is very interested in this theory, please refer to [1].

It is a problem in linear programming whether a constraints system has a solution, and there are many methods to decide whether or not a linear programming problem has a solution, such as Simplex method and Fourier-Motzkin algorithm. In [15,16,17,18], however, these methods couldn't determine efficiently and simply whether the constraints system in verification of protocols has a solution. To avoid redundant operations, we study some classic methods and give an efficient one to solve our problem in verification. The original problem is translated to the one that is to find a special cycle in the graph. It is a classic problem to find cycles in graph theory, so we can choose a more convenient method to solve it. The correctness of this translation is given by some theorems in this paper. There are also too many methods to find cycles in a graph, and we apply the method in [19] to solve our problem. And then, we give an algorithm to determine whether the constraints system has no solution. At last, we give an example in Denning-Sacco protocol to present the work of our method.

Plan of the Paper. In Section 2 we introduce our models in security protocols. In Section 3 we research our constraints system and give an efficient method to determine whether or not constraints system has a solution. In Section 4 we implement this method in our tool SPVT, and give an example how to find whether a constraints system has no solution. In Section 5 we give a conclusion to this paper, and discuss the future work.

2   Models of Protocol

In this section, we will introduce two models: the process calculus model for modeling a security protocol in section 2.1 and the Horn logic model for verifying it in section 2.2.

2.1   Protocol Representation in the Process Calculus

Security protocols are distributed parallel programs and can be described as the parallel composition of multiple role-processes using π-like calculus in [12]. We propose a π-like calculus for modeling security protocol in [1], which is Applied-π calculus [20] added some time events and is described in Table 1.

The π-like calculus is similar to the calculus in [12], and also has two parts: terms(data) and processes(programs). The identifiers x, y, z, and similar ones range over variables, and a, b, c, k range over names. f is a constructor which is used to build terms. The processes are defined similarly to [12], but we add

Table 1. The syntax of the process calculus

$M, N::=$	**Terms**
x, y, z	variable
a, b, c, k	Name
$f(M_1, \cdots, M_n)$	Constructor
$P, Q::=$	**Processes**
$\bar{c} < M > .P$	Output
$c(x).P$	Input
0	Nil
$P|Q$	Parallel
$!P$	Replication
$(\nu a)P$	Restriction
let $x = g(M_1, \cdots, M_n)$ in P	Destruction
if $M = N$ then P	Condition
$begin(M, M').P$	Begin event
$end(M, M').P$	End event
$Check(t).P$	Check event
$Mark(t).P$	Mark event
Set $d = int$ in P	Set event
$Fly(d).P$	Fly event

to it four time events: **Check, Mark, Set, Fly**. These events are used to deal with time stamps. **Check** event is used to check whether the time stamp is valid or not. The parameter d represents the lifetime of a message, that is the most allowable network delay. This event appears when the process receives a message with time stamps and need to check its freshness. **Mark** event is used to mark t as the current clock **now** when the sender wants to issue a fresh message with the time of issue, that is, t equals to the current value of **now**. **Set** event is used to assign a value of lifetime to d. **Fly** event is used to declare time rolls by, that is, the current clock **now** passes a period d.

2.2   The Model with Time Constraints on Horn Logic

Syntax [1] first introduces some special terms that can be used to represent time variables, and time functions. The syntax of the model with time constraints on Horn logic is described in Table 2.

The Horn logic uses these predicates:*attacker,begin*, and *end. attacker(M)* means that the attacker may have M, $begin(M, N)$ that the event **begin** has been executed with a parameter corresponding to M and environment N, and $end(M, N)$ that **end** has been executed in session list N with a parameter corresponding to M. [1] defines a global clock **now** which is a special place-holder, and if the constraint is **true**, then the rule will ignore this constraint, that is, $H_1 \wedge \cdots \wedge H_n \rightarrow F$ means $H_1 \wedge \cdots \wedge H_n \rightarrow F : true$. The clock **now** represents current time. The sender overprints time stamps referring to **now**. When the receiver receives this message, he will check time stamp with the current time.

Table 2. The syntax of the model with constraints on Horn logic

Term	M,N,Mes::=
time variable	$t_1, t_2, \cdots$,
time function	$tf(t_1, \cdots t_n)$ where $t_1, \cdots, t_n$ are time variables or time functions; if $n = 0$, then tf is a time constant.
variable	x, y, z
name	$a[M_1, \cdots, M_n]$,i,j
function	$f(M_1, \cdots, M_n)$
Atom, Fact	F,C::=
attacker predicate	$attacker(M)$
begin predicate	$begin(M, N)$
end predicate	$end(M, N)$
Constraint	C::=
	$f(x_1, \cdots, x_n) \# g(y_1, \cdots, y_m)|C_1, C_2,$ where f and g are n-ary and m-ary functions which return linear combination of these variables, $\# \in \{<, =, \leq\}$.
Rule	R,R'::=
logic rule	$F_1 \wedge \cdots \wedge F'_n \to F : C$ where C is constraint.

If time stamp has not expired, he will believe the message is still valid. The parameters representing network delays can be assigned in the beginning.

Translation to Horn logic. The model of roles in security protocols is a group of logic rules. [10,11,12] discuss the model on Horn logic, and we add time constraints to this model in [1] so that this model can verify time sensitive security protocols. In logic rules, d is a parameter representing network delay. If we want to do anything with some rule, we must be sure that the constraints of it can be satisfiable. The constraints of rules in Dolev-Yao model are *true* in [1], which means those constraints will be always satisfied, so we will not present them in this paper. But [1] adds a new rule in the attacker model. Suppose the lifetime of the session key k created at the clock t is $D(D$ is a big time constant), and session keys can only be leaked by accident when they have expired, as our model shows, the following attack is possible.

$$\to attacker(k(t)) : now \geq D + t$$

The honest roles are described by the process calculus in Section 2.1. The translation$[\![P]\!]\rho hC$ of a process P is a set of rules, where the environment ρ is a sequence of mappings $x \mapsto p$ and $a \mapsto p$ from (time) variables and names to patterns, h is a sequence of facts of the form $attacker(M)$ and $begin(M, M')$, and C is a list of constraints. The empty list is denoted by $\emptyset$, the concatenation of a constraint $aCons$ to the list C is denoted by $C \cup \{aCons\}$. $exp_1 \# exp_2 \# exp_3$ is short for two constraints: $exp_1 \# exp_2$ and $exp_2 \# exp_3$, where $\# \in \{<, \leq\}$. The concatenation of a mapping $x \mapsto M$ to ρ is denoted by $\rho[x \mapsto M]$, where x is a name or a variable. [1] gives abstracting rules for our process calculus as follows:"

(1) $[\![\mathbf{0}]\!]\rho hC = \emptyset;$

(2) $[\![\mathbf{P}|\mathbf{Q}]\!]\rho hC = [\![\mathbf{P}]\!]\rho hC \cup [\![\mathbf{Q}]\!]\rho hC;$

(3) $[\![!\mathbf{P}]\!]\rho hC = [\![\mathbf{P}]\!]\rho[i \mapsto i]hC$, where i is a new variable(session identifier);

(4) $[\![\nu(a)\mathbf{P}]\!]\rho hC = [\![\mathbf{P}]\!]\rho[a \mapsto a[\rho(V_0), \rho(V_s)]]hC$, where V_0 is the tuple made up with input variables, V_s is the set of session identifiers, and a becomes a new function symbol;

(5) $[\![c(x).\mathbf{P}]\!]\rho hC = [\![\mathbf{P}]\!]\rho[x \mapsto x](h \wedge \mathbf{attacker}(x))C;$

(6) $[\![\bar{c} < M > .\mathbf{P}]\!]\rho hC = [\![\mathbf{P}]\!]\rho hC \cup \{h \rightarrow \mathbf{attacker}(\rho(M)) : (\rho(C))\});$

(7) $[\![\mathbf{let}\ x{=}g(M_1, \cdots, M_n)\ \mathbf{in}\ \mathbf{P}\ \mathbf{else}\ \mathbf{Q}]\!]\rho hC = \bigcup\{[\![\mathbf{P}]\!](\sigma\rho)[x \mapsto \sigma'p'](\sigma h)(\sigma C)|$ $g(p_1, \cdots, p_n) \rightarrow p'$, where, the pair (σ, σ') is a most general unifier one, such that $\sigma\rho(M_1) = \sigma'(p_1), \cdots, \sigma\rho(M_n) = \sigma'(p_n)\} \cup [\![\mathbf{Q}]\!]\rho hC;$

(8) $[\![\mathbf{Begin}(M).\mathbf{P}]\!]\rho hC = [\![\mathbf{P}]\!]\rho(h \wedge \mathbf{begin}(\rho|_{(V_o \cup V_s)}, \rho(M)))C;$

(9) $[\![\mathbf{End}(M).\mathbf{P}]\!]\rho hC = [\![\mathbf{P}]\!]\rho hC \cup \{h \rightarrow \mathbf{end}(\rho(V_s), \rho(M)) : (\rho(C))\};$

(10) $[\![\mathbf{Check}(t).\mathbf{P}]\!]\rho hC = [\![\mathbf{P}]\!]\rho h(C \cup \{now - d \leq t \leq now\});$

(11) $[\![\mathbf{Mark}(t).\mathbf{P}]\!]\rho hC = [\![\mathbf{P}]\!]\rho h(C \cup \{t = now\});$

(12) $[\![\mathbf{Set}\ d = int\ \mathbf{in}\ \mathbf{P}]\!]\rho hC = [\![\mathbf{P}]\!](\rho[d \mapsto int])hC;$

(13) $[\![\mathbf{Fly}(d).\mathbf{P}]\!]\rho hC = [\![\mathbf{P}]\!]\rho h\{now - exp1 - d\#exp2 \mid \forall lineq \in C,$ and $lineq = now - exp1\#exp2,$ where$\# \in \{<, \leq\}, exp1, exp2$ are algebraic expressions.$\}.$

The translation of a process is a set of Horn clauses with constraints that enable us to prove that it sends certain messages or executes certain events in the constraints. The list C keeps conditions, and when they are satisfiable, the body of rule may trigger the head. From the translation, only the last four events will bring the change of constraints. The translation of a **Check** adds two constraints, meaning that the process will believe the time stamp t is new when t is in a time interval $[now - d, now]$. P can be executed after the message has been checked and these constraints are satisfiable. The translation of a **Mark** adds a constraint, meaning that P fetches the current time as its time stamp. The translation of a **Set** adds a mapping $d \mapsto int$ to the environment ρ. This event is used to assign the lifetime of messages, so the receiver will believe the message is new when the receiving time minus the lifetime is less than or equal to the time stamp. The translation of a **Fly** modifies constraints to loosen the lower bound made by **now**.

3 Constraints System

3.1 The Feather of Our Constraints System

Definition 1. *A TVPI system is a system of linear inequalities where each inequality involves at most two variables. A TVPI system is called* **monotone** *if each inequality is of the form* $ax_i - bx_j \leq c$, *where both a and b are positive. If $a = b = 1$, we call this system is a uni-TVPI system.*

In our system, inequalities can be generated by the following case:

- **Check** event: this event will check if the time stamp t is valid, so **now**$-d \leq t \leq$ **now** will be added to constraints system C. This event introduces inequalities that are of the form $x - y \leq c$.

- **Mark** event: this event only generate equality $t = \text{now}$.
- **Set,Fly** events: they only modify time constants, and don't add or minus inequalities in our system.

Therefore, our constraints system is a uni-TVPI system. we can use the Fourier-Motzkin elimination method for finding a feasible solution of a uni-TVPI system. We begin by an informal description of the method for our system.(The reader is referred to [15] for more details.)

Let the variables of our system be $x_1, \cdots, x_n$ and let the set of inequalities be denoted by E. The variables are eliminated one by one. At step i, our system will only contain variables $x_i, \cdots, x_n$; the set of inequalities at step i is denoted by E_i,where initially $E_1 = E$. To eliminate variable x_i, all the inequalities in which x_i participates are particioned into two sets, L and H, where $L = \{x_j - c | x_i \geq x_j - c \in E_i\}$, and $H = \{x_j + c | x_i \leq x_j + c \in E_i\}$. And then, $E_{i+1} = \{l \leq h | l \in L, h \in H\} \cup E_i \backslash (L \cup H)$.

Theorem 1. *The TVPI system E_{i+1} has a feasible solution if and only if the TVPI system E_i has a feasible solution.*

Hence, a feasible solution can be computed recursively for E_{i+1} and then extended to E_i. The main drawback of this method is that the number of inequalities may grow exponentially.

Lemma 1 (Farkas' Lemma). *$Ax \leq b$ has no solution if and only if some non-negative linear combination of the inequalities of the system yields $0 \leq -1$.*

In a uni-TVPI system, we can make a stronger statement:

Theorem 2. *A uni-TVPI system S has no solution if and only if there is a sequence $t_1 \geq t_2 + d_1, t_2 \geq t_3 + d_2, \cdots, t_n \geq t_1 + d_n$, and $\sum_{i=1}^{n} d_i > 0$.*

Proof. "$\Leftarrow$": A positive linear combination of the sequence is $\sum_{i=1}^{n} t_i \geq \sum_{i=1}^{n-1} (t_{i+1} + d_i) + t_1 + d_n$ which yield $0 \geq \sum_{i=1}^{n} d_i$. This result conflicts with the condition $\sum_{i=1}^{n} d_i > 0$, so $0 \leq -1$ can be derived. By **Farkas' Lemma**, this system has no solution.

"$\Rightarrow$": this system has no solution, so, by Farkas' lemma, there exists some non-negative linear combination yielding $0 \leq -1$.

(1) Suppose there is no such a sequencelike this: $t_1 \geq t_2 + d_1, t_2 \geq t_3 + d_2, \cdots, t_{n+1} \geq t_1 + d_n$, then for any non-negative linear combination of the uni-TVPI system, there is a variables x which appears in the left of the linear combination but not in the right, that is, the result of the linear combination is the form "$x \geq$ other-variables-linear-combination $+d$ ". This result couldn't yield "$0 \leq -1$", which contradicts the condition.

(2) Suppose for every sequence $t_i \geq t_{i+1} + d_i (i = 1, \cdots, n), t_n = t_1, \sum_{i=1}^{n} d_i \leq 0$. By Farkas' lemma, there are two sequences $a_i > 0$ and $e_i \in S(i = 1, \cdots, n)$ such that $\sum_{i=1}^{n} a_i e_i \Rightarrow 0 \leq -1$. From any $e \in \{e_i | 0 < i < n + 1\}$, there should be a

sequence $e_{i_1} = e, e_{i_2}, \cdots, e_{i_k}$ which are distinct and belong to $\{e_i | 0 < i < n+1\}$, and they can be written in the form $t_i \geq t_{i+1} + d_i (i = 1, \cdots, n, t_{n+1} = t_1)$.

Otherwise, $\sum_{j=1}^{k} e_{i_j} \equiv x \geq y + c$, and there isn't $y \geq z + d$ in $\mathcal{S}$, $\therefore y$ will not be eliminated, so $\sum_{i=1}^{n} a_i e_i \not\equiv 0 \leq -1$, contradiction. Therefore, we can split $\sum_{i=1}^{n} a_i e_i$ to many sequences as above. Suppose $\sum_{i=1}^{n} a_i e_i \equiv (e_{11} + e_{12} + \cdots + e_{1j_1}) + \cdots + (e_{m1} + \cdots + e_{mj_m}) \equiv \sum_{k=1}^{m} (0 \geq \sum_{i=1}^{j_k} d_{ki}) \equiv 0 \geq \sum_{k=1}^{m} \sum_{i=1}^{j_k} d_{ki}$. Because of our assumption $\sum d_i \leq 0$ in this formal sequence, $0 \geq \sum_{k=1}^{m} \sum_{i=1}^{j_k} d_{ki}$ is true, that is, $\sum_{i=1}^{n} a_i e_i \not\equiv 0 \leq -1$, contradiction. $\qquad\square$

3.2   Algorithms for the Solution of Our System

In this subsection we show how to determine whether or not a uni-TVPI system has solution. Recall from the previous subsection the idea of our algorithm is as follows:

> Suppose t_i has a greatest lower bound $t_{i+1} + d_i (i = 1, \cdots, n)$, and $t_{n+1} = t_1$. Using resolution, one obtains a sequence of inequalities $t_1 \geq t_{i+1} + \sum_{j=1}^{i} d_j (i = 1, \cdots, n)$. The last one will yield $0 \geq \sum_{i=1}^{n} d_i$. If $\sum_{i=1}^{n} d_i$ is actually greater than 0, there is a contradiction, that is, the system can't be satisfied.

An equivalent representation of the uni-TVPI system $\mathcal{S}$ is by the directed graph $G = < V, E >$. The vertex set V contains vertices $t_1, \cdots, t_n$; an edge e_{ij} from vertex t_i to $t_j (1 \leq i, j \leq n)$ represents the inequality $t_i \geq t_j + d$ and is labeled by $\max\{d | t_i \geq t_j + d \in \mathcal{S}\}$. The algorithm for constructing the graph is as follows:

ALGORITHM 1 (the construction of the graph). *Constructing a graph G corresponding to a uni-TVPI system $\mathcal{S}$ as follows:*

- *For each variable $t_i (i = 1, \cdots, n)$, generate a vertex t_i corresponding to it.*
- *For every inequality $t_i \geq t_j + c$,*
  if there is an edge $e_{ij} = < t_i, t_j >$, and the weight of it is d,
    then let the weight be $\max\{c, d\}$
    else generate an edge $e_{ij} = < t_i, t_j >$, and let the weight of it be d.

Definition 2. *A **path** from a vertex u to a vertex u' in a graph $G = < V, E >$ is a sequence $v_0 e_1 v_1 e_2 \cdots v_n$, where $u = v_0, u' = v_n, e_i = < v_{i-1}, v_i > \in E$ for $i = 1, \cdots, n$. The path is **closed** if $v_0 = v_n$. The path is **simple** if $v_0, \cdots, v_n$ are distinct. The path is a **cycle** if it is closed and the subpath from v_1 to v_n is simple. The path has a cycle if there exists $i, j (0 \leq i \leq j \leq n)$ such that $v_i e_{i+1} \cdots v_j$ is a cycle.*

We will use the adjacency-matrix representation of G in our algorithm. This representation consists of a $|V| \times |V|$ matrix Adj such that

$$Adj_{ij} = \begin{cases} d & \text{if } (v_i, v_j) \in E \text{ and its weight is } d \\ \bot & \text{otherwise} \end{cases}$$

Proposition 1. *A uni-TVPI system S has a sequence $t_i \geq t_{i+1} + d_i (1 \leq i \leq n)$ if and only if there is a path $t_1 \xrightarrow{d_1'} t_2 \xrightarrow{d_2'} \cdots \xrightarrow{d_{n-1}'} t_n$ in the corresponding graph G.*

Theorem 3. *A uni-TVPI system S has no solution if and only if there is a cycle $t_1 \xrightarrow{d_1} t_2 \xrightarrow{d_2} \cdots \xrightarrow{d_{n-1}} t_n \xrightarrow{d_n} t_1$ in the corresponding graph G, and $\sum\limits_{i=1}^{n} d_i > 0$.*

Now we need find all cycles in the graph and check whether the constraints system has no solution by theorem 3. [19] gives an algorithm to find all primary oriented circuits in a simple digraph by constructing L-matrices. We modify this method and give an on-the-fly algorithm to find a cycle which satisfies $\sum d > 0$.

Definition 3. *L-word is defined as the vertex index sequence of simple paths or cycles. The set of L-words is L-set. The matrix $M = (M_{ij})_{n \times m}$ is L-matrix if for every element M_{ij} in M, M_{ij} is an L-set.*

Definition 4. *Let $M = \{m_i | i = 1, \cdots, r\}, N = \{n_i | i = 1, \cdots, s\}$ be two L-sets. The joint of M and N is defined as follows:*

$$M * N = \begin{cases} M * N = \emptyset & \text{if } M = \emptyset \text{ or } N = \emptyset \\ \{m_i n_j | \ m_i \in M, n_j \in N \text{ and } |m_i| > 0, \\ \quad |n_j| > 0, m_i n_j \text{ is an L-word}\} & \text{otherwise} \end{cases}$$

Obviously, $M * N$ is an L-set.

Definition 5. *Let $M = (M_{ij})_{n \times k}, N = (N_{ij})_{k \times p}$ be two L-matrices. The joint of M and N is $Q = (Q_{ij})_{n \times p} = M * N$, where $Q_{ij} = \bigcup\limits_{r=1}^{k} (M_{ir} * N_{rj}), 1 \leq i \leq n, 1 \leq j \leq p$.*

Definition 6. *Let $G = < V, E >$ be a simple directed graph and $|V| = n$, then the L-matrix $M^1 = (M_{ij}^1)_{n \times n}$ is L-adjacency-matrix representation of G, where*
$$M_{ij}^1 = \begin{cases} \emptyset & , (v_i, v_j) \notin E \\ \{ij\}, & (v_i, v_j) \in E \end{cases}$$

Definition 7. *Suppose $G = < V, E >$ is a simple directed graph and $|V| = n$, and $M^1 = (M_{ij}^1)_{n \times n}$ is L-adjacency-matrix representation of G. Let $M = (M_{ij})_{n \times n}$, where $M_{ij} = \begin{cases} \emptyset & , M_{ij}^1 = \emptyset \\ \{j\}, & \text{otherwise} \end{cases}$*
Then $M^k = M^{k-1} * M (k = 2, 3, \cdots, n)$ is the k-path matrix. If $i_0 i_1 \cdots i_m \in M_{ij}^k$, then $i_0 = i, i_m = j, m = k$ and there is an edge sequence $e_1, \cdots, e_k$ such that $v_{i_0} e_1 v_{i_1} e_2 \cdots v_{i_k}$ is a simple path or a cycle in the graph G.

ALGORITHM 2. *procedure find_unsatisfication*
$Q \leftarrow M^1$
for i=2 to n do
1. $P \leftarrow Q; Q \leftarrow P * M$
2. for j=1 to n do
  2.1 for every $w \in Q_{jj}$ *do*
    2.1.1 $tmp_d \leftarrow 0; m \leftarrow |w| - 1; cur \leftarrow 0;$
    2.1.2 for l=1 to m do
      2.1.2.1 $tmp_d \leftarrow tmp_d + Adj_{w_{cur}w_l}$
      2.1.2.2 $cur \leftarrow l$
    2.1.3 if $tmp_d > 0$ *then raise an exception*
    2.1.4 for l=1 to m-1 do
      2.1.4.1 if $|Q_{w_l w_l}| = 1$ *then* $Q_{w_l w_l} = \emptyset$

Proposition 2. *Suppose the undirected graph* $G =< V, E >, |V| = n, |E| = m(n \leq m)$ *and* G *has no self-loop. The number of circles is not more than* $\frac{(m-n+1)(m-n+2)}{2}$.

This proposition can be obtained very easily by graph theory. In the algorithm 2, the operations are plus and search in nature, and the cost is in the joint of two L-matrices. $P * M$ is $O(n^3)$. So the joint operation costs $O(n^4)$. The step 2 costs $O((m - n)^2)$ by proposition 2. In fact, the graph in this paper is a directed one, so the number of loops is far less than $\frac{(m-n+1)(m-n+2)}{2}$. Therefore, the worst time of this algorithm is $O(n^4 + (m - n)^2) \approx O(n^4)$. Although [18] gives an $O(mn^2 \log m)$ algorithm for TVPI system, it doesn't discuss the time in computing the set $B_j (1 \leq j \leq d)$ of breakpoints of the edge (x_i, x_{i_j}) projected on the x_i coordinate and the implementation of our algorithms is simpler than [17][18].

4   Experiment

We implement these algorithms in our tool SPVT which is written in Objective Caml. This section will take Denning-Sacco protocol as an example for applying this method in model and verification of time sensitive security protocols verification. This protocol is described as Table 3.

Using our π-like calculus in section 2.1, Denning-Sacco protocol can be described as the following process DS:

Table 3. The formal description of Denning-Sacco protocol

① A→S: A, B
② S→A: $\{B, Kab, T_A, \{Kab, T_A, A\}Kbs\}Kas$
③ A→B: $\{Kab, T_A, A\}Kbs$
④ B→A: $\{N_B\}Kab$
⑤ A→B: $\{N_B - 1\}Kab$

processA $\triangleq$

Set $d = d1$ in $c(xB).begin(Bparam, xB).\bar{c} < (host(Kas), xB) > .c(X).$let $\{XB, XKab,$
$XT1, XTicket\} = decrypt(X, Kas)$ in if xB=XB then $Check(XT1).\bar{c} < XTicket >$
$.c(Y).$let $XN_B = decrypt(Y, XKab)$ in $Fly(d1 + d2).\bar{c} < \{XN_B - 1\}XKab >.0$

processS $\triangleq$

$c(XAB).$if $XAB = (x, y)$ then if $x = host(Kxs)$ then if $y = host(Kys)$ then $Mark(T1).$
$\bar{c} < \{y, Kab(T1), T1, \{Kab(T1), T1, x\}Kys\}Kxs >.0$

processB $\triangleq$

Set $d = d1 + d2$ in $c(XTicket).$let $\{XKab, XT1, xA\} = decrypt(XTicket, Kbs)$ in
$Check(XT1). (\nu N_B)\bar{c}<\{N_B(XT1)\}XKab>.c(xBack).$ let $xN=decrypt(xBack, Kbs)$
in if $xN = N_B(XT1) - 1$ then $Fly(d1 + d2).end(Bparam, B).0$

DS $\triangleq$

$(\nu Kas)(\nu Kbs)\bar{c} < host(Kas) > .\bar{c} < host(Kbs)> .((!processA)|(!processS)|(!processB))$

Based on the translation in Section 2.2, the logic program of Denning-Sacco
protocol can be derived as follows:

A:① $\rightarrow attacker(A) : true$

  ② $\rightarrow attacker(B) : true$

  ③ $begin(\cdots)\wedge attacker(\{B, XKab, XT1, XTicket\}Kas) \rightarrow attacker(XTicket):$
    $now - d1 \leq XT1 \leq now$

  ④ $begin(\cdots)\wedge attacker(\{B, XKab, XT1, XTicket\}Kas)\wedge attacker(\{N_B(t)\}XKab)\rightarrow$
    $attacker(\{N_B(t) - 1\}XKab) : now - d1 - (d1 + d2) \leq XT1 \leq now$

S:⑤ $attacker(x)\wedge attacker(y) \rightarrow attacker(\{y, Kab(T1), T1, \{Kab(T1), T1, x\}Kys\}Kxs:$
    $T1 = now$

B:⑥ $attacker(\{XKab, XT1, xA\}Kbs) \rightarrow attacker(\{N_B(XT1)\}XKab) : now -$
    $d1 - d2 \leq XT1 \leq now$

  ⑦ $attacker(\{XKab, XT1, xA\}Kbs)\wedge attacker(\{N_B(XT1)-1\}XKab) \rightarrow end(\cdots) :$
    $now - 2(d1 + d2) \leq XT1 \leq now$

I:⑧ $\rightarrow attacker(Kab(t)) : now \geq D + t$

By the resolution in [1], we obtain an unsatisfiable constraints in Denning-Sacco
protocol: $now-2(d1+d2) \leq t1 \leq now, t4 \geq D+t1, t3-d1-d2 \leq t1 \leq t3, t2-d1 \leq$
$t1 \leq t2 \leq t3 \leq t4, t5 \geq D + t1, t5 \geq t4$. Now we take this constraints system into
account. Because D is a very big constant, we can suppose $D > 10(d1 + d2)$. Let
the index of now be 0, and others' are their subscripts. The adjacency-matrix is

$$Adj = \begin{bmatrix} \bot & 0 & 0 & 0 & 0 & 0 \\ -2(d1 + d2) & \bot & -d1 & -(d1 + d2) & \bot & \bot \\ \bot & 0 & \bot & \bot & \bot & \bot \\ \bot & 0 & 0 & \bot & \bot & \bot \\ \bot & D & \bot & 0 & \bot & \bot \\ \bot & D & \bot & \bot & 0 & \bot \end{bmatrix}$$

The L-matrices are presented as follows:

$$M^1 = \begin{bmatrix} \{\} & \{01\} & \{02\} & \{03\} & \{04\} & \{05\} \\ \{10\} & \{\} & \{12\} & \{13\} & \{\} & \{\} \\ \{\} & \{21\} & \{\} & \{\} & \{\} & \{\} \\ \{\} & \{31\} & \{32\} & \{\} & \{\} & \{\} \\ \{\} & \{41\} & \{\} & \{43\} & \{\} & \{\} \\ \{\} & \{51\} & \{\} & \{\} & \{54\} & \{\} \end{bmatrix} \quad M = \begin{bmatrix} \{\} & \{1\} & \{2\} & \{3\} & \{4\} & \{5\} \\ \{0\} & \{\} & \{2\} & \{3\} & \{\} & \{\} \\ \{\} & \{1\} & \{\} & \{\} & \{\} & \{\} \\ \{\} & \{1\} & \{2\} & \{\} & \{\} & \{\} \\ \{\} & \{1\} & \{\} & \{3\} & \{\} & \{\} \\ \{\} & \{1\} & \{\} & \{\} & \{4\} & \{\} \end{bmatrix}$$

$$M^2 = \begin{bmatrix} \{010\} & \{051,041,031,021\} & \{032,012\} & \{043,013\} & \{054\} & \{\} \\ \{\} & \{131,121,101\} & \{132,102\} & \{103\} & \{104\} & \{105\} \\ \{210\} & \{\} & \{212\} & \{213\} & \{\} & \{\} \\ \{310\} & \{321\} & \{312\} & \{313\} & \{\} & \{\} \\ \{410\} & \{431\} & \{432,412\} & \{413\} & \{\} & \{\} \\ \{510\} & \{541\} & \{512\} & \{543,513\} & \{\} & \{\} \end{bmatrix}$$

We can obtain $M_{00}^3 = \{0210, 0310, 0410, 0510\}$. By the step 2.1 in algorithm 2, when $w = 0410$, we have $\sum d = Adj_{04} + Adj_{41} + Adj_{10} = 0 + D - 2(d1 + d2) = D - 2(d1 + d2) > 0$. So, by theorem 3, the constraints system has no solution.

5 Conclusions

Security protocol analysis is a mature field of research that has produced numerous findings and valuable tools for the correct engineering of security protocols. Despite much literature on time sensitive security protocols in constraints, most analysis methods do not take constraints solution into consideration. This paper discusses the constraints system in verification of time sensitive security protocols. Firstly, we present our models in security protocols, and describe time constraints in detail. Secondly, by analyzing the characteristic of our system, we give an effective and simple method to determine whether or not our constraints system has a solution and then implement it in our verification tool SPVT. This allows the verification of protocols that use time stamps to fulfil one or more security properties. With the work of [13,14,1], we can verify security protocols with or without time critical features and give a counter-example for an attack. The strength of the framework is that it will also allow more complicated time-dependent behavior of the protocol to be modeled naturally and verified effectively. The future work is to consider how to model sophisticate protocols, such as Kerberos V, and verifying them by our tool SPVT.

References

1. Zhou, T., Li, M.: Verification of time sensitive security protocols based on the extended horn logic model. Chinese Journal of Computer Research and Development 43(Suppl.2), 534–540 (2006)
2. Lowe, G.: Casper: A compiler for the analysis of security protocols. In: 10th IEEE Computer Security Foundations Workshop (CSFW-10), pp. 18–30 (1997)

3. Lowe, G.: A hierarchy of authentication specifications. In: 10th IEEE Computer Security Foundations Workshop (CSFW-10)
4. Evans, N., Schneider, S.: Analysing time dependent security properties in csp using pvs. In: Cuppens, F., Deswarte, Y., Gollmann, D., Waidner, M. (eds.) ESORICS 2000. LNCS, vol. 1895, pp. 222–237. Springer, Heidelberg (2000)
5. Gorrieri, R., Locatelli, E., Martinelli, F.: A simple language for real-time cryptographic protocol analysis. In: Degano, P. (ed.) ESOP 2003 and ETAPS 2003. LNCS, vol. 2618, pp. 114–128. Springer, Heidelberg (2003)
6. Gorrieri, R., Martinelli, F.: A simple framework for real-time cryptographic protocol analysis with compositional proof rules. Sci. Comput. Program. 50(1-3), 23–49 (2004)
7. Corin, R., Etalle, S., Hartel, P.H., Mader, A.: Timed model checking of security protocols. In: FMSE '04: Proceedings of the 2004 ACM workshop on Formal methods in security engineering, pp. 23–32. ACM Press, New York, USA (2004)
8. Delzanno, G., Ganty, P.: Automatic Verification of Time Sensitive Cryptographic Protocols. In: Jensen, K., Podelski, A. (eds.) TACAS 2004. LNCS, vol. 2988, pp. 342–356. Springer, Heidelberg (2004)
9. Bozga, L., Ene, C., Lakhnech, Y.: A symbolic decision procedure for cryptographic protocols with time stamps (extended abstract). In: Gardner, P., Yoshida, N. (eds.) CONCUR 2004. LNCS, vol. 3170, pp. 177–192. Springer, Heidelberg (2004)
10. Abadi, M., Blanchet, B.: Analyzing security protocols with secrecy types and logic programs. In: Symposium on Principles of Programming Languages, pp. 33–44 (2002)
11. Blanchet, B.: An efficient cryptographic protocol verifier based on prolog rules. In: 14th IEEE Computer Security Foundations Workshop (CSFW-14), pp. 82–96 (2001)
12. Blanchet, B.: From secrecy to authenticity in security protocols. In: Hermenegildo, M.V., Puebla, G. (eds.) SAS 2002. LNCS, vol. 2477, pp. 342–359. Springer, Heidelberg (2002)
13. Li, M., Li, Z., H.C.: Spvt: An efficient verification tool for security protocol. Chinese Journal of Software 17(4), 898–906 (2006)
14. Li, M., Li, Z., H.C.: Security protocol's extended horn logic model and its verification method. Chinese Journal of Computers 29(9), 1667–1678 (2006)
15. Schrijver, A.: Theory of Linear and Integer Programming. John Wiley and Sons, Chichester (1986)
16. Cormen, T.H., Leiserson, C.E., Rivest, R.L., Stein, C.: Introduction to Algorithms, 2nd edn. MIT Press, Cambridge, Massachusetts (2001)
17. Cohen, E., Megiddo, N.: Improved algorithms for linear inequalities with two variables per inequality. SIAM J. Comput. 23(6), 1313–1347 (1994)
18. Hochbaum, D.S., Naor, J.: Simple and fast algorithms for linear and integer programs with two variables per inequality. SIAM J. Comput. 23(6), 1179–1192 (1994)
19. Shuang-yan, B., Jin-mei, G., Q.M.: An algorithm and example of solution to primary oriented circuits in a digraph. Journal of Changchun post and telecom- munication institute 17(2), 41–45 (1999)
20. Sewell, P.: Applied π– a brief tutorial. Technical Report UCAM-CL-TR-498, Computer Laboratory, University of Cambridge, p. 65 (2000)

Using Bit Selection to Do Routing Table Lookup

Zhenqiang Li[1], Dongqu Zheng[2], and Yan Ma[1, 2]

[1] School of Computer Science and Technology
[2] Network Information Center
Beijing University of Posts and Telecommunications
Beijing 100876, China
{lizq,zhengdq}@buptnet.edu.cn, mayan@bupt.edu.cn

Abstract. Tree-based algorithms, such as Patricia, LC-trie, LPFST, etc, are widely used to do longest prefix matching (LPM). These algorithms use all the bits of the prefix to build the tree and the bits are used from the most significant bit to the least significant bit sequentially. Thus the tree is not balanced and the tree depth is high. In this paper, we propose bit selection tree (BS-tree) to do LPM. The bits of the prefix are selected to build BS-tree based on their costs defined in this paper. BS-tree has lower tree depth and is more balanced compared to other tree-based schemes. BS-tree has good scalability to the length of the IP address and is suitable for both IPv4 and IPv6. We evaluate the performances of BS-tree using both IPv4 and IPv6, and specially refine it for IPv6 based on the observations on IPv6 address and real IPv6 routing tables.

Keywords: Algorithm, Routing Table Lookup, Bit Selection, IPv4, IPv6.

1 Introduction

One of the most time-consuming and critical tasks in the data plane of a router is to do the longest prefix matching (LPM) to decide to which port the incoming packet should be forwarded. This is true for both IPv4 and IPv6 [1]. IPv6 is the next generation internet protocol. With 128-bit address, IPv6 provides an extremely huge address space, $3.4*10^{38}$ IPs in theory. Besides, IPv6 is expected to offer benefits for greater security, enhanced quality of service, better mobility, and new products, services, and missions for Next Generation Internet applications. Thus, IPv6 has been gaining wider acceptance to replace its predecessor, IPv4. IPv6 has already emerged out of the testing phase [2] and is seeing production deployment in Europe, Asia, and North America[1]. However, IPv6 does not change the data plane functions of the router. LPM is still needed. In fact, IPv6 with increased address length and very different prefix length distribution poses new challenges to the LPM algorithms. Major refinements to the LPM algorithms are necessary when they are applied to IPv6. LPM schemes that are suitable for both IPv4 and IPv6 are solicited.

In this paper, we use bit selection tree (BS-tree) to do LPM. Cost is defined for each bit of the prefix and only those bits with low costs are selected to construct the BS-tree. BS-tree is more balanced and the depth of BS-tree is reduced compared with

---

[1] http://www.ipv6forum.com

F.P. Preparata and Q. Fang (Eds.): FAW 2007, LNCS 4613, pp. 204–215, 2007.

other tree-based schemes. Thus, BS-tree increases the worst case lookup performance. Besides, BS-tree supports incremental updates and is suitable for both IPv4 and IPv6.

The rest of this paper is organized as follows. Section 2 presents the prior work and our contributions in this paper. The proposed algorithm is introduced in section 3. In section 4, we give the enhancements to the basic BS-tree. Section 5 shows the experimental and comparison results. Finally, the conclusion is drawn in section 6.

2 Prior Work and Main Contribution

In the past several years, LPM received great research interests and various algorithms for high performance LPM were proposed [3]. One direction to tackle this problem concentrates on partitioning routing tables in optimized tree-based data structures, such as Patricia [4], segment table [5], LC-trie [6], and tree bitmap [7].

Tree structure allows the organization of prefixes on a digital basis by using the bits of prefixes to direct the branching. Fig. 1 illustrates a binary tree representing the sample routing table shown in the right part. A prefix defines a path in the tree which is beginning from the root and ending in some node, the genuine node, in which the routing information, such as the next hop of the routing entry, is stored. The internal nodes that are not genuine nodes are called dummy nodes because there is no routing information stored in them. For example in Fig. 1, the genuine nodes are all labeled, and others are all dummy nodes. Prefix I with 1010* corresponds to the path starting from the root and ending in the leaf node I. An 8-bit IP address 10110001 has the longest prefix matching with the prefix 10110* (J) deriving from the two matching prefixes 10* (H) and 10110* (J). The worst case lookup performance of the tree-based algorithm is inversely proportional to the depth of the tree. In the worst case the depth of the tree can be up to the length of the IP address, 32 for IPv4 and 128 for IPv6. The average lookup performance of the tree-based schemes is not only relevant to the depth of the tree, but also to the shape of the tree and the traffic pattern. Detail discussions are in [8].

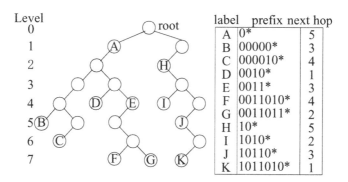

Fig. 1. A binary tree representing a sample routing table

To speed up the worst case lookup performance of the tree-based schemes, some techniques have to be used to decrease the depth of the trie. In Patricia [4], path compression is used to eliminate the one-degree dummy nodes. The segment table scheme [5] in essence is a fix-stride multi-bit trie algorithm. When searching in the multi-bit trie, each step more than one bit is inspected, thus the depth of the trie is reduced. The problem of the segment table scheme is that the stride, the number of bits to be inspected a time, is fixed other than determined by the characteristic of the specific routing table. Therefore it is not memory efficient. When it is upgraded to support IPv6 LPM, too massive memory is required to be affordable. LC-trie (Level Compressed trie) [6] overcomes the disadvantage of the segment table scheme. LC-trie replaces a large full (or nearly full) binary subtrie with a multi-bit trie iteratively to reduce the depth of the trie and finally a memory efficient variable stride multi-bit trie is constructed. The big problem of LC-trie is that it is a static algorithm, i.e. it does not support incremental updates. When the routing table changes, LC-trie has to be reconstructed from scratch. To conquer this drawback, Eatherton et al presented tree bitmap in [7]. Tree bitmap avoids leaf pushing [9], thus incremental update is supported.

In our opinion, these tree-based algorithms have three common drawbacks. First, all the bits of the prefix are used to build the tree. Thus, the depth of the tree is proportional to the length of the IP address. Second, the bits of the prefix are used from the most significant bit (MSB) to the least significant bit (LSB) in order. They are treated with no difference. Third, one and only one routing entry can be stored in the genuine node in the trie, which can result in increased memory consumption and prolonged lookup time.

Longest prefix first search tree (LPFST) [10] is a new tree-based algorithm, which puts the routing entry with longer prefix on the upper level of the tree to make the searching procedure to stop immediately as matching with some internal node while performing IP lookup. Further, some nodes may store two routing entries to minimize the number of tree nodes. LPFST is suitable for both IPv4 and IPv6. But its performance is not evaluated using IPv6 in [10]. We do this work in this paper. Although not all the bits of the prefix are used to construct LPFST, the bits are still used from MSB to LSB sequentially. Thus, LPFST can be far from balanced. For example, the first three bits of IPv6 global unicast address are 001, thus, the root of the corresponding LPFST has only left child, and so does its left child, the second generation of root node has only right child.

In this paper, we propose another tree-based LPM scheme, called BS-tree (Bit Selection tree). BS-tree has the following innovations. First, not all the bits of the prefix are used to construct the tree. Thus our scheme has good scalability to the length of the IP address. It is suitable for both IPv4 and IPv6. Second, the bits are not used from MSB to LSB sequentially. They are selected based on their costs defined in this paper. BS-tree is more balanced and the depth of BS-tree is reduced compared with other tree-based schemes. Third, all the routing entries are stored in the leaf nodes, and more than one entry can be stored in a leaf node, which is called leaf bucket. The size of the leaf bucket is determined by a tunable threshold, which is used to trade off between the memory requirement and lookup performance. Fourth, our scheme supports incremental updates.

As far as the authors' awareness, the idea of bit selection is used to solve the packet classification problem in [11]. We use it to do LPM in this paper with different bit selection criteria and discuss the incremental updates. Furthermore, the scheme is refined for both IPv4 and IPv6 based on some observations.

3 Bit Selection Algorithm

3.1 Bit Selection Criteria

The prefixes in the routing table can be viewed as an n*m matrix consisting of 0, 1, and *, where n is the number of routing entries in the routing table and m is the length of the IP address. Each bit of the prefixes in the routing table corresponds to a column in the matrix. For each column i in the matrix, we can count the number of 0, 1, and * in it, denoted by $N0[i]$, $N1[i]$, and $N*[i]$, respectively. We define the cost for column i as follows:

$$Cost[i] = \left|N0[i] - N1[i]\right| + 2 \times N*[i] \tag{1}$$

To efficiently and effectively organize the routing table in a tree, we should try to eliminate as many entries as possible from further consideration in the smallest number of steps. The finally constructed tree should be as low as possible and should be as balanced as possible. The optimal BS-tree can be constructed using dynamic programming, which is computationally expensive. Thus, we define the following simple bit selection criteria:

1) The bit with minimal cost is picked first;
2) If more than one bit has the minimal cost, the more significant one is picked.

Although BS-tree constructed using the simple bit selection criteria is not optimal because the procedure is only locally optimized at each step, it is suboptimal because $\left|N0[i] - N1[i]\right|$ is used to control the balance of the tree and $2 \times N*[i]$ is used to control the memory explosion resulting from routing entry duplication. Please note that two times of the number of the * in a column is used to calculate the cost, because such entries will be copied to the left and the right child of a node in the tree.

3.2 Node Structure and Leaf Bucket Organization

Fig. 2 shows the structure of our tree node. The node is either an internal node or a leaf bucket. The routing entries are all stored in the leaf buckets. The internal node is only used to guide the lookup procedure to reach a leaf bucket. The bit position field is also used as a flag, 0 for leaf bucket and otherwise for internal node. Internal node uses the left and right pointers to point to its children. Leaf bucket only needs one pointer points to the list of the routing entries contained in it. So, we share the right field in the internal node with the bucket field in the leaf node. In this way, we save a pointer field.

Fig. 2. The node structure

The size of the leaf bucket is determined by a threshold, which is tunable to trade off between the memory consumption and the lookup performance. In practice, it is usually less than 128. So, simple approaches can be used to finish the lookup in the leaf bucket, such as linear search, binary search on prefix intervals [12], LPFST [10], or ternary content addressable memory (TCAM). We choose the specific approach based on the specific threshold. When bucket size is not greater than 8, we use linear search and sort the bucket descendingly by the prefix length. In consequence, the first matching entry in the bucket is the most specific one that matches the IP address under lookup. Otherwise, we use binary search on prefix intervals. TCAM is used for hardware implementation.

3.3 BS-Tree Construction

The pseudo code for the BS-tree construction is recursive. Each time it is called, a new node will be created (line 2). Then the number of the routing entries in the routing table (RT) is checked. If it is less than or equals to the threshold, we organize the entries in a leaf bucket according to the criteria discussed in the previous subsection (line 3-6). Otherwise, an internal node is needed. In this case, we first calculate the cost for each column according to (1) (line 7-12). Then bit k is selected based on the bit selection criteria to partition the RT into RT0 and RT1 (line 13-15). After that, RT0 and RT1 are used to invoke the construction code iteratively to build the left and right child for the node (line 17-18).

```
1 node * constructBST(RT){
2    n = create a new node;
3    if ||RT|| <= THRESHOLD {
4       n.rightBKT =organizeBucket(RT);
5       return n;
6    }
7    for i from 1 to m {
8       N0[i] = the number of 0 in column[i];
9       N1[i] = the number of 1 in column[i];
10      N*[i] = the number of * in column[i];
11      cost[i] = |N0[i] – N1[i]| + 2 × N*[i];
12   }
13   k = the column that has minimal cost;
14   RT0 = the entries whose kth bit is 0 or *;
15   RT1 = the entries whose kth bit is 1 or *;
16   n.BitPosition = k;
17   n.left = constructBST(RT0);
18   n.rightBKT = constructBST(RT1);
19   return n;
20}
```

Fig.3 illustrates the procedure to construct the BS-tree for the sample routing table depicted in Fig.1. The threshold is 4. In step 1, bit 1 is selected and stored in the root node because it has the minimal cost, 3. Then the left child of the root node containing routing entry A, B, C, D, E, F, and G, is processed. Because the number of entries covered by this node is greater than the threshold, costs are calculated and bit 4 is picked to partition the entries (step2). Following, two leaf buckets are created for the left and right child of the prior node respectively because the number of entries in the leaf buckets equals to the threshold (step 3). Finally, in step 4, a leaf bucket including entry H, I, J, and K is created for the right child of the root node.

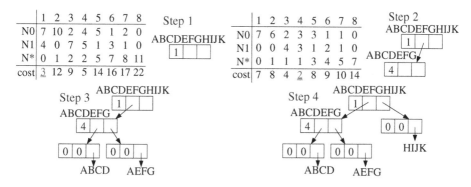

Fig. 3. A sample construction procedure of BS-tree

3.4 BS-Tree Lookup

Given the BS-tree data structure, the search procedure is straightforward. BS-tree is traversed based on the bit position stored in the internal node until a leaf bucket is identified (whose bit position is 0). Then the leaf bucket is searched according to its organization to locate the matching entry with the longest prefix.

```
1 nexthop lookupBST(root, addr){
2    node *pn = root;
3    while (pn->BitPosition){/* internal node */
4       if (the pn->BitPosition bit of addr is 0)
5          pn = pn->left;
6       else
7          pn = pn->rightBKT;
8    }
9    return lookupBucket(pn->rightBKT, addr);
10}
```

3.5 BS-Tree Update

The proposed BS-tree can be quickly updated without being reconstructed from scratch when the routing table is changed. The following pseudo code is for route insertion. Route deletion and modification have similar procedure.

```
1 void insertBST(root, p){
2    if (root->BitPosition){/* internal node */
3       if (p->length < root->BitPosition) {
4          insertBST(root->left, p);
5          insertBST(root->rightBKT, p);
6       }
7       else {
8          if (the root->BitPosition bit of p is 0)
9             insertBST(root->left, p);
10         else
11            insertBST(root->rightBKT, p);
12      }
13   }
```

```
14  else
15    insertBucket(root->rightBKT, p);
16}
```

4 Refinements

4.1 Refinements for IPv4

There is no routing entry with prefix length less than eight in the IPv4 routing tables. So, we use the first 8 bits to partition the routing entries into several groups and build a BS-tree for each group. This enhanced scheme is called MBS-tree (Multiple BS-tree). Because the height of MBS-tree is reduced, its lookup performance is increased, which we can see from the evaluation results shown in Section 5. Fig. 4 illustrates the main idea of the MBS-tree. A table is used to record the associated next-hop or BS-tree. The routing entries with the same first 8 bits are stored in the same BS-tree.

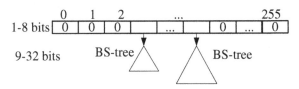

Fig. 4. The main idea of multiple BS-tree

4.2 Observations on IPv6

1) The structure of IPv6 address: The latest format of IPv6 global unicast address defined in RFC3587 has 3 parts (see Fig. 5): 45-bit global routing prefix (4-48th bits, the first three bits are '001'), 16-bit subnet ID (49-64th bits), and 64-bit interface ID (65-128th bits) in Modified EUI-64 format. Usually, only the global routing prefix and the subnet ID, totally accounting for 64 bits, are used for routing.

3	45 bits	16 bits	64 bits
001	global routing prefix	subnet ID	interface ID

Fig. 5. The format of IPv6 global unicast address

2) The policies for IPv6 address allocation & assignment: The guidelines for IPv6 address assignment to a site defined in RFC3177 are as follows: /48 in the general case, except for very large subscribers; /64 when it is known that one and only one subnet is needed by design; /128 when it is absolutely known that one and only one device is connecting. IANA (the Internet Assigned Numbers Authority) issued the memo titled "IPv6 Address Allocation and Assignment Policy"[2] on June 26, 2002 to govern the distribution of IPv6 address space. In consequence, RIRs (Regional Internet Registries) jointly published the document with the same title on January 22,

---

[2] http://www.iana.org/ipaddress/ipv6-allocation-policy-26jun02

2003 to define registry policies for the assignment and allocation of globally unique IPv6 addresses to Internet Service Providers and other organizations[3]. One of their goals is to distribute IPv6 address space in a hierarchical manner according to the topology of network infrastructure so as to permit the aggregation of routing information by ISPs, and to limit the expansion of Internet routing tables. Address allocation hierarchy is shown in Fig.6. IANA, the top level of the hierarchy, allocates /23 from 2000::/3 to the RIRs. RIRs in turn allocate /32 to subordinate address agencies, ISPs or LIRs (Local Internet Registries). The bottom level of the hierarchy, EUs (End Users), are generally given /48, /64, or /128 according to RFC3177.

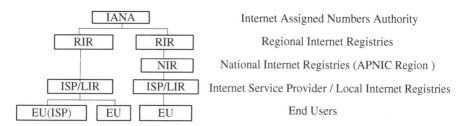

IANA	Internet Assigned Numbers Authority
RIR RIR	Regional Internet Registries
NIR	National Internet Registries (APNIC Region)
ISP/LIR ISP/LIR	Internet Service Provider / Local Internet Registries
EU(ISP) EU EU	End Users

Fig. 6. Address allocation hierarchy

3) Characteristics of real IPv6 backbone routing tables: We collected 6 IPv6 BGP routing tables from the backbone network on July 3, 2006. Four of them are from four different peers (Vatican, United Kingdom, USA, and Japan) in the RouteViews project[4] and the other two are from Tilab[5] and Potaroo[6] respectively. The prefix length distribution is listed in Table 1, from which we can see that: 1) the number of route entries in a routing table is less than 1000 now; 2) most of the prefix lengths are less than 64 bits (only 1 exception in Japan table with 128 bits); 3) there is no routing entry with prefix length less than 16 (The length 0 route in Japan table is the default route); 4) the entries peak at prefix lengths of 32, 35, and 48, the bolded rows in the table. Routing entries with length less than 33 amount to 80% of the total entries.

The characteristics of the real IPv6 backbone routing tables reflect the IPv6 address structure and the policies for IPv6 address allocation and assignment.

4.3 Refinements for IPv6

The observations can be used to enhance the performance of the BS-tree when it is applied to do IPv6 LPM.

1) Because there is no routing entry with prefix length less than 16 and the number of distinct values of the first 16 bits in a routing table is no more than 10 (The statistic results are omitted duo to space limitation), we use the first 16 bits to

---

[3] http://www.ripe.net/ripe/docs/ripe-267.html

[4] http://www.routeviews.org/

[5] http://net-stats.ipv6.tilab.com/bgp/bgp-table-snapshot.txt

[6] http://bgp.potaroo.net/v6/as1221/bgptable.txt

Table 1. Prefix Length Distributions of real IPv6 backbone routing tables

	Tilab	Potaroo	VA	UK	USA	Japan
0	0	0	0	0	0	1
16	0	1	1	1	1	1
19	1	1	1	1	1	1
20	3	3	3	3	3	3
21	2	2	2	2	2	2
24	5	4	0	0	4	4
27	1	1	1	1	1	1
28	2	2	2	1	3	4
29	1	1	1	1	1	1
30	1	2	1	1	1	1
32	514	526	515	513	516	508
33	5	5	5	5	5	5
34	5	5	5	5	5	5
35	31	33	31	31	31	31
39	0	1	1	0	1	0
40	3	15	8	6	8	3
42	0	10	1	1	1	0
45	1	1	1	1	1	0
48	50	76	66	71	66	18
64	6	0	6	2	0	5
128	0	0	0	0	0	1

partition the routing entries into several groups and construct a small BS-tree for each group. The depth of such small BS-tree is reduced, and the overall lookup performance of our scheme is improved.

2) We only calculate the costs for the 17th-48th bits and only use these 32 bits to construct the BS-tree, because very few routing entries with prefix length more than 48. N*[j], 49<=j<=128, is very large, thus bit 49-128 should not be selected to build the BS-tree. This improvement has two advantages. First, it greatly reduces the task to calculate the costs. Thus it significantly speeds up the initial construction of the BS-tree. Second, we need only 32 bits of the destination IPv6 address to traverse the BS-tree. This facilitates the lookup procedure, because 128-bit IPv6 address has to be organized with a structure and to fetch a specific bit from a 128-bit number is not convenient on a 32-bit machine.

The improved BS-tree for IPv6 LPM, also called MBS-tree (Multiple BS-tree), is illustrated in Fig. 7. With MBS-Tree, at the beginning of the lookup, the first 16 bits of the IPv6 address are used to choose an appropriate BS-tree to start the traverse using the following 32 bits of the IPv6 address.

5   Performance Evaluation and Comparison

In the section we present the evaluation and comparison results. We do the evaluations using both real routing tables and synthetically created IPv6 routing tables. The synthetically created IPv6 routing tables with different sizes are generated

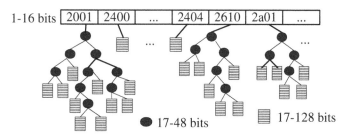

1-16 bits | 2001 | 2400 | ... | 2404 | 2610 | 2a01 | ...

● 17-48 bits

▤ 17-128 bits

Fig. 7. The improved BS-tree for IPv6 longest prefix matching

to test the scalability of our scheme because the current IPv6 routing tables are small. Our test application measures the average lookup time using generated IPs. For each test run, we generate one million IP addresses, 80% of which are generated based on the routing entries randomly selected from the corresponding routing table, others are generated purely randomly.

All the tests are done on a PC with one Pentium 4 2.4 GHz CPU, 512M DDR333 memory, and Linux OS. The sample implementations are all written in C language.

Due to page limitation, we can only present some of the evaluation results in this paper. Table 2 shows the comparison results using the rrc00 IPv4 routing table[7] collected on Sep. 1, 2006. From Table 2, we can see that BS-tree has the lowest tree depth (14 for basic BS-tree and 9 for MBS-tree), thus BS-tree is the best for the worst case lookup time.

Table 2. Comparison results for IPv4 using rrc00 with 198687 entries

	Patricia	LC-trie[*]	LPFST	BS-tree[#]	MBS-tree[#]
Tree depth	30	15	31	14	9
Memory(MB)	7.40	4.32	4.54	7.46	6.96
Total node	369895	279955	189053	10173	9483
1-degree node	7478	NA	48232	0	0
2-degree node	181208	NA	70410	5086	4667
Leaf node	181209	181209	70411	5087	4816
Building (s)	0.142	0.25	0.114	0.562	1.59
AVG time(ns)	896	510	651	905	728

*fill factor = 1.0, root branch = 0.
#bucket size = 64, bucket organized in binary search on prefix interval.

The evaluation results using the Potaroo IPv6 routing table collected on Sep. 10, 2006 are listed in Table 3, which shows that BS-tree has the best lookup time in both worst case and average case. The memory consumption of BS-tree is relatively large. In fact, most of the memories are consumed by the leaf buckets which are organized as binary search on prefix interval [12]. The memories occupied by BS-tree itself are small, which we can derive from the number of nodes in the BS-tree.

To test the scalability of our scheme and its adaptability to the future IPv6 network, we synthetically creates several IPv6 routing tables with different sizes by v6Gen [13] and use them to do the experiments. The prefix length distribution of the

---

[7] http://www.ripe.net/projects/ris/

Table 3. Comparison results for IPv6 using potaroo with 815 entries

	Patricia	LPFST	BS-tree[#]	MBS-tree[#]
Tree depth	30	44	6	6
Memory(KB)	43.8	29.0	53.3	47.6
Total node	1563	806	35	43
1-degree node	40	297	0	0
2-degree node	761	254	17	17
Leaf node	762	255	18	26
Building (ms)	1.32	0.54	2.53	1.53
AVG time(ns)	375	291	201	174

bucket size = 64, bucket organized in binary search on prefix interval.

IPv6 routing table with 100K routing entries is depicted in Fig. 8. Other generated IPv6 routing tables with different sizes have the similar prefix length distribution. The size of the current backbone IPv4 routing table is about 200K. So, we display the comparison results using synthesized IPv6 routing table with 200K routing entries in Table 4. We can see that BS-tree has the best lookup time in both worst and average case. Under our test circumstance, BS-tree can complete almost one million IPv6 LPM in one second even when the number of entries in the routing table reaches 200K.

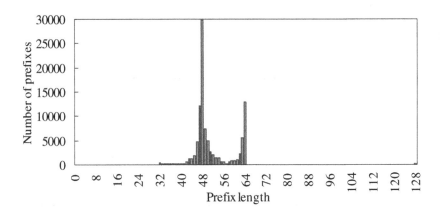

Fig. 8. Prefix length distribution of the synthesized IPv6 routing table with 100K entries

Table 4. Comparison results for IPv6 using v6gen with 200k entries

	Patricia	LPFST	BS-tree[#]	MBS-tree[#]
Tree depth	44	58	20	19
Memory(MB)	10.8	7.01	22.6	19.9
Total node	386885	194819	20139	18271
1-degree node	4396	51288	0	0
2-degree node	191244	71765	10069	9134
Leaf node	191245	71766	10070	9137
Building (s)	0.74	0.20	3.92	2.35
AVG time(us)	1.34	1.30	1.18	1.04

bucket size = 64, bucket organized in binary search on prefix interval.

6 Conclusion

In this paper, we propose bit selection tree (BS-tree) to do longest prefix matching (LPM) for both IPv4 and IPv6. Only the bits with low costs are selected to build the tree. Both the balance and the depth of the tree are controlled by the bit selection criteria. The evaluation results show BS-tree has the best worst case lookup performance.

IPv6 has been gaining wider acceptance to replace IPv4 and IPv6 production deployment has been seeing in Europe, Asia, and North America. Therefore, we specially refine BS-tree for IPv6 LPM based on the observations on the IPv6 addressing architecture, the IPv6 address allocation policies, and the real IPv6 backbone routing tables. Furthermore, its performance is evaluated with real IPv6 routing tables and synthetically created large IPv6 routing tables. For IPv6, BS-tree has the best lookup performance in both average case and worst case. The sample software implementation of BS-tree only needs several tens of kilobytes memory for the current real IPv6 backbone routing tables and can achieve 6 million lookups per second (3.8Gbps for 80-byte minimal IPv6 packet) on a PC with one Pentium 4 2.4GHz CPU, 512M DDR333 memory, and Linux operating system.

References

1. Deering, S., Hinden, R.: Internet protocol version 6 specification. RFC2460 (December 1998)
2. Fink, R., Hinden, R.: 6bone (IPv6 testing address allocation) phaseout. RFC3701 (March 2004)
3. Ruiz-Sbnches, M., Biersack, E., Dabbous, W.: Survey and taxonomy of IP address lookup algorithms. IEEE Network 15(2), 8–23 (2001)
4. Sklower, K.: A tree-based packet routing table for Berkeley Unix. In: 1991 Winter USENIX Conf., pp. 93–99 (1991)
5. Gupta, P., Lin, S., McKeown, N.: Routing lookups in hardware at memory access speed. IEEE INFOCOM, 1240–1247 (1998)
6. Nilsson, S., Karlsson, G.: IP-address lookup using LC-tries. IEEE JSAS 17(6), 1083–1092 (1999)
7. Eatherton, W., Dittia, Z., Varghese, G.: Tree bitmap: hardware/ software IP lookups with incremental updates. ACM SIGCOMM Computer Communication Review 34(2), 97–122 (2004)
8. Li, Z., Ma, Y.: Trie-based observations on the routing tables. In: Japan-China Joint Workshop on Frontier of Computer Science and Technology, pp. 157–163 (November 2006)
9. Srinivasan, V., Varghese, G.: Fast address lookups using controlled prefix expansion. ACM Sigmetrics, 1–11 (June 1998)
10. Wuu, L., Chen, K., Liu, T.: A longest prefix first search tree for IP lookup. IEEE ICC, 989–993 (May 2005)
11. Woo, T.Y.C.: A modular approach to packet classification: algorithms and results. IEEE INFOCOM, 1213–1222 (March 2000)
12. Lampson, B., Srinivasan, V., Varghese, G.: IP lookups using multiway and multicolumn search. IEEE INFOCOM, 1248–1256 (March 1998)
13. Zheng, K., Liu, B.: V6Gene: a scalable IPv6 prefix generator for route lookup algorithm benchmark. IEEE AINA, 147–152 (April 2006)

A New Fuzzy Decision Tree Classification Method for Mining High-Speed Data Streams Based on Binary Search Trees*

Zhoujun Li[1], Tao Wang[2], Ruoxue Wang[3], Yuejin Yan[2], and Huowang Chen[2]

[1] School of Computer Science & Engineering, Beihang University, Beijing, 100083, China
[2] Computer School, National University of Defense Technology, Changsha, 410073, China
[3] Journal of Computer Research and Development, Beijing, 100080, China
lizj@buaa.edu.cn

Abstract. Decision tree construction is a well-studied problem in data mining. Recently, there has been much interest in mining data streams. Domingos and Hulten have presented a one-pass algorithm for decision tree constructions. Their system using Hoeffding inequality to achieve a probabilistic bound on the accuracy of the tree constructed. Gama et al. have extended VFDT in two directions. Their system VFDTc can deal with continuous data and use more powerful classification techniques at tree leaves. Peng et al. present soft discretization method to solve continuous attributes in data mining. In this paper, we revisit these problems and implemented a system sVFDT for data stream mining. We make the following contributions: 1) we present a binary search trees (BST) approach for efficiently handling continuous attributes. Its processing time for values inserting is $O(nlogn)$, while VFDT's processing time is $O(n^2)$. 2) We improve the method of getting the best split-test point of a given continuous attribute. Comparing to the method used in VFDTc, it decreases from $O(nlogn)$ to $O(n)$ in processing time. 3) Comparing to VFDTc, sVFDT's candidate split-test number decrease from $O(n)$ to $O(logn)$.4)Improve the soft discretization method to increase classification accuracy in data stream mining.

Keywords: Data Streams, Fuzzy, VFDT, Continuous Attribute, Binary Search Tree.

1 Introduction

Decision trees are one of the most used classification techniques for data mining. Its induction offers a highly practical method for generalizing from instances whose class membership is known. The most common approach to induce a decision tree is to partition the labeled instances recursively until a stopping criterion is met. The partition is defined by way of selecting a test that has a manageable set of outcomes,

---

* This work was supported by the National Science Foundation of China under Grants No. 60573057, 60473057 and 90604007.

F.P. Preparata and Q. Fang (Eds.): FAW 2007, LNCS 4613, pp. 216–227, 2007.

creating a branch for each possible outcome, passing each instance down the corresponding branch, and treating each block of the partition as a subprbolem, for which a subtree is built recursively[20].

Continuous data stream poses new problems to traditional classification methods of data mining. The common approach of the traditional methods is to store and process the entire set of training examples. The growing amounts of training data increase the processing requirements of data mining systems up to a point, where they either run out of memory, or their computation time becomes prohibitively long. The key issue of data stream mining is that only one pass is allowed over the entire data. Moreover, there is a real-time constraint, i.e. the processing time is limited by the rate of arrival of instances in the data stream, and the memory and disk available to store any summary information may be bounded. For most data mining problems, a one-pass algorithm cannot be very accurate. The existing algorithms typically achieve either a deterministic bound on the accuracy or a probabilistic bound [21, 23].

Domingos and Hulten [2, 6] have addressed the problem of decision tree construction on data streams. Their algorithm guarantees a probabilistic bound on the accuracy of the decision tree that is constructed. Gama et al. [5] have extended VFDT in two directions: the ability to deal with continuous data and the use of more powerful classification techniques at tree leaves. Wang et al. [28,29] use binary search trees to handle continuous attributes based on top of VFDT and VFDTc.

Peng et al.[30]propose the soft discretization method in traditional data mining field, it solves the problem of noise data and improve the classification accuracy.

The rest of the paper is organized as follows. Section 2 describes the related works that is the basis for this paper. Section 3 presents the technical details of sVFDT. The system has been implemented and evaluated, and experimental evaluation is done in Section 4. Last section concludes the paper, resuming the main contributions of this work.

2 Related Work

In this section we analyze the related works that our sVFDT bases on.

2.1 VFDT

VFDT(Very Fast Decision Tree) system[2], which is able to learn from abundant data within practical time and memory constraints. In VFDT a decision tree is learned by recursively replacing leaves with decision nodes. Each leaf stores the sufficient statistics about attribute-values. The sufficient statistics are those needed by a heuristic evaluation function that evaluates the merit of split-tests based on attribute-values. When an example is available, it traverses the tree from the root to a leaf, evaluating the appropriate attribute at each node, and following the branch corresponding to the attribute's value in the example. When the example reaches a leaf, the sufficient statistics are updated. Then, each possible condition based on attribute-values is evaluated. If there is enough statistical support in favor of one test over the others, the leaf is changed to a decision node. The new decision node will have as many descendant leaves as the number of possible values for the chosen attribute (therefore this tree is not necessarily binary). The decision nodes only maintain the information about the split-test installed in this node. The initial state of the tree consists of a single leaf: the root of the tree. The heuristic evaluation function is the Information Gain (denoted by G($\cdot$)). The sufficient statistics for estimating the merit of a discrete attribute are the counts n_{ijk}, representing the number of examples of class k that reach the leaf, where the attribute j takes the value i. The Information Gain measures the amount of information that is necessary to classify an example that reach the node: $G(A_j)=info(examples)-info(A_j)$. The information of the attribute j is given by: $\inf o(A_j) = \sum_i P_i(\sum_k -P_{ik} \log(P_{ik}))$ where $P_{ik} = n_{ijk} / \sum_a n_{ajk}$, is the probability of observing the value of the attribute i given class k and $P_i = \sum_a n_{ija} / \sum_a \sum_b n_{ajb}$ is the probabilities of observing the value of attribute i.

As mentioned in Catlett and others [23], that it may be sufficient to use a small sample of the available examples when choosing the split attribute at any given node. To determine the number of examples needed for each decision, VFDT uses a statistical result known as Hoeffding bounds or additive Chernoff bounds. After n independent observations of a real-valued random variable r with range R, the Hoeffding bound ensures that, with confidence $1-\delta$, the true mean of r is at least $\overline{r}-\varepsilon$, where $\overline{r}$ is the observed mean of samples and $\varepsilon = \sqrt{\dfrac{R^2 \ln(1/\delta)}{2n}}$. This is true irrespective of the probability distribution that generated the observations.

Let G($\cdot$) be the evaluation function of an attribute. For the information gain, the range R, of G($\cdot$) is $log_2 \#classes$. Let x_a be the attribute with the highest G($\cdot$), x_b the attribute with second-highest G($\cdot$) and $\Delta\overline{G} = \overline{G}(x_a) - \overline{G}(x_b)$, the difference between the two better attributes. Then if $\Delta\overline{G} > \varepsilon$ with n examples observed in the leaf, the Hoeffding bound states with probability $1-\delta$ that x_a is really the attribute with highest value in the evaluation function. In this case the leaf must be transformed into a decision node that splits on x_a.

For continuous attribute, whenever VFDT starts a new leaf, it collects up to M distinct values for each continuous attribute from the first examples that arrive at it. These are maintained in sorted order as they arrive, and a candidate test threshold is maintained midway between adjacent values with different classes, as in the traditional method. Once VFDT has M values for an attribute, it stops adding new candidate thresholds and uses additional data only to evaluate the existing ones. Every leaf uses a different value of M, based on its level in the tree and the amount of RAM available when it is started. For example, M can be very large when choosing the split for the root of the tree, but must be very small once there is a large partially induced tree, and many leaves are competing for limited memory resources. Notice that even when M is very large (and especially when it is small) VFDT may miss the locally optimal split point. This is not a serious problem here for two reasons. First, if data is an independent, identically distributed sample, VFDT should end up with a value near (or an empirical gain close to) the correct one simply by chance. And second, VFDT will be learning very large trees from massive data streams and can correct early mistakes later in the learning process by adding additional splits to the tree.

Thinking of each continuous attribute, we will find that the processing time for the insertion of new examples is O (n^2), where n represents the number of distinct values for the attribute seen so far.

2.2 VFDTc

VFDTc is implemented on top of VFDT, and it extends VFDT in two directions: the ability to deal with continuous attributes and the use of more powerful classification techniques at tree leaves. Here, we just focus on the handling of continuous attributes.

In VFDTc a decision node that contains a split-test based on a continuous attribute has two descendant branches. The split-test is a condition of the form $attrib_j \leq T$. The two descendant branches correspond to the values *TRUE* and *FALSE* for the split-test. The cut point is chosen from all the possible observed values for that attribute. In order to evaluate the goodness of a split, it needs to compute the class distribution of the examples at which the attribute-value is less than or greater than the cut point. The counts n_{ijk} are fundamental for computing all necessary statistics. They are kept with the use of the following data structure: In each leaf of the decision tree it maintains a vector of the classes' distribution of the examples that reach this leaf. For each continuous attribute j, the system maintains a binary attribute tree structure. A node in the binary tree is identified with a value i(that is the value of the attribute j seen in an example), and two vectors (of dimension k) used to count the values that go through that node. Two vectors, *VE* and *VH* contain the counts of values respectively $\leq i$ and $> i$ for the examples labeled with class k. When an example reaches leaf, all the binary trees are updated. In [5], an algorithm of inserting a value in the binary tree is presented. Insertion of a new value in this structure is *O(nlogn)* where n represents the number of distinct values for the attribute seen so far.

To obtain the Information Gain of a given attribute, VFDTc uses an exhaustive method to evaluate the merit of all possible cut points. Here, any value observed in the examples seen so far can be used as cut point. For each possible one, the information of the two partitions is computed using equation 1.

$$\inf o(A_j(i)) = P(A_j \leq i) \ * \ iLow(A_j(i)) + P(A_j > i) \ *i \ High(A_j(i)) \quad (1)$$

Where i is the cut point, $iLow(A_j(i))$ the information of $A_j \leq i$ (equation 2) and $iHigh(A_j(i))$ the information of $A_j > i$ (equation 3).

$$iLow(A_j(i)) = -\sum_K P(K = k \mid A_j \leq i) \ * \ \log(P(K{=}k|A_j \leq i)) \quad (2)$$

$$iHigh(A_j(i)) = -\sum_K P(K = k \mid A_j > i) \ * \ \log(P(K{=}k|A_j{>}i)) \quad (3)$$

VFDTc only considers a possible cut_point if and only if the number of examples in each of subsets is higher than P_{min} (a user defined constant) percentage of the total number of examples seen in the node. [5] Presents the algorithm to compute #($A_j \leq i$) for a given attribute j and class k. The algorithm's processing time is *O(logn)*, so the best split-test point calculating time is $O(nlogn)$. Here, n represents the number of distinct values for the attribute seen so far at that leaf.

2.3 Soft Discretization

Soft discretization could be viewed as an extension of hard discretization, and the classical information measures defined in the probability domain have been extended to new definitions in the possibility domain based on fuzzy set theory [13]. A crisp set A_c is expressed with a sharp characterization function $A_c(a) : \Omega \to \{0,1\} : a \in \Omega$, alternatively a fuzzy set A is characterized with a membership function $A(a) : \Omega \to [0,1] : a \in \Omega$. The membership $A(a)$is called the possibility of A to take a value $a \in \Omega$[14]. The probability of fuzzy set A is defined, according to Zadeh [15], by, $P_F(A) = \int_\Omega A(a)dP$, where dP is a probability measure on Ω, and the subscript F is used to denote the associated fuzzy terms. Specially, if A is defined on discrete domain $\Omega = \{a_1, .. a_i, ..., a_m\}$, and the probability of $P(a_i) = P_i$ then its probability is $P_F(A) = \sum_{i=1}^{m} A(a_i) p_i$.

Let $Q = \{A_1, ..., A_k\}$be a family of fuzzy sets on Ω. Q is called a fuzzy partition of Ω [16] when $\sum_{r=1}^{k} A_r(a) = 1, \forall a \in \Omega$.

A hard discretization is defined with a threshold T, which generates the boundary between two crisp sets. Alternatively, a soft discretization is defined by a fuzzy set pair which forms a fuzzy partition. In contrast to the classical method of non-overlapping partitioning, the soft discretization is overlapped. The soft discretization is defined with three parameters/functions, one is the cross point T, the other two are the membership functions of the fuzzy set pair A_1 and A_2: $A_1(a)+A_2(a)=1$. The cross point T, i.e. the localization of soft discretization, is determined based on whether it can maximize the information gain in classification, and the membership functions of the fuzzy set pair are determined according to the characteristics of attribute data, such as the uncertainty of the associated attribute.

3   Technique Details

Improving soft discretization method, we implement a system named sVFDT on top of VFDT and VFDTc. It handles continuous attributes based on binary search trees, and uses a more efficient best split-test point calculating method.

3.1 sVFDT Framework

[2] lists the details of the Hoffding tree algorithm just for processing discrete attributes. Although sVFDT bases on the top of VFDT and VFDTc, its framework is much different. For the reason of simplicity, we just list the detail framework of sVFDT for handling continuous attributes in Table 1.

Table 1. The framework of sVFDT

Inputs:	D	is a sequence of examples,
	X	is a set of continuous attributes,
	G (.)	is a split evaluation function,
	δ	is one minus the desired probability of choosing the correct attributes at any given node.

Output: SDT is a soft decision tree.

Procedure SDT (D, X, G, δ)

    Let SDT be a tree just with root r

    Let $X_1 = X \cup \{X_\varnothing\}$

    Let $\overline{G}_1(X_\varnothing)$ be the $\overline{G}$ obtained by predicting the most frequent class in D.

    For each continuous attribute $X_i \in X$ builds an empty binary search tree BST_i.

    For each example (x, c_k) in D

        Sort (x, c_k) into a leaf l using SDT.

        Inserts the example values into each BSTi of leaf l using the inserting algorithm in Figure 1.

        Compute $\overline{G}_l(X_j)$ of all possible split point for each attribute X_i using its corresponding BST as represented in Figure 2.

        Replace l by an internal decision node splits on the best split-point basing on VFDT's theory and soft discretization method.

    Return SDT

3.2 Updates the Binary Search Binary Tree When New Examples Arrives

One of the key problems in decision tree construction on streaming data is that the memory and computational cost of storing and processing the information required to obtain the best split gain can be very high. For discrete attributes, the number of distinct values is typically small, and therefore, the class histogram does not require much memory. Similarly, searching for the best split predicate is not expensive if number of candidate split conditions is relatively small.

However, for continuous attributes with a large number of distinct values, both memory and computational costs can be very high. Many of the existing approaches for scalable, but multi-pass, decision tree construction requires a preprocessing phase in which attribute value lists for continuous attributes are sorted [20]. Preprocessing of data, in comparison, is not an option with streaming datasets, and sorting during execution can be very expensive. Domingos and Hulten [2] have described and evaluated their one-pass algorithm focusing only on discrete attributes, and in later version they uses sorted array to handle continuous attribute. This implies a very high memory and computational overhead for inserting new examples and determining the best split point for a continuous attribute.

Our sVFDT manages a binary search tree for each continuous attribute. The binary search tree data structure will benefit the process of inserting new example and getting the best split point. For each continuous attribute i, the system maintains a binary search tree structure. A node in the binary search tree is identified with a value *keyValue* (that is the value of the attribute i seen in the example), and a vector(of dimension k) used to count the values that go through that node. This vector *classTotals[k]* contains the counts of examples which value is *keyValue* and class labeled with k. A node manages *left* and *right* pointers for its left and right child, where its left child corresponds to $\leqslant keyValue$, while its right child corresponds to >*keyValue*. To get the best split point, each continuous attribute manages a *head* pointer.

In sVFDT a Hoeffding tree node manages a binary search tree for each continuous attribute before it becomes a decision node.

```
Procedure InsertValueBSTTree(x, k, BSTTree)
Begin
  while (BSTTree ->right != NULL || BSTTree ->left != NULL )
        if (BSTTree ->keyValue = = x )   then break;
        Elseif (BSTTree ->keyValue > x ) then BSTTree = BSTTree ->left;
        else BSTTree = BSTTree ->right;
  Creates a new node curr based on x and k;
  If ( BSTTree.keyValue = = x )   then   BSTTree.classTotals[k]++;
  Elesif (BSTTree.keyValue > x)   then   BSTTree.left = curr;
  else       BSTTree.right = curr;
  Threads the tree ;( The details of threading is in figure2)
End
```

Fig. 1. Algorithm to insert value x of an example labeled with class k into a binary search tree corresponding to the continuous attribute i

In the induction of decision trees from continuous-valued data, a suitable threshold T, which discretizes the continuous attribute i into two intervals $atrr_i \leqslant T$ and $atrr_i > T$, is determined based on the classification information gain generated by the corresponding discretization. Given a threshold, the test $atrr_i \leqslant T$ is assigned to the left branch of the decision node while $atrr_i > T$ is assigned to the right branch. As a new example (x,c) arrives, the binary search tree corresponding to the continuous attribute i is updated as Figure1. In [5], when a new example arrives, $O(logn)$ binary search tree nodes need be updated, but sVFDT just need update a necessary node here.

VFDT will cost $O(n^2)$, and our system sVFDT will just cost $O(nlogn)$ (as presented in Figure 1)in execution time for values inserting, where n represents the number of distinct values for the given attribute seen so far.

3.3 Soft Discretization for Continuous Attributes

Taking advantage of binary search tree, we use a more efficient method to obtain the fuzzy information gain of a given attribute.

Assuming we are to select an attribute for a node having a set S of N examples arrived, these examples are managed by a binary tree according to the values of the continuous attribute i ; and an ordered sequence of distinct values a_1, a_2 ... a_n is formed. Every pair of adjacent data points suggests a potential threshold $T = (a_i + a_{i+1})/2$ to create a split test point and generate a corresponding partition of attribute i. In order to calculate the goodness of a split, we need to compute the class distribution of the examples at which the attribute value is less than or greater than threshold T. The counts *BSTTree.classTotals[k]* are fundamental for computing all necessary statistics.

As we know, inorder traversal of a binary search tree will produce a list of sorted values, so we take the advantage of this property to make best split-point selecting more efficient. As presented in Figure 2, traversing from the root pointer, we can easily compute the fuzzy information of all the potential thresholds. sVFDT implies soft discretization by managing Max/Min value and example numbers.

Procedure BSTInorderAttributeSplit(BSTtreePtr ptr,int *belowPrev[])
    BSTInorderAttributeSplit(ptr->next,int *belowPrev[]);
    For (k = 0 ; k < count ; k++)
     *belowPrev[k] += ptr->classTotals[k];
    Calculates the information gain using *belowPrev[];
    BSTInorderAttributeSplit(ptr->next,int *belowPrev[]);

Fig. 2. Algorithm to compute the information gain of each possible split point

Here, VFDTc will cost $O(nlogn)$, and our system sVFDT will just cost $O(n)$ in processing time, where n represents the number of distinct values for the given continuous attribute seen so far.

3.4 Classify a New Example

The classification for a given unknown object is obtained from the matching degrees of the object to each node from root to leaf. The possibility of an object belonging to class C_i is calculated by a fuzzy product operation $\otimes$. In the same way, the possibility of the object belonging to each class can be calculated, $\{\Pi_i\}_{i=1...k}$. If more than one leaf are associated with a same class C_i, say, the value of $\Pi_i = \oplus (\Pi_j)$ will be considered as the possibility of the corresponding class, where the maximum operation is used as the fuzzy sum operation $\oplus$ In the end, if one possibility value, such as Π_k , is much higher than others, that is $\Pi_k \gg \Pi_{i...k}$, then the class will be assigned as the class of the object, otherwise the decision tree predicts a distribution over all the classes.

4 Evaluation

In this section we empirically evaluate sVFDT. The main goal of this section is to provide evidence that the use of binary search tree decreases the processing time of VFDT, while keeps the same error rate and tree size. The soft discretization method will increase accuracy for the reason that it can solve the problem of noise data.

We first describe the data streams used for our experiments. We use a tool named *treeData* mentioned in [2] to create synthetic data .It creates a synthetic data set by sampling from a randomly generated decision tree. They were created by randomly generating decision trees and then using these trees to assign classes to randomly generated examples. It produced the random decision trees as follows. Starting from a tree with a single leaf node (the root) it repeatedly replaced leaf nodes with nodes that tested a randomly selected attribute which had not yet been tested on the path from the root of the tree to that selected leaf. After the first three levels of the tree each selected leaf had a probability of f of being pre-pruned instead of replaced by a split (and thus of remaining a leaf in the final tree). Additionally, any branch that reached a depth of 18 was pruned at that depth. Whenever a leaf was pruned it was randomly (with uniform probability) assigned a class label. A tree was completed as soon as all of its leaves were pruned.

VFDTc`s goal is to show that using stronger classification strategies at tree leaves will improve classifier's performance. With respect to the processing time, the use of naïve Bayes classifier will introduce an overhead [5], VFDTc is slower than VFDT. In order to compare the VFDTc and sVFDT , we implement the continuous attributes solving part of VFDTc ourselves.

We ran our experiments on a Pentium IV/2GH machine with 512MB of RAM, which running Linux RedHat 9.0.

Table 2 shows the processing (excluding I/O) time of learners as a function of the number of training examples averaged over nine runs. VFDT and sVFDT run with

parameters $\delta = 10^{-7}, \tau = 5\%, n_{\min} = 300, example\ number = 100000K$, no

leaf reactivation, and no rescan. Averagely, comparing to VFDT, sVFDT`s average reduction of processing time is 16.65%, and comparing to VFDTc, sVFDT`s average reduction is 6.87%.

Table 2. The comparing result of processing time

time(seconds) / example numbers	VFDT	VFDTc	sVFDT
10000	4.66	4.21	3.65
20736	9.96	8.83	8.01
42996	22.88	20.59	18.37
89156	48.51	43.57	40.47
184872	103.61	93.25	86.62
383349	215.83	187.77	174.12
794911	522.69	475.65	439.28
1648326	1123.51	1022.39	936.26
3417968	2090.31	1839.45	1751.17
7087498	3392.94	3053.65	2872.12
14696636	5209.47	4688.53	4369.26
30474845	8203.05	7382.75	6829.03
43883922	13431.02	11953.61	11038.15
90997707	17593.46	15834.12	14826.38
100000000	18902.06	16822.86	15683.12

Figure 3 shows the error rate curves of VFDT and sVFDT. Both algorithms have 10% noise data, VFDT`s error rate trends to 13.3%, while the sVFDT`s error rate trends to 8.2%. Experiment results show that sVFDT get better accuracy by using soft discretization, and it solves the problem of noise data.

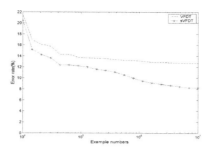

Fig. 3. Error rate as a function of the examples numbers

5   Conclusions and Future Work

On top of VFDT and VFDTc, we propose a system sVFDT to improve the soft discretization method. Focusing on continuous attribute, we have developed and evaluated a new technique named BST to insert new example and calculate best split-test point efficiently. It builds binary search trees, and its processing time for values insertion is $O(nlogn)$. Comparing to the method used in VFDTc, it improves from $O(nlogn)$ to $O(n)$ in processing time for best split-test point calculating. As for noise data, we improve the soft discretization method in traditional data mining field, so the sVFDT can deal with noise data efficiently and improve the classification accuracy.

In the future, we would like to expand our work in some directions. First, we do not discuss the problem of time changing concept here, and we will apply our method to those strategies that take into account concept drift [4, 6, 10, 14, 15, 19, 24, 25]. Second, we want to apply other new fuzzy decision tree methods in data streams classification [8, 13, 17, 18, 26].

References

[1] Babcock, B., Babu, S., Datar, M., Motawani, R., Widom, J.: Models and Issues in Data Stream Systems. In: PODS (2002)

[2] Domingos, P., Hulten, G.: Mining High-Speed Data Streams. In: Proceedings of the Association for Computing Machinery Sixth International Conference on Knowledge Discovery and Data Mining, pp. 71–80 (2000)

[3] Mehta, M., Agrawal, A., Rissanen, J.: SLIQ: A Fast Scalable Classifier for Data Mining. In: Proceedings of The Fifth International Conference on Extending Database Technology, Avignon, France, pp. 18–32 (1996)

[4] Fan, W.: StreamMiner: A Classifier Ensemble-based Engine to Mine Concept Drifting Data Streams. In: VLDB'2004 (2004)

[5] Gama, J., Rocha, R., Medas, P.: Accurate Decision Trees for Mining High-Speed Data Streams. In: Domingos, P., Faloutsos, C. (eds.) Proceedings of the Ninth International Conference on Knowledge Discovery and Data Mining, ACM Press, New York (2003)

[6] Hulten, G., Spencer, L., Domingos, P.: Mining Time-Changing Data Streams. In: ACM SIGKDD, ACM Press, New York (2001)

[7] Jin, R., Agrawal, G.: Efficient Decision Tree Construction on Streaming Data. In: Proceedings of ACM SIGKDD, ACM Press, New York (2003)

[8] Last, M.: Online Classification of Nonstationary Data Streams. Intelligent Data Analysis 6(2), 129–147 (2002)

[9] Muthukrishnan, S.: Data streams: Algorithms and Applications. In: Proceedings of the fourteenth annual ACM-SIAM symposium on discrete algorithms, ACM Press, New York (2003)

[10] Wang, H., Fan, W., Yu, P., Han, J.: Mining Concept-Drifting Data Streams using Ensemble Classifiers. In: The 9th ACM International Conference on Knowledge Discovery and Data Mining, Washington DC, USA. SIGKDD, ACM Press, New York (2003)

[11] Arasu, A., Babcock, B., Babu, S., Datar, M., Ito, K., Nishizawa, I., Rosenstein, J., Widom, J.: STREAM: The Stanford Stream Data Manager Demonstration Description – Short Overview of System Status and Plans. In: Proc. of the ACM Intl Conf. on Management of Data (SIGMOD 2003), ACM Press, New York (2003)

[12] Aggarwal, C., Han, J., Wang, J., Yu, P.S.: On Demand Classification of Data Streams. In: Proc. 2004 Int. Conf. on Knowledge Discovery and Data Mining (KDD'04), Seattle, WA (2004)

[13] Guetova, M., Holldobter, Storr, H.-P.: Incremental Fuzzy Decision Trees. In: 25th German conference on Artificial Intelligence (2002)

[14] Ben-David, S., Gehrke, J., Kifer, D.: Detecting Change in Data Streams. In: Proceedings of VLDB 2004 (2004)

[15] Aggarwal, C.: A Framework for Diagnosing Changes in Evolving Data Streams. In: Proceedings of the ACM SIGMOD Conference, ACM Press, New York (2003)

[16] Gaber, M.M., Zaslavskey, A., Krishnaswamy, S.: Mining Data Streams: a Review. SIGMOD Record 34(2) (June 2005)

[17] Cezary, Janikow, Z.: Fuzzy Decision Trees: Issues and Methods. IEEE Transactions on Systems, Man, and Cybernetics 28(1), 1–14 (1998)

[18] Utgoff, P.E.: Incremental Induction of Decision Trees. Machine Learning 4(2), 161–186 (1989)

[19] Xie, Q.H.: An Efficient Approach for Mining Concept-Drifting Data Streams, Master Thesis

[20] Quinlan, J.R.: C4.5: Programs for Machine Learning. Morgan Kaufmann, San Mateo, CA (1993)

[21] Hoeffding, W.: Probability Inequalities for Sums of Bounded Random Variables. Journal of the American Statistical Association 58, 13–30 (1963)

[22] Breiman, L., Friedman, J.H., Olshen, R.A., Stone, C.J.: Classification and Regression Trees. Wadsworth, Belmont, CA (1984)

[23] Maron, O., Moore, A.: Hoeffding Races: Accelerating Model Selection Search for Classification and Function Approximation. In: Cowan, J.D., Tesauro, G., Alspector, J. (eds.) Advances in Neural Information Processing System (1994)

[24] Kelly, M.G., Hand, D.J., Adams, N.M.: The Impact of Changing Populations on Classifier Performance. In: Proc. of KDD-99, pp. 367–371 (1999)

[25] Black, M., Hickey, R.J.: Maintaining the Performance of a Learned Classifier under Concept Drift. Intelligent Data Analysis 3, 453–474 (1999)

[26] Maimon, O., Last, M.: Knowledge Discovery and Data Mining,the Info-Fuzzy Network(IFN) Methodology. Kluwer Academic Publishers, Dordrecht (2000)

[27] Fayyad, U.M., Irani, K.B.: On the Handling of Continuous-valued Attributes in Decision Tree Generation. Machine Learning 8, 87–102 (1992)

[28] Wang, T., Li, Z., Yan, Y., Chen, H.: An Efficient Classification System Based on Binary Search Trees for Data Streams Mining, ICONS (2007)

[29] Wang, T., Li, Z., Hu, X., Yan, Y., Chen, H.: A New Decision Tree Classification Method for Mining High-Speed Data Streams Based on Threaded Binary Search Trees. In: Workshop on High Performance Data Mining and Application, PAKDD (2007)

[30] Peng, Y.H., Flach, P.A.: Soft Discretization to Enhance the Continuous Decision Tree Induction. In: Proceedings of ECML/PKDD-2001 Workshop IDDM-2001, Freiburg, Germany (2001)

Hamiltonian Property on Binary Recursive Networks

Yun Sun[1], Zhoujun Li[2], and Deqiang Wang[3]

[1] School of Computer, National University of Defense Technology, China
cloud_sun76@126.com
[2] School of Computer Science and Engineering, Beihang University, China
lizj@buaa.edu.cn
[3] Institute of Nautical Science and Technology, Dalian Maritime University, China
dqwang@dlmu.edu.cn

Abstract. By means of analysis and generalization of the hypercube and its variations of the same topological properties and network parameters, a family of interconnection networks, referred to as binary recursive networks, is introduced in this paper. This kind of networks not only provides a powerful method to investigate the hypercube and its variations on the whole, but also puts forth an effective tool to explore new network structure. A constructive proof is presented to show that binary recursive networks are Hamiltonian based on their recursive structures, and thus a universal searching algorithm for Hamiltonian cycle in binary recursive networks is derived.

Keywords: interconnection network; hypercube; binary recursive networks; Hamiltonian cycle.

1 Introduction

Network topology is a crucial factor for interconnection networks since it determines the performance of networks. Many interconnection network topologies have been proposed for the purpose of connecting thousands of processing elements. We find that there are some interconnection network topologies with the following attractive topological properties and good parameters: they are n-regular graphs with 2^n vertices and $n \times 2^{n-1}$ edges, their shortest cycles are 4-length (when $n \geq 2$), and their structures are strictly recursive (every n-dimensional network is constructed by two $(n-1)$-dimensional networks). These networks include the hypercube [11] and lots of its variations (such as the crossed cube [4, 5], the Möbius cube [3], the generalized twisted cube [2], the twisted n-cube [6] and the twisted-cube connected network [14]). The family of these networks is referred to as binary recursive networks in this paper.

The topology of an interconnection network is usually represented as a graph. A ring (called a separating cycle) structure, which is a linear array with wraparound [8, 13], is widely used in interconnection networks, owing to its good properties such as low connectivity, simplicity, extensibility, as well as its feasible implementation. The embedding problem, which maps a source graph into a host graph, is an important topic of recent studies. The embedding of rings into various networks has been discussed in

F.P. Preparata and Q. Fang (Eds.): FAW 2007, LNCS 4613, pp. 228–235, 2007.

[10, 12]. A cycle which visits each vertex in a graph exactly once is called a Hamiltonian cycle. The Hamiltonian cycle is the most important of all kinds of cycles.

A graph G is Hamiltonian if there exists a Hamiltonian cycle in it. The Hamiltonian problem has long been fundamental in graph theory [7]. Hamiltonian is an important property for networks, which means that some Hamiltonian cycles exist in the networks. Hamiltonian cycles have been found in some binary recursive networks [2, 3, 4, 5, 6, 14, 15]. However, for different topologies, the methods of searching for Hamiltonian cycles are quite different and only suit their own special structures. There being no unified perspective to these variants, it is difficult to extend the results from the individual topology to the whole family. Park et al. [9] tried to use the induction principle to prove the existence of Hamiltonian cycle, but he failed to find out the uniform method for searching the Hamiltonian cycle of the family. The reason for the failure lies in the limitation of the definition they used for the family. Their definition is not algebraical, but descriptive. It cannot reflect the topology nature of the family. In this paper, however, we first redefine the family using an algebraic method, and then use this definition to prove the fact that the family of binary recursive networks is Hamiltonian graph. The proof is constructive, which is based on the structure recursive property of binary recursive networks. We can prove that all members of the family are Hamiltonian graphs. Based on the proof, we present a routing algorithm to find out a Hamiltonian cycle. This algorithm can begin with any vertex, and get a Hamiltonian cycle.

Let $G = (V, E)$ be an undirected simple graph. We adopt the fundamental graph terminology in [1, 5, 11] when using undirected graph to model interconnection networks. Given a graph G, the vertex set and the edge set of G are denoted by $V(G) = V$ and $E(G) = E$, respectively. A path $P(v_0, v_t) = \langle v_0, v_1, \cdots, v_t \rangle$ is a sequence of nodes where two consecutive nodes are adjacent. A cycle (denoted as C_v hereafter) is a path with at least three vertices where the first vertex is the same as the last vertex. For $x = x_n x_{n-1} \cdots x_{i-1} x_i x_{i+1} \cdots x_1 \in V$, denote $\text{bit}_i(x) = x_i$ and $\text{pre}_i(x) = x_n x_{n-1} \cdots x_{n-i+1}$.

The rest of this paper is organized as follows: Section 2 explains the basic definitions of the binary recursive networks. The main theorem and algorithm are proved in Section 3 and Section 4. We release our conclusion in Section 5.

2 Binary Recursive Networks

Definition 1. Let $n \geq 1$ be an integer. The graph G is n-labeled, i.e. each of its vertex is labeled by an n-bit binary string. $*^{n-i}$ is a $(n-i)$-length binary string ($* \in \{0,1\}$, $1 \leq i \leq n$), $\Gamma(*^{n-i})$ is the set of vertices whose first $n-i$ bits are $*^{n-i}$. If R_i is a one-to-one mapping from $\Gamma(*^{n-i}0)$ to $\Gamma(*^{n-i}1)$, we call it an i-dimensional binary recursive adjacent function (adjacent function for short).

Let R_i be an i-dimensional adjacent function of the n-labeled graph G, then according to the property of one-to-one mapping, we can obtain

(1) $\forall x \in \Gamma_G(*^{n-i}0)$, $\exists y \in \Gamma_G(*^{n-i}1)$ such that $y = R_i(x)$, or

(2) $\forall y \in \Gamma_G(*^{n-i}1)$, $\exists x \in \Gamma_G(*^{n-i}0)$ such that $y = R_i(x)$, i.e. $x = R_i^{-1}(y)$.

(3) $\forall x_1, x_2$, if $x_1 \neq x_2$, then $R_i(x_1) \neq R_i(x_2)$.

We denote the relationship of vertices x and y as $y = R_i^*(x)$.

Definition 2. Let G be an n-labeled graph and R_i ($1 \leq i \leq n$) an i-dimensional adjacent function of G. $\forall x, y \in V(G)$, if and only if $(x, y) \in E(G)$, there exists an integer k ($\in \{1, 2, \cdots, n\}$) such that $y = R_k^*(x)$, then we call G an n-dimensional binary recursive network determined by $R_1, R_2, \cdots, R_n$, denoted as RN_n. If $y = R_k^*(x)$ ($k \in \{1, 2, \cdots, n\}$), we call $(x, y) \in E(G)$ a kth-dimensional conjunction edge (k-conjunction edge for short), where x and y are the k-adjacent vertex of each other.

According to the definitions of the hypercube, the crossed cube, the Möbius cube, the generalized twisted cube, the twisted n-cube and the twisted-cube connected network, we obtain their adjacent function (see Table 1).

According to Definition 2, we know the networks above are all binary recursive networks.

Table 1. Adjacent functions of typical binary recursive networks

Type of networks	Adjacent functions of networks ($i = 1, 2, \cdots, n$)
The hypercube	$H_i(x) = x + \varepsilon_i$
The crossed cube	$C_i(x) = x + \varepsilon_i + \sum\limits_{k=1}^{\lceil i/2-1 \rceil} \text{bit}_{2k-1}(x)\varepsilon_{2k}$
The Möbius cube	$M_i(x) = \begin{cases} x + \varepsilon_i, & \text{bit}_{i+1}(x) = 0 \\ x + \sum\limits_{k=1}^{i} \varepsilon_k, & \text{bit}_{i+1}(x) = 1 \end{cases}$
The generalized twisted cube	$G_i(x) = \begin{cases} x + \varepsilon_i, & \mod(i, 3) = 1, 2, \\ x + \varepsilon_i + \text{bit}_{i-1}(x)\varepsilon_{i-2}, & \mod(i, 3) = 0 \end{cases}$
The twisted n-cube	$T_i(x) = \begin{cases} x + \varepsilon_1 + \varepsilon_2, & i = 2, x \in \Gamma(0^{n-2}), \\ x + \varepsilon_i, & i = 1, 3, \cdots, n \end{cases}$
The twisted-cube connected network	$N_i(x) = x + \varepsilon_i + \sum\limits_{k=1}^{\lceil i/2-1 \rceil} \text{bit}_{2k-1}(x)\varepsilon_{2k} + \sum\limits_{k=3}^{i-1} \text{bit}_k(x)\varepsilon_2$

3 Hamiltonian Cycles in RN_n

Let α and β be binary strings and $\Gamma_{RN_n}(\alpha) = (V_\alpha, E_\alpha)$ ($\Gamma(\alpha)$ for short), where

$$V_\alpha = \left\{ x \mid x \in V(RN_n) \text{ and } \text{pre}_{|\alpha|}(x) = \alpha \right\},$$

$$E_\alpha = \left\{ (x, y) \mid x, y \in V_\alpha \text{ and } (x, y) \in E(RN_n) \right\}.$$

And denote $\Gamma^{(\alpha)}(\beta) = \Gamma(\alpha\beta)$.

Theorem 1. RN_n ($n \geq 2$) are Hamiltonian graphs.

Proof. $\forall x \in \Gamma(00)$, let

$$P_x = \left\langle x, R_n^*(x), R_{n-1}^*(R_n^*(x)), R_n^*(R_{n-1}^*(R_n^*(x))) \right\rangle ,$$

where $R_n^*(R_{n-1}^*(R_n^*(x))) \in \Gamma(01)$. So P_x is a 3-length path ($3 = 2^2 - 1$) (see Fig. 1).

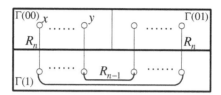

Fig. 1. ($2^2 - 1$)-length paths

R_i ($i = 1, 2, \cdots, n$) are one-to-one mappings, so if $x \neq y$, there are no same vertices in P_x and P_y for any $x, y \in \Gamma(00)$. Thus, we gain $|\Gamma(00)| = 2^{n-2}$ disjoint 3-length paths, their endvertices (the first vertex and the last vertex) are in $\Gamma(00)$ and $\Gamma(01)$ respectively.

From the presentation above, we know that the $2 \times 2^{n-2} = 2^{n-1}$ endvertices of the 2^{n-2} disjoint paths $\{P_x | x \in \Gamma(00)\}$ are all in $\Gamma(0)$. Whereas, each vertex in the $\Gamma(0)$ is a endvertex of some disjoint path because $|\Gamma(0)| = 2^{n-1}$. Furthermore, both $\Gamma^{(0)}(10)$ and $\Gamma^{(0)}(11)$ have 2^{n-3} vertices (other 2^{n-2} vertices in $\Gamma^{(0)}(0) = \Gamma(00)$). According to

$$R_{n-2}(\Gamma^{(0)}(10)) = \Gamma^{(0)}(11) ,$$

we connect the ($n-2$)-adjacent vertices between $\Gamma^{(0)}(10)$ and $\Gamma^{(0)}(11)$ (i.e. select every ($n-2$)-conjunction edge in $\Gamma^{(0)}(1)$). Then the original 2^{n-2} disjoint ($2^2 - 1$)-length paths become 2^{n-3} disjoint ($2^3 - 1$)-length paths (see Fig.2).

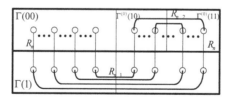

Fig. 2. ($2^3 - 1$)-length paths

We deal with the 2^{n-2} endvertices in $\Gamma(00)$ in the same way. There are 2^{n-4} vertices in $\Gamma^{(00)}(10)$ and $\Gamma^{(00)}(11)$ respectively, after connecting the ($n-3$)-adjacent vertices between $\Gamma^{(00)}(10)$ and $\Gamma^{(00)}(11)$, we gain 2^{n-4} disjoint ($2^4 - 1$)-length paths (see Fig.3).

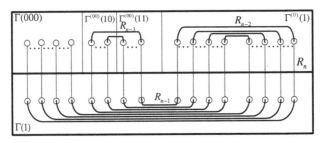

Fig. 3. ($2^4 - 1$)-length paths

Generally, there are 2^{n-k} endvertices of 2^{n-k-1} disjoint $(2^{k+1}-1)$ -length paths in $\Gamma(0^k)$. Both $\Gamma^{(0^k)}(10)$ and $\Gamma^{(0^k)}(11)$ have 2^{n-k-2} endvertices. After connecting those $(n-k-1)$ -adjacent vertices, we gain 2^{n-k-2} disjoint $(2^{k+2}-1)$ -length paths ($k=3$, $4,...,n-3$).

Finally, only $(2^2 =) 4$ endvertices are left in $\Gamma(0^{n-2})$, and there are 2 endvertices in $\Gamma^{(0^{n-2})}(0)$ and $\Gamma^{(0^{n-2})}(1)$ respectively. Here $(2^0 =) 1$ endvertex $u = 0^{n-2}10$ in $\Gamma^{(0^{n-2})}(10)$ and $(2^0 =) 1$ endvertex $v = 0^{n-2}11$ in $\Gamma^{(0^{n-2})}(11)$. According to $u = R_1(v) = v + \varepsilon_1$, we connect them, and gain $(2^0 =) 1$ disjoint $(2^n - 1)$ -length path. Its endvertices are $s = 0^{n-1}0$ and $t = 0^{n-1}1$. Obviously they are 1-adjacent vertices. After connecting s and t, the path we gained becomes a cycle which is 2^n -length and its vertices are different to one another. There are exactly 2^n vertices in RN_n, so this cycle is a Hamiltonian cycle of the RN_n.

In a word, all RN_n ($n \geq 2$) are Hamiltonian graphs.

Example 1. Finding a Hamiltonian cycle in 5-crossed cube CQ_5.

Solution: First, constructing $2^{5-2} = 8$ vertex-disjoint $(2^2 - 1 =) 3$ - length paths,

$P_{00000} = \langle 00000, 10000, 11000, 01000 \rangle$, $P_{00001} = \langle 00001, 10011, 11001, 01011 \rangle$,

$P_{00010} = \langle 00010, 10010, 11010, 01010 \rangle$, $P_{00011} = \langle 00011, 10001, 11011, 01001 \rangle$,

$P_{00100} = \langle 00100, 11100, 10100, 01100 \rangle$, $P_{00101} = \langle 00101, 11111, 10101, 01111 \rangle$,

$P_{00110} = \langle 00110, 11110, 10110, 01110 \rangle$, $P_{00111} = \langle 00111, 11101, 10111, 01101 \rangle$.

Each of these paths has one endvertex in $\Gamma_{CQ_5}(00)$. Since $R_3(\Gamma^{(0)}(10)) = \Gamma^{(0)}(11)$, connecting 3-adjacent vertices which are in $\Gamma^{(0)}(10)$ and $\Gamma^{(0)}(11)$ respectively, we gain $2^{5-3} = 4$ vertex-disjoint $(2^3 - 1 =)$ 7-length paths. Their endvertices are in $\Gamma_{CQ_5}(00)$:

$\langle 00000,10000,11000,01000,0110010100,11100,00100 \rangle$,

$\langle 00001,10011,11001,01011,01101,10111,11101,00111 \rangle$,

$\langle 00010,10010,11010,01010,01110,10110,11110,00110 \rangle$,

$\langle 00011,10001,11011,01001,01111,10101,11111,00101 \rangle$.

Similarly, connecting these 2-adjacent vertices 00100 and 00110, 00101 and 00111 (which are in $\Gamma^{(00)}(10)$ and $\Gamma^{(00)}(11)$ respectively), we gain 2 vertex-disjoint $(2^4 - 1) = 15$-length paths. Their endvertices are in $\Gamma(000)$:

$$P_0 = \langle 00000,10000,11000,01000,01100,10100,\ 11100,00100,$$
$$00110,11110,10110,\ 01110,01010,11010,10010,00010 \rangle,$$
$$P_1 = \langle 00001,10011,11001,01011,01101,10111,11101,00111,$$
$$00101,11111,10101,\ 01111,01001,11011,10001,00011 \rangle.$$

Connecting the endvertices 00000 of P_0 and 00001 of P_1, 00010 of P_0 and 00011 of P_1, we gain one vertex-disjoint $(2^5 =) 32$-length cycle -- a Hamiltonian cycle in CQ_5 (see Fig 4).

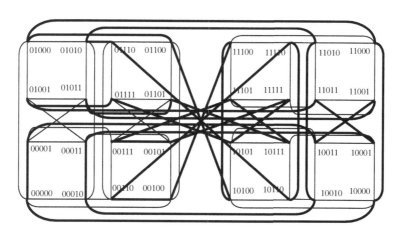

Fig. 4. Hamiltonian cycle in CQ_5

4 A Routing Algorithm for Hamiltonian Cycle in RN_n

Algorithm 1. Computing Hamiltonian cycle in RN_n .

```
Input: Integer  n,   R₁*(x),R₂*(x),⋯,Rₙ*(x) , and the beginning
  vertex  x_o .
Output: A Hamiltonian cycle  C_x^{2ⁿ} .
```

Step 1: Reason about $P^{2^{n-1}-1}R_1^*P^{2^{n-1}-1}(x)$ with $R_1^*(x),R_2^*(x),\cdots,R_n^*(x)$, where

$$P^{2^{n-i}-1} = P^{2^{n-i-1}-1}R_{i+1}^*P^{2^{n-i-1}-1} \quad (1\le i \le n-1) \ .$$

Step 2: Count $C_x^{2^n} = R_1^*(P^{2^{n-1}-1}R_1^*P^{2^{n-1}-1})(x_0)$ step by step, and record the value of every step.

Step 3: Return $C_x^{2^n}$.

Validity of this algorithm can be directly verified by Theorem 1.

Example 2. Finding a Hamiltonian cycle in a 5-crossed cube CQ_5 .

Input: $n=5$, $x_0 = 00000$ and $R_i(x)$ $(i=1,2,\cdots,n)$,

$$R_i(x) = x+\varepsilon_i + \sum_{k=1}^{\lceil i/2-1\rceil} \mathrm{bit}_{2k-1}(x)\varepsilon_{2k} \ .$$

Ste 1: $C_x^{2^5} = R_1^*(P^{2^{5-1}-1}R_1^*P^{2^{5-1}-1})(x)$

$= R_1^*(R_5^*R_4^*R_5^*R_3^*R_5^*R_4^*R_5^*R_2^*R_5^*R_4^*R_5^*R_3^*R_5^*R_4^*R_5^*R_1^*R_5^*R_4^*R_5^*R_3^*R_5^*R_4^*R_5^*R_2^*R_5^*R_4^*R_5^*R_3^*R_5^*R_4^*R_5^*)(x)$

Step 2: $R_5^*(00000) = 10000$, $R_4^*(10000) = 11000$, $R_5^*(11000) = 01000$,

$R_3^*(01000) = 01100$, $R_5^*(01100) = 10100$, $R_4^*(10100) = 11100$,

$R_5^*(11100) = 00100$, $R_2^*(00100) = 00110$, $R_5^*(00110) = 11110$,

$R_4^*(11110) = 10110$, $R_5^*(10110) = 01110$, $R_3^*(01110) = 01010$,

$R_5^*(01010) = 11010$, $R_4^*(11010) = 10010$, $R_5^*(10010) = 00010$,

$R_1^*(00010) = 00011$, $R_5^*(00011) = 10001$, $R_4^*(10001) = 11011$,

$R_5^*(11011) = 01001$, $R_3^*(01001) = 01111$, $R_5^*(01111) = 10101$,

$R_4^*(10101) = 11111$, $R_5^*(11111) = 00101$, $R_2^*(00101) = 00111$,

$R_5^*(00111) = 11101$, $R_4^*(11101) = 10111$, $R_5^*(10111) = 01101$,

$R_3^*(01101) = 01011$, $R_5^*(01011) = 11001$, $R_4^*(11001) = 10011$,

$R_5^*(10011) = 00001$, $R_1^*(00001) = 00000$.

Step 3: return $C_{00000}^{2^5}$:

$$C_{00000}^{2^5} = \langle 00000, 10000, 11000, 01000, 01100, 10100, 11100, 00100,$$
$$00110, 11110, 10110, 01110, 01010, 11010, 10010, 00010,$$
$$00011, 10001, 11011, 01001, 01111, 10101, 11111, 00101,$$
$$00111, 11101, 10111, 01101, 01011, 11001, 10011, 00001, 00000\rangle .$$

5   Conclusion

In this paper, we named a family of interconnection networks as Binary Recursive Networks. The network structures are variations of hypercube. Then we get a general algebra expression of the adjacent functions of the binary recursive networks.

Using the adjacent functions, we give a constructive proof to show that all the binary recursive networks are Hamiltonian. This proof provides a method of constructing Hamiltonian cycle in the binary recursive networks. An example is given to illustrate the method. At last, an algorithm for rapidly searching Hamiltonian cycles is given.

In the future, we are going on to study Hamiltonian-like properties (such as almost-, odd-, even-pancyclicity) and the Hamiltonian connectivity of $RN_n, (n \geq 2)$.

References

1 Bondy, J.A., Murty, U.S.R.: Graph theory with applications. The Macmillan Press Ltd, NYC (1976)
2 Chedid, F.B., Chedid, R.B.: A new variation on hypercubes with smaller diameter. Infromation Processing Letters 46, 275–280 (1993)
3 Cull, P., Larson, S.M.: The Möbius cubes. IEEE Trans. Computers 44(5), 647–659 (1995)
4 Efe, K.: A Variation on the Hypercube with Lower Diameter. IEEE Trans. On Computers 40(11), 1312–1316 (1991)
5 Efe, K.: The crossed cube architecture for parallel computation. IEEE Trans. Parallel and Distributed Systems 3(5), 513–524 (1992)
6 Esfahanian, A.H., Ni, L.M., Sagan, B.E.: The twisted n-cube with application to multiprocessing. IEEE Trans. Computers 40(1), 88–93 (1991)
7 Gould, R.J.: Updating the hamiltonian problem – a survey. J. Graph Theory 15(2), 121–157 (1991)
8 Leighton, F.T.: Introduction to Parallel Algorithms and Architectures: Array, Tree, Hypercubes. Morgan Kaufmann, San Mateo, Calif. (1992)
9 Park, C.D., Chwa, K.Y.: Hamiltonian properties on the class of hypercube-like networks. Information Processing Letters, 11–17 (2004)
10 Rowley, R.A., Bose, B.: Fault-tolerant ring embedding in de Bruijn Networks. IEEE Trans. Comput. 42(12), 1480–1486 (1993)
11 Saad, Y., Schultz, M.H.: Topological properties of hypercube. IEEE Trans. Computers 37(7), 867–872 (1988)
12 Hsieh, S.Y., Chen, G.H., Ho, C.W.: Fault-free Hamiltonian cycles in fauly arrangement graphs. IEEE Trans. Parallel and Distributed Systems 10(3), 223–237 (1999)
13 West, D.B.: Introduction to graph theory. PrenticeHall, Englewood Cliffs (1996)
14 Wang, D., Zhao, L.: The twisted-cube connected network. J. of Computer Science and Technology 14(2), 181–187 (1999)
15 Wang, D., An, T., Pan, M., et al.: Hamiltonian-like properties of k-ary n-cubes. In: The Proceeding of International Conference on Parallel and Distributed Computing, pp. 1002–1007 (2005)

A Performance Guaranteed New Algorithm for Fault-Tolerant Routing in Folded Cubes

Liu Hongmei

College of Science, Three Gorges University
Hubei Yichang 443002, PRC
liuhm@ctgu.edu.cn

Abstract. In folded cube computer network, we developed a new algorithm for fault-tolerant routing which is based on detour and backtracking techniques. Its performance is analyzed in detail in the presence of an arbitrary number of components being damaged and derived some exact expressions for the probability of routing messages via optimal paths from the source to obstructed node. The probability of routing messages via an optimal path between any two nodes is a special case of our results, and can be obtained by replacing the obstructed node with the destination node. It is also showed that in the presence of component failures this algorithm can route messages via an optimal path to its destination with a very high probability.

1 Introduction

Data distribution among processors and internode message routing are central issues in the implementation of a parallel algorithm on a multiprocessor computing system([1][2]).The the n-dimensional hypercube Q_n is a well-known interconnection topology, and it meets several computational demands of a multiprocessors of computing system. However among a few drawbacks of this architecture, the connectivity and diameter are not small when its dimension is big enough and the complete binary tree B_n on $2^n - 1$ vertices can not be embedded into the hypercube Q_n with dilation one([3]). Hence, there are many improved structures such as $m-$ary $n-$cube $Q_n(m)$ and the folded cube FQ_n. Recent years $Q_n(m)$ multi-computer networks have received considerable attention mainly focusing on its features, routing algorithm ,fault-tolerant and transmission delay ([4] [5] [6] [7]). Folded cube FQ_n is a popular topological structure of multi-computers network. To make FQ_n multi-computer useful for reliable critical applications, significant research efforts have been made on it. For example, [8] investigated the properties and performance of FQ_n in detail, paper [9] discussed the embedding binary trees into folded cube and [10] found that there are hamilton cycles in FQ_n with faulty links. Meanwhile [11] made a good job by constructing method and proposed that there exists an extendible embedding of complete binary tree B_n into FQ_n for every even $n \geq 2$. The hardware cost in designing $Q_n(m)$ and FQ_n are greater when compared to hypercube Q_n. However, the overhead is negligible when n is large. These networks achieve considerable improvement in

F.P. Preparata and Q. Fang (Eds.): FAW 2007, LNCS 4613, pp. 236–243, 2007.
© Springer-Verlag Berlin Heidelberg 2007

the running time of message routing, specially in FQ_n because of its smaller diameter and smaller internode distance. Owing to their higher node degree, these networks also have higher connectivity, so have better fault tolerance and better diagnostic capabilities compared to hypercubes.

This paper will make use of probabilistic idea and detour and backtracking techniques to investigate the fault tolerant routing based on the depth-first search approach in FQ_n multi-computer. Similar discussion has been made for hypercube Q_n ([12]).

A connected folded cube with fault components is called an injured folded cube. In order to enable non-fault nodes (called normal nodes) in an injured folded cube to communicate with each other, enough network information must be either kept at each node or added to the message to be routed. The first approach requires each node to keep a certain amount of global information for its routing decisions. This however necessities the propagation of updated network information when the network condition (fault or not) of its adjacent links or nodes changed. For the second approach, each node is required to know only the condition (faulty or not)of its adjacent components. In this paper, we develop a new fault-tolerant routing algorithm scheme for FQ_n multi-computers, in which each message is accompanied with a stack which keeps track of the history of the path travelled, and tries to avoid visiting a node more than once unless a backtracking is enforced.

We shall first develop the routing scheme, in which every node is only required to know the status of its adjacent components. The path that a message has traversed is kept track of by the message as it is routed toward its destination. Then the performance of this routing algorithm is analyzed rigorously.

An optimal path between a pair of nodes is a path of length equal to the distance between these two nodes. Under the proposed routing scheme, the first node in the message's route that is aware of the nonexistence of an optimal path from itself to the destination is called obstructed node. At the obstructed node, the message has to take a detour, that is, non-optimal path will be chosen. In this paper, we derive exact expressions for the probabilities of optimal path routing from the source node to a given obstructed node in the presence of both link and node failures. Note that determination of the probability for optimal path routing between any two nodes can be viewed as an obstructed node that is 0 hop away from the destination node.

This paper is organized as follows: necessary notations and definitions are introduced in Section 2. A fault-tolerant routing algorithm for folded cube FQ_n multi-computers is presented in Section 3. The performance of this routing scheme is analyzed rigorously in Section 4. The paper concludes with Section 5.

2   Preliminaries

Suppose $G(V, E)$ is a simple graph. The set of vertices and the set of edges of G are denoted by $V(G) = V$, and $E(G) = E$, respectively. If $xy \in E$, then x and y are neighbor vertices. G is $k-$regular if the degrees of all vertices is k. computer network topology usually can be represented by a graph $G(V, E)$, where vertices (or nodes) represent processors and edges represent links between processors.

The n-dimensional hypercube Q_n has vertex set consisting of all the n-bit, that is, $V(Q_n) = \{x_1 x_2 \ldots x_n : x_i = 0, 1\}$, with two vertices x and y adjacent iff they differ in exactly one bit. Folded cube is defined as graph FQ_n which vertex set is $V(FQ_n) = V(Q_n)$ and edge set is $E(FQ_n) = E(Q_n) \bigcup \{x\overline{x} : x \in V(FQ_n)\}$, where we defined $x = x_1 x_2 \ldots x_n$, $\overline{x} = \overline{x}_1 \overline{x}_2 \ldots \overline{x}_n$ and $\overline{x}_i = 1 - x_i$. We use $x = x_1 x_2 x_3 \ldots x_n$ to express vertex x, the leftmost coordinate of the address will be referred to as dimension-1, and the second to the leftmost coordinate as dimension-2, and so on.

Suppose $x = x_1 x_2 x_3 \ldots x_n$, $x_i \in \{0, 1\}$ and $y = y_1 y_2 y_3 \ldots y_n$, $y_i \in \{0, 1\}$ be two nodes of FQ_n, xy is an edge of dimension$-i$ if $x_1 = y_1$, $x_2 = y_2$, $\ldots$, $x_{i-1} = y_{i-1}$, $x_i \neq y_i$, $x_{i+1} = y_{i+1}$, $\ldots$, $x_n = y_n$. xy is an edge of opposite-dimension if $y_i = \overline{x}_i$, for all $i \in \{1, 2, \ldots, n\}$. From the definition, FQ_n contains 2^n vertices and $2^{n-1}(n+1)$ edges, and FQ_n is a $(n+1)$-regular graph with diameter $\lceil \frac{n}{2} \rceil$. If $n = 1$, it is a complete graph K_2. If $n = 2$, it is a complete graph K_4.

The Hamming distance between two nodes $x = x_1 x_2 x_3 \ldots x_n$ and $y = y_1 y_2 y_3 \ldots y_n$ in FQ_n is defined as

$$H(x, y) = \sum_{i=1}^{n} |x_i - y_i|$$

Let $d(x, y)$ denote the distance between vertices x and y. Then with the definition of Hamming distance, we have the following lemma.

Lemma 1. If $H(x, y) \leq \lceil \frac{n}{2} \rceil$, then $d(x, y) = H(x, y)$; If $H(x, y) > \lceil \frac{n}{2} \rceil$, then $d(x, y) = n - H(x, y) + 1$.

For the sake of convenience, we now introduce two concepts.

A path in FQ_n can be represented by a sequence of n-bit strings such that every two consecutive n-bit strings differ by exactly one bit or differ at all bits. An optimal path is a path whose length is equal to the distance between the source and destination. We call the routing via an optimal path the optimal routing. A link of node x is said to be toward another node y if the link belongs to one of the optimal path from x to y and call y the forward node of x.

Suppose $x = x_1 x_2 \ldots x_n$ and $y = y_1 y_2 \ldots y_n$ be two vertices, and $H(x, y) = k$. We define $diff(x, y) = \{d_1, d_2, \ldots, d_k\}$ be the dimension set in which x and y differ and let $d_1 < d_2 < \ldots < d_k$ and define $same(x, y) = \{s_1, s_2, \ldots, s_{n-k}\}$ the dimension set in which x and y have the same value and $s_1 < s_2 < \ldots < s_{n-k}$, that is, $x_{d_j} \neq y_{d_j}$ for $j \in \{1, 2, \ldots, k\}$ and $x_{s_i} = y_{s_i}$ for $i \in \{1, 2, \ldots, n-k\}$. For instance, in FQ_6, if $x = 100100$ and $y = 111001$, then $d_1 = 2, d_2 = 3, d_3 = 4, d_4 = 6$ and $s_1 = 1, s_2 = 5$. A given path of length k between x and y in FQ_n can be described by a coordinate sequence $C = [c_1, c_2, \ldots, c_k]$ where $1 \leq c_i \leq n$, the coordinate sequence is a sequence of ordered dimensions. A coordinate sequence is said to be simple if any dimension does not occur more than once in that sequence. It is easy to see that a path is optimal only if its coordinate sequence is simple. For example, $[000001, 001001, 001101, 101101]$ is an optimal path from 000001 to 101101, and can be represented by a coordinate sequence $[3, 4, 1]$. If $H(x, y) \leq \lceil \frac{n}{2} \rceil$, then the optimal path from x to y can be expressed the coordinate sequence $C = [d_{j_1}, d_{j_2}, \ldots, d_{j_k}]$, where $j_1, j_2, \ldots, j_k$ is a permutation of $1, 2, \ldots, k$. Otherwise, we will transmit message from x to $\bar{x}$ and then by the coordinate sequence $C = [s_{j_1}, s_{j_2}, \ldots, s_{j_{n-k}}]$ message can be transmitted from $\bar{x}$ to y, which is an optimal path between x and y.

Definition 1. *The exclusive operation between binary strings $x = x_1 x_2 \ldots x_n$ and $y = y_1 y_2 \ldots y_n$, denoted by $x \oplus y = r_1 r_2 \ldots r_n$, is defined as $r_i = 0$ if $x_i = y_i$ and $r_i = 1$ if $x_i = \bar{y}_i$*

Let $e^i = e_1 e_2 \ldots e_i \ldots e_n$ where $e_i = 1$ and $e_j = 0$ for all $j \neq i$. For example, $000001 \oplus e^3 = 001001$

Definition 2. *The number of inversions of a simple coordinate sequence $C = [c_1, c_2, \ldots, c_k]$ denoted by $V(C)$, is the number of pairs (c_i, c_j) such that $1 \leq i < j \leq k$ but $c_i > c_j$. For example $V([3, 4, 1]) = 2$.*

3 Routing Algorithm

An adaptive routing algorithm is proposed here, which requires every node to know only the condition (fault or not)of its own links. The case that a node is faulty is treated as that all neighbors of the node are faulty. This algorithm can successfully route messages between any pair of connected non-fault nodes.

The algorithm will attempt to avoid visiting the same node more than once except when backtracking is forced. Thus, those nodes that message traversed so far are recorded in a set TD and will be delivered together with the message to the next node. When the source node begins routing a message, TD is set to be empty set ϕ. Therefore, information to be phased on to the next node can be represented as (message, TD). The source node is denoted by s, destination node by d, $H(s, d) = k$, and the current node by u. A message reaches its destination when $u = d$.

When a node has received a message, it will check the value of u to see if the destination is reached. If not, the intermediate node will try to send the message along an optimal path to the destination. however, if all the optimal paths are blocked by fault components and those nodes visited before, the node will route the message via an alternative path, which is called detour. When there is no alternative path available, backtracking is enforced, that is, the message is returned to the original node.

Algorithm A: Fault-tolerant Routing Algorithm
Step 1. If $diff(u, d) = \phi$, the message is reached destination, Stop.
    Set $u := s$
Step 2. Compute $H(u, d)$, Set $k = H(u, d)$
Step 3. If $k \leq \lceil \frac{n}{2} \rceil$,

Step 4. For $j := 1, k$ do,

    if $d_j \in diff(u,d)$, $u \oplus e^{d_j}$ and $u(u \oplus e^{d_j})$ are normal and $u \oplus e^{d_j} \notin TD$,

    send (message, TD) to $u \oplus e^{d_j}$, $TD = TD \cup \{u \oplus e^{d_j}\}$, $u = u \oplus e^{d_j}$

Step 5. If $k > \lceil \frac{n}{2} \rceil$, then send message to $\bar{s}$, set $u := \bar{s}$

step 6. For $j := 1, n - k$ do,

    if $s_j \in same(u,d)$, $u \oplus e^{s_j}$ and $u(u \oplus e^{s_j})$ are normal, and $u \oplus e^{s_j} \notin TD$,

    send (message, TD) to $u \oplus e^{s_j}$, $TD = TD \cup \{u \oplus e^{s_j}\}$, $u = u \oplus e^{s_j}$.

/*If the algorithm is not terminated yet, all optimal paths to destination are blocked by fault components and nodes traversed before*/

Step 7. For $j := 1, n - k$ do,

    if $s_j \in same(u,d)$, $u \oplus e^{s_j}$ and $u(u \oplus e^{s_j})$ are normal, and $u \oplus e^{s_j} \notin TD$,

    send (message, TD) to $u \oplus e^{s_j}$, $TD = TD \cup \{u \oplus e^{s_j}\}$, $u = u \oplus e^{s_j}$. Go to Step 2.

/*A detour is taken*/

Step 8. If the algorithm is not terminated yet, then Backtracking is taken, the message must be returned to the node from which this message was originally received.

To route a message to its destination in folded cube FQ_n networks allowing some fault components, those nodes traversed before by the message must be made known to the intermediate nodes so as to avoid message looping. This is the very reason that under A every intermediate nodes has to append to the message a tag TD.

4 Performance Analysis of Routing Algorithm

We consider a folded cube FQ_n network with a given number of fault nodes and fault links, and all possible distributions of fault components are assumed to be evenly.

For convenience, we introduce a definition of weight of a set of dimensions to be $W(d_{j_1}, d_{j_2}, \ldots, d_{j_t}) = \sum_{i=1}^{t} j_i$.

Theorem 1. *Suppose x and y are respectively the source and destination nodes in a folded cube FQ_n network, and $H(x,y) = \lceil \frac{n}{2} \rceil$. Then the number of fault components required for the simple coordinate sequence $C = [d_{j_1}, d_{j_2}, \ldots, d_{j_t}]$ to be the path chosen by algorithm A to an obstructed node located j hops away from y is $V(C) + W(d_{j_1}, d_{j_2}, \ldots, d_{j_t}) - \sum_{i=1}^{t} i + j$, where $t = \lceil \frac{n}{2} \rceil - j$.*

Proof. Due to the symmetric structure of FQ_n, without loss of generality, we suppose that the source $x = x_1 x_2 x_3 \ldots x_n$, $y = \bar{x}_1 \bar{x}_2 \ldots \bar{x}_{\lceil \frac{n}{2} \rceil} x_{\lceil \frac{n}{2} \rceil + 1} \ldots x_n$. With algorithm A, the lowest dimension among those dimensions not traversed before will be chosen first. Since $H(x,y) = \lceil \frac{n}{2} \rceil$, the selection of dimension d_{j_1} to the first hop implies that $j_1 - 1$ fault components have been encountered. Also, the selection of dimension j_2 to the second hop means there are another $j_2 - 1$ fault components encountered if $d_{j_2} < d_{j_1}$, or there are another $j_2 - 2$ fault components encountered if $d_{j_2} > d_{j_1}$. Similarly, we know that up to the obstructed node (i.e., the first t hops), the message must have encountered $\sum_{i=1}^{t}(j_i - 1) - V(C^R)$. But $V(C) + V(C^R) = t(t-1)/2$. Since $C = [d_{j_1}, d_{j_2}, \ldots, d_{j_t}]$ is a simple coordinate sequence, either $d_{j_i} < d_{j_r}$ or $d_{j_i} > d_{j_r}$, every pair of (d_{j_i}, d_{j_r}) must be counted in $V(C)$ or $V(C^R)$. However, there are $C_t^2 = t(t-1)/2$ different ways to choose pairs $((d_{j_i}, d_{j_r}))$ from $(d_{j_1}, d_{j_2}, \ldots, d_{j_t})$. Therefore $\sum_{i=1}^{t}(j_i - 1) - V(C^R) = \sum_{i=1}^{t} j_i - t - t(t-1)/2 + V(C) = V(C) + \sum_{i=1}^{t} j_i - t(t+1)/2$ fault components. This completes the proof.

Denote the set of combinations of r different numbers out of the set $\{1, 2, \ldots, n\}$ by $S(n, r)$, clearly $|S(n, r)| = C_n^r$ is the number of combinations of r objects out of n different objects. Let $I_n(r)$ denote the number of permutations of n numbers with exactly r inversions. we discuss $I_n(r)$.

Let $i_1 i_2 \ldots i_n$ be a permutation of the set $\{1, 2, 3, \ldots, n\}$. The pair (i_k, i_l) is called an inversion if $k < l$ and $i_k > i_l$. For example, the permutation 31542 has five inversions, namely (3,1), (3,2), (5,4), (5,2), (4,2). For a permutation $i_1 i_2 \ldots i_n$, we let a_j denote the number of inversions in permutation which precede j but are greater than j. The sequence of numbers $a_1, a_2, \ldots, a_n$ is called the inversion sequence of the permutation $i_1 i_2 \ldots i_n$. [13] affords the following theorem.

Theorem 2. *Let $b_1, b_2, \ldots, b_n$ be a set of sequence of integers with $0 \leq b_1 \leq n - 1$, $0 \leq b_2 \leq n - 2$, $\ldots$, $0 \leq b_{n-1} \leq 1$, $b_n = 0$. Then there exists a unique permutation of $\{1, 2, \ldots, n\}$ whose inversion sequence is $b_1, b_2, \ldots, b_n$.*

By Theorem 2, we know that $I_n(t)$ equals the number of non-negative integer solutions of the equation:

$$b_1 z_1 + b_2 z_2 + \ldots + b_n z_n = t, \quad 0 \leq z_i \leq 1, 1 \leq i \leq n. \tag{1}$$

This equation can be solved by using generation function.

Theorem 3. *Suppose there are f fault links in a folded cube FQ_n, and a message is routed by A from node x to y where $H(x, y) = \lceil \frac{n}{2} \rceil$. Let h_L be the Hamming distance between obstructed node and the destination node. Then*

$$P(h_L = j) = \frac{1}{C_L^f} \sum_{\sigma \in S(\lceil \frac{n}{2} \rceil, t)} \sum_{i=0}^{\min\{\frac{\lceil \frac{n}{2} \rceil (\lceil \frac{n}{2} \rceil - 1)}{2}, f-j\}} I_t(\alpha) C_{L - \lceil \frac{n}{2} \rceil - i}^{f - j - i}$$

where $\alpha = i - W(\sigma) + \frac{t(t+1)}{2}$ and $P(A)$ is the probability of event A, $L = (n+1)2^{n-1}$ and $t + j = \lceil \frac{n}{2} \rceil$.

Proof. There are C_L^f different configurations of fault links. The problem of obtaining $P(h_L = j)$ is then reduced to that of counting the number of configurations which lead to the case of $h_L = j$. Since the message traverses $t = \lceil \frac{n}{2} \rceil - j$

hops before it reaches the obstructed node, there are $|S(\lceil\frac{n}{2}\rceil, t)| = C^t_{\lceil\frac{n}{2}\rceil}$ possible locations of the obstructed node. No loss of generality, let $x = 00\ldots0$ be source and $y = \underbrace{11\ldots1}_{\lceil n/2 \rceil}00\ldots0$ destination. Then there is a one-to one correspondence between each element in $S(\lceil\frac{n}{2}\rceil, t)$ and each possible location of the obstructed node.

Consider an obstructed node location u which is determined by an element $\sigma \in S(\lceil\frac{n}{2}\rceil, t)$. Let C be the coordinate sequence from x to u. From Theorem 1, we know that the message has encountered $V(C) + W(\sigma) + t(t+1)/2$ fault links before reaching u. Thus, the number of different paths from x to u while traversing the dimensions in σ and encountering i fault links can be expressed as $I_t(i - W(\sigma) + t(t+1)/2)$. For each given coordinate sequence to u, the location of these k fault links encountered before reaching u are determined. Moreover, there are additional j fault links adjacent to u. also, note that t links in the path from x to u are non-fault. Therefore, the number of different configurations for a given coordinate sequence or path to a certain obstructed node location u is $C^{f-j-i}_{L-j-i-t} = C^{f-j-i}_{L-\lceil\frac{n}{2}\rceil-i}$. Thus, this theorem finishes.

The probability of an optimal path routing can be viewed as a special case of Theorem 3 by setting the obstructed node to the destination node, namely, $P(h_L = 0)$. So we have the following corollary.

Corollary 1. *The probability for a message to be routed in an FQ_n with f fault links via an optimal path to a destination node which is n hops away can be expressed as*

$$P(h_L = 0) = \beta \sum_{i=0}^{\min\{\frac{\lceil\frac{n}{2}\rceil(\lceil\frac{n}{2}\rceil-1)}{2}, f\}} I_{\lceil\frac{n}{2}\rceil}(i) C^{f-i}_{(n+1)2^{n-1}}$$

where $\beta = \dfrac{1}{C^f_{(n+1)2^{n-1}}}$.

Note that if $j = 0$, then $t = \lceil\frac{n}{2}\rceil$. $S(\lceil\frac{n}{2}\rceil, t) = S(\lceil\frac{n}{2}\rceil, \lceil\frac{n}{2}\rceil)$ is $(1, 2, \ldots, \lceil\frac{n}{2}\rceil)$ and $W(1, 2, \ldots, \lceil\frac{n}{2}\rceil) = \lceil\frac{n}{2}\rceil(\lceil\frac{n}{2}\rceil + 1)/2$. The Corollary is true.

5   Conclusion

Computer network safety is an important issue in the research of computer science. When there are many components suffering damage, transmitting data safely and efficiently is an essential work. Folded cube FQ_n is a popular network topology for parallel processing computer systems. This paper has proposed a new fault-tolerant routing algorithm for FQ_n networks. Its performance has been analyzed in the presence of arbitrary number of damaged components. Some exact expressions for the probability of routing messages via optimal paths from the source node to obstructed node have been derived. The existing result of the probability of routing messages via an optimal path between any two nodes is a special case of our results, and can be obtained by replacing the obstructed node with the destination node.

Acknowledgment. This project is supported by NSFC (10671081) and The Science Foundation of Hubei Province(2006AA412C27).

References

1. Leighton, F.T.: Introduction to parallel algorithms and architectures: Arrays, tree, hypercubes. Morgan Kaufmann, San Mateo,CA (1992)
2. Quinn, M.J.: Parallel computing: Theory and practice. McGraw-Hill, Singapore (1994)
3. Junming, X.: Analysis of Topological Structure of Interconnection Networks. Kluwer Academic publishers, Dordrecht/Boston/London (2001)
4. Liu, H.: The restricted connectivity and restricted fault diameter in m-ary n-cube systems. In: The 3rd international conference on impulsive dynamic systems and applications, vol. 4, pp. 1368–1371 (July 21-23, 2006)
5. Hongmei, L.: Topological properties for m-ary n-cube. Journal of Wuhan University of Technology Transportation Science and Engineering 30, 340–343 (2006)
6. Hongmei, L.: The routing algorithm for generalized hypercube. Mathematics in Practice and Theory 36, 258–261 (2006)
7. Junming, X.: Fault tolerance and transmission delay of generalized hypercube networks. Journal of China University of Science and Technology 31, 16–20 (2001)
8. EI-Amawy, A., Latifi, S.: Performance of folded hypercubes. IEEE Transact Parallel distributed Syst. 2, 31–42 (1991)
9. Wang, D.: On embedding binary trees into folded hypercubes. Congressus Numeranbtium 134, 89–97 (1998)
10. Wang, D.: Embedding Hamilton cycles into folded hypercubes with faulty links. J. Parallel Distributed Comput. 61, 545–564 (2001)
11. Choudum, S.A., Usha nandini, R.: Complete Binary Trees in Folded and Enhanced Cube. NETWORKS 43, 266–272 (2004)
12. Chen, M.-S., Shin, K.G.: Depth-first search approach for fault-tolerant routing in hypercube multicomputers. IEEE Transactions on parallel and distributed systems 1, 152–129 (1990)
13. Brualdi, R.A.: Introductory Combinatorics. Elsevier Science Publishing Co., Amsterdam (1977)

Pathologic Region Detection Algorithm for Prostate Ultrasonic Image Based on PCNN

Beidou Zhang[1], Yide Ma[1], Dongmei Lin[1], and Liwen Zhang[2]

[1] School of Information Science & Engineering, Lanzhou University,
730000 Lanzhou, China
[2] B-Ultrasonic room of People's Hospital of Gansu Province,
730000 Lanzhou, China

Abstract. It is quite important and difficult for doctors to detect pathologic regions of prostate ultrasonic images. An automated region detection algorithm is proposed to solve this problem, especially for ultrasonic images containing all kinds of noise and speckle. First, all the pixels of an ultrasonic image are fired by Pulse Coupled Neural Network (PCNN). Then after being processed by morphological closing, binary reversing and region labeling, the seeds are detected automatically using PCNN, by which the region of interest (ROI) of the ultrasonic image is detected by Region Growing. In the end, we code the ROI by pseudo-color. Detected pathologic regions can be used for further clinical inspection and quantitative analysis of ultrasonic images.

Keywords: Prostate Ultrasonic Image, Pulse Coupled Neural Network, Image Segmentation, Pseudo-color.

1 Introduction

In the field of modern clinical diagnosis, medical imaging technologies, such as US, CT, MRI, PET, have been playing an important role in detecting and treating of numerous diseases. The radiologists present 2D or 3D images, giving patients a detailed view of their anatomies. Because of the diverse physiological properties, tissues would display kinds of medical images by different medical imaging equipments. Appropriate equipments should be selected to detect different tissues because single medical imaging equipment is not suitable for all kinds of disease diagnosis. Ultrasonic imaging is a common modality in current medical practice. It is used to image soft tissues, such as lungs, prostate, liver, spleen, thyroid or the neonatal brain. The advantages of ultrasound imaging are its rapid speed, high security, cost effectiveness and portability of the equipment, which make it more suitable than CT or MRI in many situations [1].

In medical imaging, ultrasonic image analyzing remains a difficult task. And for the same image, the opinions of different doctors are not consistent [2]. Along with the improvement of image acquisitions, more and more image data are obtained from various imaging modalities. Especially for video stream, doctors need to process large numbers of data every day, and the manual or semiautomatic processing technologies can not satisfy

F.P. Preparata and Q. Fang (Eds.): FAW 2007, LNCS 4613, pp. 244–251, 2007.

their requirements. In this instance, Computer-Aided Diagnosis (CAD) technology provides doctors with automated and impersonal processing methods. However, most medical images have lots of shortcomings, such as complexity, variability or blur, requiring users to operate the CAD system manually sometimes [3~5].

PCNN is a new artificial neural network which comes from the research of small mammals' visual properties [6]. It has an excellent ability for segmentation because of its synchronous pulse burst, changeable threshold and controllable parameters. Combining PCNN with mathematical morphology, we propose an automated pathologic region detection algorithm.

2 PCNN and Mathematical Morphology

In 1990, Eckhorn proposed the model of Pulse-Coupled Neural Network after researching the synchronous pulse burst phenomenon of the cat visual cortex [7]. PCNN has predominance in image processing, image recognition, moving object recognition and so forth [8].

Fig.1 shows a single neural model of PCNN. It is composed of three elements: Dendritic Tree, Linking Modulation and Pulse Generator Element. The Dendritic tree includes two parts of the neuron element, the linking and the feeding. The linking region incorporates neighborhood information, namely other neurons' outputs, with the internal activity of the neuron element. The feeding region incorporates the input signal information and also neighbor information. Dendritic Tree receives the inputs from other neurons, and then transmits them through two channels, one is F Channel and the other is L Channel. L_j is added to a constant positive bias and then multiplied by F_j which comes from F Channel. Pulse Generator Element is composed of a pulse generator and a comparator, whose threshold is changeable. It compares the internal activity with a dynamic threshold to decide whether the neuron fires or not. When the threshold θ_j is greater than U_j, the pulse generator is turned off and the pulses are stopped to put out. Otherwise, the pulse generator is opened, the neuron fires, and a

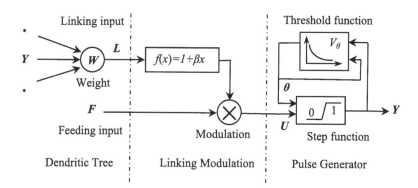

Fig. 1. A neuron model of PCNN

pulse or pulse sequence is emitted. The whole Pulse-Coupled Neural Networks work as follows: If the neuron has a pulse to put out, the changeable threshold would increase abruptly. Therefore, the second firing is forbidden, and the threshold decays exponentially. After that, θ_j would be less than U_j again, the neuron is activated secondly. Distinctly, these pulses are input to other neurons to affect their firing states. Every neuron carries their iterative computations as follows:

① The Feeding and Linking compartments receive inputs from stimulus and previous states, and communicate with neighboring neurons through the synaptic weights M and W respectively. The values of these two compartments are determined by,

$$F_{ij}[n]=\exp(-\alpha_F)F_{ij}[n-1]+V_F\sum m_{ijkl}Y_{kl}[n-1]+S_{ij} \tag{1}$$

$$L_{ij}[n]=\exp(-\alpha_L)L_{ij}[n-1]+V_L\sum w_{ijkl}Y_{kl}[n-1] \tag{2}$$

② The states of these two compartments are combined in a second order fashion to create the internal state of the neuron, U. The combination is controlled by the linking strength, β. The internal activity is calculated by,

$$U_{ij}[n]=F_{ij}[n]\left(1+\beta L_{ij}[n]\right) \tag{3}$$

③ In pulse generate compartment, U is compared with dynamic threshold, θ, to produce the output, Y. If U_{ij} is greater than θ_{ij}, the Pulse Generator would output 1, otherwise 0, by

$$Y_{ij}[n]=1 \quad if \ U_{ij}[n]>\theta_{ij}[n] \quad or \ 0 \ otherwise \tag{4}$$

④ When the neuron fires, the dynamic threshold, θ, increases its value abruptly, and then decays until the neuron fires again by,

$$\theta_{ij}[n]=\exp(-\alpha_\theta)\theta_{ij}[n-1]+V_\theta Y_{ij}[n-1] \tag{5}$$

When PCNN is used for image processing, it is a monolayer two-dimensional array of laterally linked neurons. The number of neurons in the network is equal to the number of pixels in the input image. One-to-one correspondence exists between image pixels and neurons. Each pixel is connected to a unique neuron and each neuron is connected with the surrounding neurons. The intensities of pixels are put into neurons correspondingly, and the neuron firing equals to the pixel firing.

Mathematical Morphology has been enriched and developed continuously since it was put forward for the first time by G. Matheron and J. Serra in 1964. This subject is based on strict mathematical theories such as integral geometry, set algebra and topology theory, and refers to modern probability theory, neo-mathematics, etc. Though its theories foundation is very complex, the basic principle is relatively simple. When Mathematical Morphology is applied to image processing, the pixels are regarded as the sets of points, a structural element is used to do the operations as follows: intersection, union, and shift. Accordingly, other mathematical morphology processing algorithms come into being in terms of the basal set operations [9].

In practical image processing applications, dilation and erosion are in common use and can make up of opening, closing, hitting, thinning or thicking in various combinations.

PCNN is the research result of biology, so it has the biological background and has an equivalence relation with mathematical morphology in image processing. A neuron's firing in some time would bring on the neurons besides it firing and these neurons would also bring on their surrounding neurons' firing. The pulses would spread like automatic waves. Inverse the fired regions, we will get the shrunk regions which are caused by spreading pulse. So, the parallel pulse spreading is equal to the dilation operation in mathematical morphology. Using this property, we can construct other mathematical morphology operators.

3   Ultrasonic Image Processing Using PCNN

Ultrasonic images are usually characterized by complexity, low contrast and all kinds of speckle, making ultrasonic image processing very difficult. PCNN was introduced into the field of ultrasonic image processing, and edge detection algorithm and image enhancement algorithm were proposed by us for the first time.

We use PCNN to segment the original ultrasonic image and one area of the image is obtained after each iteration. Detect this area's edge, judge whether the current pixel is an edge pixel and mark it in another matrix. Also mark the current area, and place it to another matrix. After several segmentation and region detection, all pixels are detected. Combine the results, and we can get the image's edges. PCNN model is the simulation of visual behavior, and its output reflects some human visual properties. All of the parameters of PCNN, threshold θ_{ij} is one of the most active parameters. It determines neurons' firing time. After neurons' firing, θ_{ij} increases to a great value, and then decays exponentially. When U_{ij} is greater than θ_{ij} again, the neuron fires once more. Now θ_{ij} is greater than the corresponding pixel's gray level, and has been stretched exponentially. Here, we set Y as equation (6):

$$Y_{ij}[n] = \theta_{ij} \quad if \ U_{ij}[n] \geq \theta_{ij}[n] \quad or \ 0 \ otherwise \qquad (6)$$

Superpose the firing images of different time, and we get the primary enhanced image. Fig. 2 and Fig. 3 depict the processing outcome of edge detection algorithm and image enhancement algorithm using PCNN.

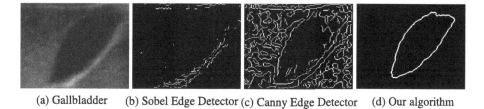

(a) Gallbladder   (b) Sobel Edge Detector (c) Canny Edge Detector   (d) Our algorithm

Fig. 2. Gallbladder ultrasonic image edge detection

4   Pathologic Region Detection Algorithm

Aiming at the more and more medical data and noise of ultrasonic images, we propose a pathologic region automatic detection algorithm based on PCNN. Prostate ultrasonic images are selected as experiment images, which contain small sick regions

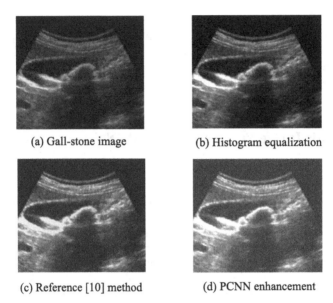

(a) Gall-stone image

(b) Histogram equalization

(c) Reference [10] method

(d) PCNN enhancement

Fig. 3. Gall-stone ultrasonic image enhancement

depicted in Fig. 4. The concrete algorithm is depicted as follows, in which the pre-processing part contains taking out irrespective regions, mean filtering and smoothing.

(1) Acquire the original ultrasonic image I, and pre-process it;

(2) Segment I by simplified PCNN:

    a. Initial PCNN's parameters:

$\beta=0.4$, $\alpha_\theta=0.3$, $V_\theta=240$,

$$W_1 = \begin{bmatrix} 0.5 & 1 & 0.5 \\ 1 & 0 & 1 \\ 0.5 & 1 & 0.5 \end{bmatrix},$$

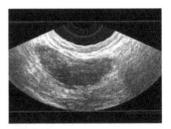

Fig. 4. Prostate ultrasonic image

define threshold matrix T, storing matrix Y for fired pixels, storing matrix Y_1 for pixels fired the first time, edge storing matrix E, fired regions storing matrix Z;

    b. $T(i,j)=V_\theta$, n=1;

    c. Scan the image matrix I, and calculate them one by one, $F(i,j)=S(i,j)$, L=W1*Y[n-1], U=F(1+β*L) ;

If U(i,j)> T(i,j) Y(i,j)=1, isolate the (i,j) pixel so as to ensure it will never fire again;

else Y(i,j)=0 ;

d. Detect the edge of the region fired this time, and note it in E; label the region, note it in Z;

e. All the neurons unfired are decayed by multiplying exp(-α_θ) ;

f. n=n+1, if all neurons are fired, end calculation, output E and Z, else go back to step c.

(3) Do closing operation for Y_1;

(4) Inverse the outcome of step (3), and label it;

(5) Take the pixels of small region as seed pixels, use region growing to extract ROI;

(6) Code the ROI using pseudo-color to enhance it;

(7) Integrate the outcomes, and output them.

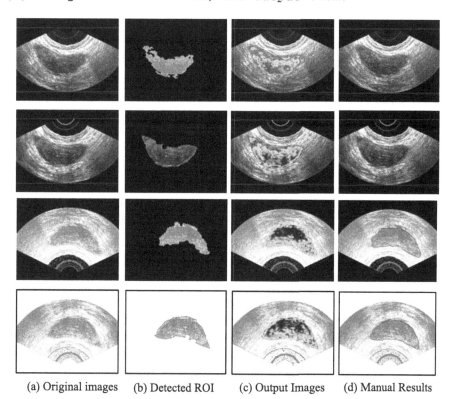

(a) Original images   (b) Detected ROI   (c) Output Images   (d) Manual Results

Fig. 5. Detection results by the automated region detection algorithm

We test the algorithm under the platform of MATLAB 7.0. Fig.5(b) are extracted outcomes detected by our algorithm. From Fig.5, we can see that the extracted ROI have good region integrality, uniformity and connectivity, which make the segmentation much more robust and exact. Using pseudo-color to enhance the ROI, we get the result shown in Fig.5(c), and Fig.5(d) are manual results. Hereto, the image details have been enhanced according to their intensities, by which quantitive analysis can be done further. The disadvantage of Region Growing is that the operator needs to select seed pixels manually. In order to extract useful regions, operators must select a seed among them. In this paper, the algorithm of seed pixels automatic extraction based on PCNN successfully solves these problem of selecting seed pixels though ultrasonic images are badly degraded.

5 Conclusion

The disadvantage of Region Growing is that operator needs to select seed pixels manually. In order to extract useful regions, operators must select a seed among them. In this paper, an algorithm of seed pixels automatic extraction based on PCNN is proposed, which solves the problem of selecting seed pixels though ultrasonic images are badly degraded due to much noise. This is the successful application of PCNN for ultrasonic image detection. Medical images are characterized by complexity and variety, so the automatic detection algorithms for all kinds of medical images need to be researched further.

Acknowledgments. The authors thank Dr. Aboul Ella Hassanien for providing the original ultrasonic images and his help for us. This work was supported by National Natural Science Fund of China (NO.60572011) and 985 Special Study Project (LZ85-231-582627).

References

1 Jensen, J.A.: Medical ultrasound imaging. Biophysics and Molecular Biology, 1–13 (2006)
2 Stippel, G., Philips, W., Govaert, P.: A tissue-specific adaptive texture for medical ultrasound images. Ultrasound in Med. & Biol. 1211–1223 (2005)
3 Van Stralen, Marijn, Bosch, Johan, G., et al.: A semi-automatic endocardial border detection method for 4D ultrasound data. In: Medical Image Computing and Computer-Assisted Intervention, 7th International Conference, Proceedings, pp. 43–50 (2004)
4 Levienaise-Obadia, Barbara, Gee, Andrew: Adaptive segmentation of ultrasound images. Image and Vision Computing, 583–588 (1999)
5 Hiransakolwong, Nualsawat: Automated liver detection in ultrasound images. In: 4th International Conference on Image and Video Retrieval, pp. 619–628 (2005)
6 Johnson, J.L., Padgett, M.L.: PCNN Models and Applications. IEEE Trans. Neural Networks, 480–498 (1999)
7 Eckhorn, R., Reitboechk, H.J., Arndt, M., et al.: Feature linking via correlated synchronization among distributed assemblies: Simulation of results form cat cortex. Neural Compu. 293–307 (1990)

8 Lindblad, T., Kinser, J.M.: Image Processing using Pulse-Coupled Neural Networks. Springer, Heidelberg (2005)

9 Gonzalez, R.C., Woods, R.E.: Digital Image Processing. Addison-Wesley Longman Publishing Co., Redwood City, CA, USA (1992)

10 Meihong, S., Junying, Z., Yonggang, L., et al.: A New Method of Low Contrast Image Enhancement. Application Research of Computers, 235–238 (2005)

A Novel Heuristic Approach for Job Shop Scheduling Problem

Yong-Ming Wang[1,3], Nan-Feng Xiao[1], Hong-Li Yin[2], and En-Liang Hu[4]

[1] School of Computer Science and Engineering, South China University of Technology,
Guangzhou 510640, China
[2] School of Computer Science and Information Technology, Yunnan Normal University,
Kunming 650092, China
[3] Department of Computer Science, Qujing Normal University, Qujing 655011, China
[4] School of mathematics, Yunnan Normal University, Kunming 650092, China
{yymm_wang,honli_yin}@163.com

Abstract. Job shop scheduling problem has earned a reputation for being difficult to solve. Varieties of algorithms are employed to obtain optimal or near optimal schedules. Optimization algorithms provide optimal results if the problems to be solved are not large. But most scheduling problems are NP-hard, hence optimization algorithms are ruled out in practice. The quality of solutions using branch and bound algorithms depends upon the good bound that requires a substantial amount of computation. Local search-based heuristics are known to produce decent results in short running times, but they are susceptible to being stuck in local minima. Therefore, in this paper, we presented a brand-new heuristic approach for job shop scheduling. The performance of the proposed method was validated based on some benchmark problems of job shop scheduling, with regard to both solution quality and computational time.

Keywords: Job shop scheduling; Heuristic algorithm; Simulated annealing algorithm; Production scheduling.

1 Introduction

In the production environment, scheduling allocates resources over time in order to perform a number of tasks. Typically, resources are limited and tasks are assigned to resources in a temporal order. The jobs are input to the scheduling engine of a production planning system. It determines the sequences and periods for processing jobs on dedicated machines. Jobs often follow technological constraints that define a certain type of shop floor. Typically, the objectives are the reduction of makespan of an entire production program, the minimization of mean tardiness, the maximization of machine load or some criteria [1,2]. Within the great variety of production scheduling problems, the general job shop problem is probably the most studied one during the last decade. It illustrates some of the demands required by a wide range of real-world problems.

F.P. Preparata and Q. Fang (Eds.): FAW 2007, LNCS 4613, pp. 252–260, 2007.
© Springer-Verlag Berlin Heidelberg 2007

The classical job shop problem can be stated as follows: there are m different machines and n different jobs to be scheduled. Each job is composed of a set of operations and the operation order on machines is pre-specified. Each operation is characterized by the required machine and the fixed processing time. The objective is to determine the operation sequences on the machines in order to minimize the makespan, i.e., the time required to complete all jobs [3-5].

There are many researches focused upon the development of exact methods to solve relatively small scheduling problems. Several heuristics have also been developed to solve larger, more practical, job shop scheduling problems [6]. Priority based rules and heuristics were some of the earliest techniques that were developed. More recent works included the development of heuristic approaches that provided improved solutions. One such method is shifting bottleneck heuristic [7-9]. This heuristic has been prominently documented as a high quality job shop scheduling method. Shifting Bottleneck Heuristic (SBH) attempts to minimize the makespan of jobs. SBH employs disjunctive arc graphs where the disjunctive arcs represent constraints that occur in scheduling. SBH handles the scheduling problem one machine at a time. Each one machine-scheduling problem is iteratively solved until one machine with the maximum lateness is identified as the bottleneck. This bottleneck is sequenced to minimize the maximum lateness. The process continues with additional iterations until all machines have been scheduled [10-13]. With some iterative search, enumeration, and comparison process, the computational effort using the SBH in solving job shop scheduling problem may be substantial. Furthermore the quality of solution obtained by SBH is not guaranteed. There clearly exist tradeoffs among computational time and solution quality. Thus, it is desirable to design a scheduling method that can yield high quality solutions with less computational time.

In this paper, we presented a brand-new heuristic approach. It evaluates machine availability and job requirements at each incremental point in a time horizon. A sequence matrix (indicating the routing of each job through its processing machine) and a process time matrix (for each job at each machine in its routing sequence) are utilized in determining the candidate jobs for scheduling. If more than one candidate job is available, the processing time matrix is partitioned and computations are performed to determine which job to schedule next. The new approach is derived from a classical scheduling approach.

2 A Classical Scheduling Approach

A scheduling concept proven useful in solving scheduling problems, in there, makespan is the performance measure, is proposed by Johnson [14]. This method was presented for minimizing the makespan in a two-machine flow shop scheduling problem. In Johnson's problem, there are n jobs, which must go through one machine and then through a second machine of a flow shop. Only one job can be processed on one machine at a given time. He denoted A_i as the processing time of the ith job on the first machine and B_i the corresponding time on the second machine. The objective is to the makespan.

Johnson proved a schedule with a minimum makespan resulted if the jobs were sequenced by the following rules [14]:

1) Consider all jobs A_i and B_i for $i=1, 2,\ldots, n$.

2) Identify the minimum processing time among all A_i's and B_i's. If this time is on the first machine (i.e. A_i) then schedule job i first. If the time is on the second machine (i.e. B_i), schedule job i at last.

3) Eliminate scheduled jobs from further consideration. Repeat step 2 until all jobs are scheduled, working from both ends toward the middle.

4) If a tie exists in the minimum tunes between A_i and B_i or A_i and B_j then schedule job i.

There are fundamental differences exist between flow shop and job shop scheduling problems. However, Johnson's work provides an important reference in the development of a new job shop scheduling heuristic. In the follows, a novel heuristic approach (NHA) has been developed for job shop scheduling. This scheduling approach employs a concept similar to Johnson's rules in its job sequencing strategy.

3 The Novel Heuristic Approach for Job Shop Scheduling

A job shop scheduling problem with n jobs and m machines can be described by the following two matrices:

1) A sequence matrix S indicating the operation routing of each job.
2) A processing time matrix P showing the processing times for each job at each operation in it's routing sequence.

Sequence matrix S					Processing time matrix P				
Job Number	Routing Sequence				Job Number	Processing time			
1	M_{11}	M_{12}	$\cdots$	M_{1m}	1	P_{11}	P_{12}	$\cdots$	P_{1m}
2	M_{21}	M_{22}	$\cdots$	M_{2m}	2	P_{21}	P_{22}	$\cdots$	P_{2m}
$\vdots$	$\vdots$	$\vdots$	$\vdots$	$\vdots$	$\vdots$	$\vdots$	$\vdots$	$\vdots$	$\vdots$
n	M_{n1}	M_{n2}	$\cdots$	M_{nm}	n	P_{n1}	P_{n2}	$\cdots$	P_{nm}

The new heuristic is designed to construct a schedule by applying decision criteria to a partitioned processing time matrix for the selection among the particular jobs in contention for a particular resource at a particular time. If no contention exists among jobs at a particular instant, the heuristic schedules the only available candidate. When determining the next job to schedule among multiple contending candidates, job-scheduling priority is resolved through the use of a partitioned processing time submatrix [X|Y]. This partitioned submatrix is established according to the following method.

For each row (job i), the processing time can be computed as

$$R_i = \sum_{j=1}^{m} p_{ij} \qquad \text{for each } i=1, 2,\ldots,\ n. \tag{1}$$

And the total processing time for all jobs is

$$TR = \sum_{i=1}^{n} R_i \tag{2}$$

Similarly, the sum of the processing times of all operations at each column, can be computed as

$$C_j = \sum_{i=1}^{n} p_{ij} \qquad \text{for each } j=1, 2,\ldots,\ m. \tag{3}$$

And the total processing times for all columns is

$$TC = \sum_{j=1}^{m} C_j \tag{4}$$

If the total processing time was equally balanced in two elements (B_1 and B_2), each element could be defined as:

$$B_1=B_2=TR/2 \tag{5}$$

Applying the element B_1, the original processing time matrix P can be partitioned as shown below:

Let D_j represent the following differences:

$$D_1=B_1-C_1 \quad \text{for } j=1.$$
and $\qquad\qquad D_j=D_{j-1}-C_j \quad \text{for } j=2,3,\ldots,\ m.$ $\tag{6}$

The residual for column r-1($i.e., D_{r-1}$) can be applied to partition P when $D_{r-1} \le C_r$.

By letting $D_p=D_{r-1}/C_r$ and by restricting D_p to the range $0 \le D_p \le 1$ for values of r ranging $1<r<m$,the fractional value, D_p, can be used in partitioning decisions. By selecting column r as the partitioning column, the original processing time matrix P is partitioned as $[X|Y]$ for each job i:

$$X_i = \sum_{j=1}^{r-1} p_{ij} + D_p (p_{ir}) \tag{7}$$

and

$$Y_i = (1 - D_p)(p_{ir}) + \sum_{j=r+1}^{m} p_{ij} \tag{8}$$

This submatrix of partitioned processing times for the jobs is structured as an h-job ($1<h \le n$) two-machine problem similar to those addressed in Johnson's research [15-17]. According to Johnson's method scheduling decisions are made on the basis of the shortest processing time in each machine. Jobs with processing times of shortest duration at machine one are scheduled firstly; those jobs with short processing time duration at machine two are scheduled at last. This scheduling strategy has been

shown by Johnson to produce the optimal makespan for two-machine flow shop problems. As such, a strategy similar to Johnson's rule has been incorporated, in various forms, into scheduling heuristics for a variety of problem types.

Extending this rationale to the partitioned processing times submatrix provides the following procedure for determining which job should be scheduled at a particular time t from among the h jobs in queue at this instant. The procedural steps are as listed as the following:

Step 1: For each job i, determine $m_i = \min\{X_i, Y_i\}$.

Step 2: If for some job k, $X_k = Y_k = m_k$, go to step 4.

Step 3: Let $N = \{\text{job}| m_i = X_i\}$, Let $m_N = \min\{m_i\}$, for job $i \in N$, If there exists at least two jobs, say job r and job s, with $m_r = m_s = m_N$, compute partitioned submatrix and go to step 4; Otherwise job k is scheduled next, where $m_k = m_N$.

Step 4: Certain tie breaking and decision rules are employed in these instances noted in Steps 2 and 3. These rules are presented and discussed in literature [14], in this paper the random decision method is adopted.

When one machine is available and more than one jobs that requires the machine is awaiting, the above decision method is invoked.

Therefore, procedurally, the novel heuristic approach can be listed as the following:

Resources

Step 1: For machine j, check availability at the current time (i.e., at time=t).

Step 2: If machine j is available, proceed to Jobs Step 1.

Step 3: If machine j is unavailable and $j < m$, set $j = j+1$ and return to Resources Step 1.

Step 4: If machine j is unavailable and $j = m$ (i.e., the last machine); advance time to $t=t+1$, set $j = 1$; return to Resources Step 1.

Jobs

Step 1: For machine j, check the status and precedence relationships of each job to eliminate inactive jobs and those jobs with unsatisfied precedence constraints from further consideration.

Step 2: If no jobs are available to schedule at the current time machine j is idle. Return to Resources and consider the next available machine.

Step 3: If only one job is available, then schedule this job. Return to Resources and consider the next available machine.

Step 4: If two or more jobs are candidates, compute the partitioned processing time submatrix and apply a rule similar to Johnson's rule to schedule the appropriate job. Return to Resources module and consider the next available machine.

Step 5: If all jobs have been scheduled, then terminate the process. Otherwise, advance time to $t=t+1$ and return to the Resources module to consider resource at time $t=t+1$.

When some operation is scheduled the processing time matrix P must be updated and the result is denoted by P'.

For instance, if p_{21} is scheduled, we get P' as the following:

$$P = \begin{vmatrix} p_{11} & p_{12} & \cdots & p_{1m} \\ p_{21} & p_{21} & \cdots & p_{2m} \\ \vdots & \vdots & \cdots & \vdots \\ p_{n1} & p_{n1} & \cdots & p_{nm} \end{vmatrix} \qquad P' = \begin{vmatrix} p_{11} & p_{12} & \cdots & p_{1m} \\ p_{22} & p_{23} & \cdots & 0 \\ \vdots & \vdots & \cdots & \vdots \\ p_{n1} & p_{n1} & \cdots & p_{nm} \end{vmatrix}$$

And the next scheduling operation is selected based on P', this repetition running until all the elements of P' equal 0. This gives a termination condition to the novel heuristic approach.

4 Experiments and Results

Our experiments were conducted on a Dell personal computer, operating at 2.66G Hz. Since the SBH is a prominently documented and well-accepted performance standard for job shop scheduling method evaluation, it will be used as a benchmark reference in assessing the performance of the new heuristic approach. The makespan is an important performance measure in scheduling applications since a reduction in the makespan generally implies an increasing throughput and a reduction in machine idle time. Computational times are important considerations concerning the cost and practical application of the scheduling procedure. If the scheduling procedure requires an excessive amount of time and computing resources, its application may be limited to only rather small problems. Therefore, the new heuristic approach is validated with considering the quality of obtained makespan and computational time.

From Table 1, we can see that NHA got 5 optimal scheduling results out of 7 problems. In the LA10 and LA16, NHA got better outcomes than BNH. Computational times used to solve a particular instance by SBH were substantially longer than the time required by the NHA for the same problem, which can be seen from Table 2. To sum up, NHA outperform BNH, it can obtain better scheduling results with a less computational time. Fig. 1 and Fig. 2 are a benchmark problem instance and the Gantt chart of scheduling results scheduled by NHA.

Sequence matrix S

Processing time matrix P

Job Number	Routing Sequence					Job Number	Processing time				
1	1	4	2	5	3	1	20	87	31	76	17
2	5	3	1	2	4	2	25	32	24	18	81
3	2	3	5	1	4	3	72	23	28	58	99
4	3	2	5	1	4	4	86	76	97	45	90
5	5	1	4	3	2	5	27	42	48	17	46
6	2	1	5	4	3	6	67	98	48	27	62
7	5	2	4	1	3	7	28	12	19	80	50
8	2	1	3	4	5	8	63	94	98	50	80
9	5	1	3	2	4	9	14	75	50	41	55
10	5	3	2	4	1	10	72	18	37	79	61

Fig. 1. The benchmark LA2 represented by S and P

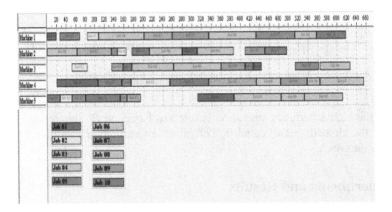

Fig. 2. The Gantt Chart of LA2's scheduling results

Table 1. The makspan comparison obtained by NHA and SBH

Problem Name	Jobs	Machines	The best known solution	Makespan obtained by new heuristic approach	Makespan obtained by shift bottleneck heuristic
LA2(F1)	10	5	655	655	740
LA5(F5)	10	5	593	593	632
LA9(G4)	15	5	951	651	714
LA10(G5)	15	5	958	973	1023
LA16(A1)	10	10	945	982	996
LA18(A3)	10	10	848	848	927
LA25(B5)	15	10	977	977	1084

Table 2. The computational time comparison consumed by NHA and SBH

Problem Name	Jobs	Machines	The best known solution	Computational time (and makespan) of new heuristic approach	Computational time (and makespan) of shift bottleneck heuristic
SWV3	20	10	1398	67s (1491)	91s (1797)
SWV4	20	10	1483	52s (1562)	123s (1635)
SWV6	20	15	1678	93s (1783)	182s (1825)
SWV9	20	15	1663	104s (1722)	176s (1903)
SWV10	50	10	1667	232s (1754)	483s (1966)
SWV12	50	10	3003	289s (3177)	528s (3326)
SWV15	50	10	2904	359s (3054)	497s (3420)

5   Conclusion

Job shop scheduling problems belong to a class of problems, which are NP-Complete. As such, these types of problems are the most intractable to solve. In the passed decades, such as integer programming, branch and bound techniques and enumerative based methods have been applied to determine scheduling solutions. Unfortunately, even if possible solutions could be generated and evaluated at the rate of one per second, this method could still take centuries to solve problems involving five or more jobs with five or more machines. Heuristic approaches are known to produce good results in a acceptable running times. Therefore, in this paper, we presented a brand-new heuristic approach based on Johnson's classical flow shop scheduling method for job shop scheduling. The performance of the proposed method was accessed based on some benchmark problems of job shop scheduling, with regard to both solution quality and computational time. The experiments results indicate that the presented new heuristic approach can produce higher quality solution with considerably less computational time when compared with the outcomes achieved by shifting bottleneck heuristic method.

Acknowledgments. The authors wish to express their sincere gratitude to the Editor, the Associate Editor, and the three referees, for their valuable comments and suggestions to help improve the paper.

References

1 Blazewicz, J., Ecker, K., Schmidt, G., Weglarz, J.: Scheduling in Computer and Manufacturing Systems, pp. 35–52. Springer, Heidelberg (1993)
2 Gen, M., Cheng, R.: Genetic Algorithm and Engineering Design, pp. 60–63. John Wiley, New York (1997)
3 Brucker, P., Jurisch, B.: A New Lower Bound for the Job Shop Scheduling Problem. European Journal of Operations research 64, 156–167 (1993)
4 Batas, E., Lenstra, J., Vazacopoulos, A.: One Machine Scheduling with Delayed Precedence Constraints. Management Science 41, 94–109 (1995)
5 Brucker, P., Jurisch, B., Sievers, B.: A Branch and Bound Algorithm for the Job Shop Scheduling Problem. Discrete Applied Mathematics 49, 107–127 (1994)
6 Carlier, J., Pinson, E.: Adjustments of Heads and Tails for the Job Shop Problem. European Journal of Operations Research 78, 146–161 (1994)
7 Ramudhin, A.: The Generalized Shifting Bottleneck Procedure. European Journal of Operations Research 93, 34–48 (1996)
8 Lourenco, H.: Job Shop: Computational Study of Local Search and Large Step Optimization Methods. European Journal of Operations Research 83, 347–364 (1995)
9 Ivens, P., Lambrecht, M.: Extending the Shifting Bottleneck Procedure to Real-Life Applications. European Journal of Operations Research 90, 252–268 (1996)
10 Droubouchevitch, I.G., Strusevich, V.A.: A polynomial algorithm for the three - machine open shop with a bottleneck machine. Annals of Operations Research 92, 111–137 (1999)
11 Demirkol, E., Mehta, S., Uzsoy, R.: A Computational Study of Shifting Bottleneck Procedures for Shop Scheduling Problems. Journal of Heuristics 3, 185–211 (1997)

12 Mukherjee, S., Chatterjee, A.K.: On the representation of the one machine sequencing problem in the shifting bottleneck heuristic. European Journal of Operational Research 166, 1–5 (2006)

13 Mönch, L., Drießel, R.: A distributed shifting bottleneck heuristic for complex job shops. Computers & Industrial Engineering 49, 363–380 (2005)

14 Parks, D.R.: A new job shop scheduling heuristic. Dissertation, University of Houston, pp. 60–81 (1998)

15 Janiak, A., Kozan, E.: Metaheuristic approaches to the hybrid flow shop scheduling problem with a cost-related criterion. International Journal of Production Econ omics 105 (2), 407–424 (2007)

16 Akrami, B., Karimi, B.: Two metaheuristic methods for the common cycle economic lot sizing and scheduling in flexible flow shops with limited intermediate buffers: The finite horizon case. Applied Mathematics and Computation 183(1), 634–645 (2006)

17 Yavuz, M., Tufekci, S.: Dynamic programming solution to the batching problem in just-in-time flow-shops. Computers & Industrial Engineering 51(3), 416–432 (2006)

18 Qiu, R.G., Tang, Y., Xu, Q.: Integration design of material flow management in an e-business manufacturing environment. Decision Support Systems 42(2), 1104–1115 (2006)

An Efficient Physically-Based Model for Chinese Brush

Bendu Bai[1,2], Kam-Wah Wong[2], and Yanning Zhang[1]

[1] School of Computer Science,
Northwestern Polytech University, Xi'an 710072, China
[2] School of Creative Media,
City University of Hong Kong, Kowloon, Hong Kong China
baibendu@cityu.edu.hk

Abstract. This paper presents a novel physically-based model for virtual Chinese brush. Compared with previous works, the main advantage of our method lies in the use of physically based modeling methods that describe the behavior of the real brush's deformation in terms of the interaction of the external and internal forces with the virtual writing paper. Instead of simulating the brush using bristles, we use points to simulate the whole brush bundle, which can drastically decrease the complexity inherent in the conventional bristle-level approach. A spring network is derived to calculate the physical deflection of brush according to the force exerted on it. With this model, we can get a more effective simulation of real brush painting.

1 Introduction

Chinese calligraphy was thought to be the highest and purest form of Chinese painting, and Chinese have been painting with hair brushes and ink on paper over two thousands years. Traditional Chinese hair brushes are made from animal hairs and the ink are made from pine soot and animal glue. With the development of modern computers, to design and develop a digital painting environment that simulates Chinese brush has attracted a lot of researchers.

Chinese hair brush model includes a model for the ink and the paper,covers the various stages of the brush going through a painting or calligraphy process [2]. Some existing models for Chinese brush simulate the process of a traditional brush moving. Because of the computational complexity and the randomness of the brush dynamic variations, building an accurate model is a challenge. Other models attempt to exploit a synthesize approach by using realistically Chinese calligraphic writings simulate the different brush stroke detail and various brush effect [1]. Obviously, the synthetic results depend on the library of realistic Chinese calligraphic writings. In this paper we have only considered the purely physically-based modeling techniques.

F.P. Preparata and Q. Fang (Eds.): FAW 2007, LNCS 4613, pp. 261–270, 2007.

1.1 Related Work

Several physically-based 3D brush models have been proposed recently and we will review only those are directly related to our work. The first purely physically based 3D brush model was proposed by Saito [10]. Saito used energy optimization to determine the brush deformation. His model accounted for brush stiffness, friction and kinetic energy. However, the model cannot simulate brush spreading. Wong et al. [1] modeled a calligraphy brush as an inverted cone, with some ellipses to synthesize Chinese calligraphic writings. Since their model pay no attention to the brush tip while drawing stroke, it fails to produce the biased-tip strokes. Baxter et al. [5] modeled the western brush as spring-mass systems, but the model was still lacking of the effect of brush spreading or splitting. Xu et al. [2][3] presented a more complex geometry which can split into smaller tufts, but the bulk of the brush must penetrate the paper in the drawing process, which leads to unrealistic brush footprints. Chu and Tai [6][7][8] delivered a very convincing model for Chinese calligraphy that included factors such as plasticity, tip spreading, and pore resistance. Like Saito, Chu and Tai also used energy optimization for the brush dynamics, but their approach had to solve a static constrained minimization problem by using local sequential quadratic programming(SQP). Although the method of SQP can avoid solving stiff differential equations, it is still complex in computation due to the iterations of SQP. Moreover, In their design, Chu and Tai deal with the effect of split brush by applying a method named the split map[7]. The advantage of split map is that many lines or dots can be achieved by a single brush tuft. However, the method lacks variety as expressive as a real brush due to the effect achieved relying on large amount of map texture, and they did not describe how to generate enough map texture to simulate the effect of split bristle as a real-life brush.

1.2 Contributions

Our aim is to provide a simple but efficient brush model for interactively creating oriental artworks with computer. Since a typical real brush may consist of thousands or even tens of thousands of individual bristles, physically simulating each and every bristle is not practical based on the present hardware capability. Inspired by the fact that artists use real brush with elastic bristles to draw strokes, and based on the physical characteristics of the motion of brush, we convert simulating the complex physical deformation of bristles to simulating the footprint variation of brush, since the final effect is determined by the latter. For this purpose, we simplify the geometric representation of brush to be a collection of points instead of hair threads and use it as the basic unit to construct a 3D brush bundle. To better simulate the brush plasticity, we formulate the brush dynamics as a set of energy function according to the force exerted on it. A spring network is derived to calculate those energy functions. This method significantly improves the brush's appearance, producing the plastic effect of brush that user expect. Considering the self-similar of the split brush cluster, we propose a simple split algorithm. Since our method generate complex split brush bundle only by applying some affine transformations on a single brush bundle, it can efficiently decrease the complexity of physical simulation.

2   Brush Geometry

Our brush geometry model (BGM) is inspired by Wong's BGM [1]. Wong's BGM used an inverted cone approximately represented a normal state brush bundle(see Fig.1a). Our BGM also uses an inverted cone to represent the initial state of the brush bundle as shown in Fig.1b.

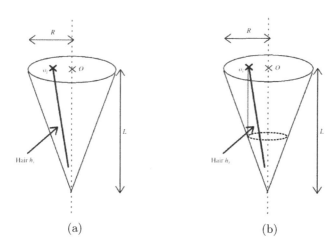

(a)   (b)

Fig. 1. (a) H.T.F. Wong's BGM. (b) our geometry model.

It should be noted that the main difference between our BGM and Wong's BGM lies in the composition of hairs. Wong's BGM was composed of a bundle of slant lines relative to the cone's main axis(see Fig.1a)which started from hair root(o_i) to the tip. While our BGM is composed of a bundle of parallel lines parallel to the cone's main axis(see the red line in Fig.1b). In this way, we can easily build the whole brush cone using a series of circles along the cone's main axis. And the tip of each piece of hair is determined by the root(o_i). The length of the piece of hair l_i can be got directly by model's initialization, which does not need additional computation of trigonometric function as Wong's BGM did. Given the height of the brush stem from the paper at any writing instance, the height of the intersection of the brush bundle and the paper can be computed as

$$Height_t = L - P_t.z \qquad (1)$$

where t=0,1,2... is the discretized instances of writing, L is the length of brush bundle, $P_t(x, y, z)$ is the coordinate of the footprint of brush stroke at instance t, and we define $P_t.z$ as the vertical distance from the root of brush bundle to the paper at instance t. $P_t.z$ is obtained from an pressure-sensitive input device (e.g. tablet) if it is an interactive drawing session; or it is specified in a file if it is an off-line rendering session. Similar to a real brush, the hairs distribute uniformity alone the radius. The density of hairs in the center area is much thicker than that near the boundary. Crowded hairs are obviously found in the center, and loser around the brush boundary. According to this role, we design a single circular disk as shown in Fig.2. A 3D virtual brush in static state can be built using a

Fig. 2. The distribution of hair alone the radius

series of circular disks (in Fig.2) with different sizes, starting from the brush tip to the brush root. Obviously, the smallest disk may only have one dot. However, for hair brush, a small brush tip including only one dot does not have a practical meaning. Our experiments show that the tip of the brush including four dots is a minimum.

3   Brush Dynamics

Brush dynamic model describes the deformation of the brush bundle at every instance of brush movement. In real-life Chinese calligraphy, a typical stroke usually includes three stages: Qibi (press), Yunbi (move) and Tibi (lift). During the different stage, the brush dynamic model should have the different footprint when the brush is pressed onto or lifted from the paper.

Qibi. Fig3.(a) shows a horizontal stroke. As the artist writes the stroke, firstly he presses the brush vertically against the paper until the brush tip reaches the writing paper (see arrow A of Fig3.(b)). Starting from the moment when the brush touched the paper, with the brush continually moving downward(i.e., brush only moves along z-axis), to the moment when the brush is beginning to move on the paper surface (i.e. brush moves along x-axis or y-axis), is the stage of Qibi.

During this stage, the brush is dipped with ink. The attractive force between ink and hair molecules holds the hair together. Moreover, in real-life calligraphy writing, usually only the tip of the brush is used to paint in the Qibi stage. Although the force that the brush suffers in the stage is rather complex, such as the

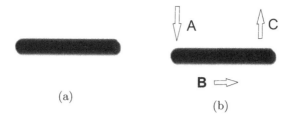

Fig. 3. (a) A stroke generated by our model. (b) The moving direction of the brush during a single stroke.

external pressure and the internal friction between wet bristles, the footprint in a real-life writing process looks like a circular disk according to our observations. Therefore, our brush model in Qibi stage is designed as a series of concentric circular disks, which are laid to lap over each other onto the cross-sectional plane (as shown in Fig.4. A). This simple representation is computationally efficient and quite similar to the observed reality.

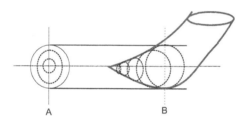

Fig. 4. The dynamic deformation of our brush during the different stage of a single stroke

Yunbi. The stage of Yunbi starts from which the artist drugs the brush along the path of the stroke (see arrow B of Fig3.(b)). At the very beginning of the stage, the artist will continually press the brush against the paper surface, as far as the stroke width reaches certain value which the painter wants to stroke to be.

During this stage, the brush suffers from the pressure and friction with the paper, so that the brush will alter its shape from the invert cone to a bent invert cone (see Fig.4 red line). Moreover, the pressure and the friction also make the circles change as ellipses (see Fig.5. B). Here, we simulate the bent brush through moving the circles with a horizontal displacement along x-axis of the paper plane(see Fig.4 B). In order to maintain the smoothness of the stroke boundaries, the displacement between two adjacent circle should meet the constrain as follows

$$r_i - r_{i-1} \leqslant d_i \leqslant 2(r_i - r_{i-1}) \tag{2}$$

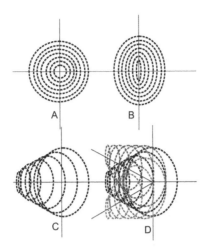

Fig. 5. The dynamic deformation of our brush. A is the platform of our model. B illustrates brush anamorphosis from circle to ellipse. C illustrates the deformation during the stage of Yunbi . D illustrates the brush deformation of split.

where d_i represent the displacement between the ith circle and the $(i-1)$th circle, r_i represents the radii of the ith circle, and r_{i-1} is the radii of the adjacent of the ith circle nearer to the brush tip.

The major axis and minor axis of the ith ellipses can be computed by follow equations

$$majoraxis(u_i) = r_i k_u, \quad minoraxis(v_i) = r_i k_v \tag{3}$$

where k_u are the deformation constant of major axis, and k_v is the deformation constant of minor axis. Typically, k_u and k_v are from 1.0 to 1.5 and from 1.0 to 0.5, respectively.

Tibi. The stage of Tibi describes the process that the brush was lifted from the writing paper, which can be found at the end of the stroke (see arrow C of Fig.3.(b)). During this stage, a moist brush is deformed. In order to return the bent brush to its original shape, the flexible force has to overcome the resistance of molecular friction. Due to the effect of this internal frication, the bent brush does not entirety revert to its original shape. We formulate the energy function based on the principle of conservation of energy when the brush was pressed or lifted as follows:

$$E_{press} = E_{bend} + E_{fric} \tag{4}$$

$$E_{lift} = E_{bend} - E_{fric} \tag{5}$$

$$E_{press} = E_{lift} \tag{6}$$

where E_{bend} is the brush bent energy, and E_{fric} is the internal energy of molecular friction when the brush was pressed or lifted. Then, we deal with the problems by using a spring network to approximate the behavior of the deformations. We use bend springs at each joint point between cone major axis and circles to model the bending deformation of the brush. The relationship between the amount of bent displacement and the spring force can be closely approximated by using Hooke's law [11]:

$$F_s = -F_x = -kx \tag{7}$$

where F_s is the equal and opposite restoring force of the spring on the stretch node, k is the spring constant, and x is the displacement of a node position under the influence of a force F_x. Taking the definition of the Hooke's law (Equation7) into Equation 4 and 5, we thus have our energy functions as:

$$E_{press} = E_{bend} + E_{fric} = -k_b d_i - k_f d_i \tag{8}$$

$$E_{lift} = E_{press} - E_{fric} = -k_b d_i + k_f d_i \tag{9}$$

where k_b and k_f are the bent spring constant and the internal friction constant, d_i is the displacement of spring node between the ith circle and the $(i-1)$th circle. The value of d_i should satisfy the constraint of formula (2).

Split Algorithm. The effect of brush splitting results from the brush deformation, the decrease of the volume of the ink, and rapid moving of the brush. With a single brush bundle, it is impossible to achieve the effect of brush splitting. To deal with the problem, Xu [2] proposed a two-level hierarchical brush geometry model. The main advantage of Xu's model lies in modeling the complex geometry of a realistic split brush as several similar cluster using the comparability of the hair macros. We cope with the problem by using a similar approach, with some necessary modifications due to a different primitive model adopted in our framework. According to our geometry model, we use a simple algorithm to achieve the splitting effect. Since the amount of ink plays an important role in the process of brush splitting, we firstly set an initial ink value for entire brush bundle(the ink model will be discussed in next section). For every instant t, we check the decrease of the amount of ink and the deformation of the brush tip. When the ellipse of the brush tip reaches its limit($k_u = 1.5$) and the amount of ink reduce to a threshold, a new primitive brush generates. Then, we use the affine transformations composing the parent brush bundle and the new subbrush bundle together. Each brush bundle can be classified as a collection of vertices as shown in Fig.2. The split brush bundle can be made from the same model which reduces the number of models that need to be generated. This is helpful to build a complex model by a single model, as which allows Graphical Processing Unit(GPU) to work on a smallest set of vertices at a time. Fig.5. D shows a model with one parent brush bundle and two subbrush bundles. By this method, with less than 6 subbrush bundles, we have not only modeled the complex geometry of a split brush, but also diminish computation.

Ink model. The ink model describes the process of ink depositing in which brush geometry model is instanced and transformed into a real screen presentation. During the painting process, the brush sweeps paper plane along a stroke path which transfers the brush footprint to the paper plane on each pass. To simulate the decrease of the amount of the ink, we define a parameter c_i (1.0-0.0) for the ink consumption at unit interval of the brush bundle's ith ellipse. When the single brush bundle split into several bundle, each ellipse of the new subbrush bundles have the same value as the ellipse on the same contour line of main brush bundle. Therefore, the amount of the ink of ith ellipse at instant t, i.e. $e_{i,t}$, is

$$e_{i,t} = 1 - \sum_{j=1,\ldots,t} c_{i,j}. \tag{10}$$

Suppose we are drawing a stroke, we want to have a brush that gradually adds color so that each points of a footprint contributes a little more ink with whatever is currently in the image. To do so, We simulate the process using alpha blending with OpenGL. The blending functions that use the source and the destination blending factor as GL_SRC_ALPHA and GL_ON_MINUS_SRC_ALPHA. To avoid the alias brush shape, we vary the alphas across the brush to make the brush add more of its ink in the middle and less on the edges. The process of the ink deposition is optimized by GPU's hardware implementation. As a result, the CPU can offload the rendering processing to do physics simulation, and therefore increase system performance.

4   Results

We have implemented a prototype system based on our brush model. Our system was written in c++ using Microsoft Visual C++6.0. All graphical rendering operations are accelerated by using OpenGL on the GPU. All experiments are done on a 2GHZ Pentium 4 with NVIDIA GeForce FX 5200 display card, and it runs in real-time at 60 frames per second.

Figures 6 and 7 show some sample calligraphy and strokes obtained using our prototype system. A calligraphy horizontal stoke shown in Fig 6 (a) was written with a dipped sufficiently ink brush. Fig 6 (b) and (c) illustrate the effect of ink diffusion due to the decrease of the amount of ink and the amount of time that the brush spent in contact with the paper. Fig 6 (d) illustrates the effect of brush split generated by a dry and multi-tuft brush. Fig 7 shows a comparison between a digitized real calligraphy sample and the imitation artwork created by our prototype system. It proves that very realistic-looking calligraphy artwork can be generated by our brush model.

5   Further Work

NPR. One of the most important motivations of this work is the hope to create a system that will automatically paint or write Chinese brush strokes by simulating the artistic processes. Therefore, based on our brush model, developing a Chinese-painting style non-photorealistic rending system will be one of the major tasks in the near future.

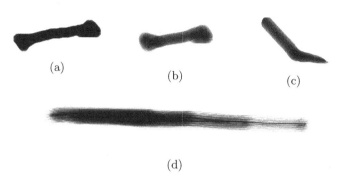

Fig. 6. Sample calligraphy stoke from our prototype system. (a) A horizontal stroke with full-inking. (b) A horizontal stroke with medium-inking. (c) Blade-like calligraphic stroke with medium-inking. (d) Made with a drying splitting brush.

Fig. 7. (a) Digitized image of a real art work. (b) The image generated by our prototype system.

Better input device. In the current implementation, our prototype system uses input file for specifying the brushes parameters, or uses mouse and keyboard as input devices for manual painting. This can work fine for automatic painting, but is not user friendly for real-time artist's painting. For a real-time interactive painting system, a user friendly input device as expressive as a real brush would give significant improvement to the final appearance. We are exploring a pressure sensitive touch pad and table brush. Other possible input devices include Wacom 6D Art Pen (see http://www.cgvisual.com/headlines/Moxi/ CGVheadlines_Moxi_p2.htm)

6   Conclusion

We have presented a novel model for virtual brushes that can produce the effects needed by the digital Chinese calligraphy. The model is very simple but efficient. With the proposed model, realistic Chinese calligraphy artwork can be generated effectively with direct manipulation. Major advantages of our model over previous works are

- We use a collection of points instead of hair threads as the basic unit to construct a 3D brush bundle, so that we convert simulating the complex physical deformation of bristles (as did in most previous models) to simulating the footprint variation of brush, which can efficiently decrease the complexity of physical simulation.
- We formulate the brush dynamics as a set of energy function according to the force exerted on it. A spring network is derived to calculate those energy function. This method significantly improves the brush's appearance, producing the plastic effect of brush that user expect.

Up to now, our research mainly focused on modeling a medium stiff hair brush, it is expected that our model can also mimic other kinds of brushes by modifying the bent spring parameter. Moreover, only a vertical brush dynamic model was simulated presently. In order to simulate a hairy brush in Chinese painting, a slant angle brush dynamic model should also be studied.

Acknowledgments. The work described in this paper was fully supported by a grant from the Research Grants Council of the Hong Kong Special Administrative Region, China [Project No. CityU 121205]

References

1. Wong, H.T.F., Ip, H.H.S.: Virtual brush: a model-based synthesis of Chinese calligraphy. Computers and Graphics 24(1), 99–113 (2000)
2. Xu, S., et al.: A Solid Model Based Virtual Hairy Brush. In: Eurographics 2002 Proceedings Saarbrucken, Germany, pp. 299–308. Blackwell Publishers, Oxford (2002)
3. Xu, S., et al.: Advanced Design for a Realistic Virtual Brush. Computer Graphics Forum 22(3), 533–542 (2003)
4. Baraff, D., Witkin, A.: Large steps in cloth simulation. In: SIGGRAPH98 Proceedings, pp. 43–54 (1998)
5. Baxter, B., Scheib, V., Lin, M., Manocha, D.: DAB: Interactive haptic painting with 3D virtual brushes. In: SIGGRAPH 2001 Proceedings (2001)
6. Chu, N.S., Tai, C.L.: An efficient brush model for physically-based 3D painting. In: Proc. of Pacific Graphics (2002)
7. Chu, N.S., Tai, C.L.: Real-Time Painting with an Expressive Virtual Chinese Brush. IEEE Computer Graphics and Applications 24(5), 76–85 (2004)
8. Chu, N.S., Tai, C.L.: MoXi: Real-Time Ink Dispersion in Absorbent paper. In: SIGGRAPH 2005 Proceedings (2005)
9. Desbrun, M., Schroder, P., Barr, A.: Interactive animation of structured deformable objects. In: Proc. of Graphics Interface 99 (1999)
10. Saito, S., Nakajima, M.: 3D physically based brush model for painting. In: SIGGRAPH99 Conference Abstracts and Applications, p. 226 (1999)
11. Landau, L.D., Lifshitz, E.M.: Theory of Elasticity, Course of Theoretical Physics, vol. 7. Pergamon Press, Oxford (1986)

A Trigram Statistical Language Model Algorithm for Chinese Word Segmentation

Jun Mao[1], Gang Cheng[1], Yanxiang He[1], and Zehuan Xing[2]

[1] Computer School, Wuhan University, Wuhan 430072, P. R. China
[2] Department of Linguistics, Central China Normal University
Wuhan 430079, P. R. China

Abstract. We address the problem of segmenting a Chinese text into words. In this paper, we propose a trigram model algorithm for segmenting a Chinese text. We also discuss why statistical language model is appropriate to be applied to Chinese word segmentation and give an algorithm for segmenting a Chinese text into words. In particular, we solve the problem of searching which often leads to low performance brought by trigram model. Finally, the issue of OOV word identification is discussed and merged to trigram model based method in order to improve the accuracy of segmentation.

1 Introduction

In many applications of natural language processing, we intend to obtain and analyze basic linguistic units, usually words. For example, counting and indexing the frequency of every word is often used in information retrieval. For English and other western languages, the segmentation of texts is not necessary at all. Sentences in those languages are always naturally segmented into independent words by using spaces and punctuations which are called word delimiters. But for Asian languages like Chinese and Japanese, things are quite different. As an ideographic, Chinese sentences are composed of characters without any spaces, and even none of any punctuation exists in ancient Chinese texts. In Chinese tradition, each character corresponds to a single syllable. Most words in all modern varieties of Chinese are polysyllabic and thus they are usually made up of two or more characters. Thus, Chinese word segmentation which is to find word boundaries is a crucial task for applications in naturally language processing like machine translation, information retrieval, etc.

Researches on Chinese word segmentation have been conducted for many years. Many methods aiming to resolve the problem have been proposed. Generally, these can be classified into heuristic dictionary-based methods, statistical machine learning methods, and hybrid methods. In some sense, these methods are practical to segment Chinese texts. However, dictionary-based, statistical-based approaches are both limited in the performance and suffer from some innate difficulties. Hybrid approaches, although imperfect, outperform the two traditional ones and have been applied more widely.

F.P. Preparata and Q. Fang (Eds.): FAW 2007, LNCS 4613, pp. 271–280, 2007.

One of the most important difficulties is resolution of ambiguity. As a Chinese character can occur in different word internal positions in different words, it is often difficult to determine word boundaries in such conditions. To resolve this problem, a lot of machine learning techniques have been applied. In this paper, we propose an algorithm based on statistical language models, which is to collect linguistic information in a statistical way so that to resolve the ambiguity problem and improve higher accuracy compared with traditional methods. OOV (out-of-vocabulary) word identification is another essential problem for Chinese word segmentation. We combine N-gram model based approach with the linguistic rules for OOV word identification. Our experimental results show that our algorithm outperforms the traditional algorithms by dealing with the problem of ambiguity and unknown word discovery effectively.

This paper is organized as follows. In the following section 2, we review traditional segmentation methods and related work. In section 3, we employ statistical language models for Chinese word segmentation. In section 4, we discuss OOV (out-of-vocabulary) word identification. Our experimental results and analysis are presented in section 5. Finally, we give in section 6 a conclusion and future work.

2   Traditional Segmentation Approaches and Related Work

2.1   Dictionary-Based Approaches

The dictionary-based approaches are the most straightforward approaches for Chinese segmentation. They segment Chinese sentences by matching words in a large machine-readable dictionary.

Cheng, Young, and Wong[1] described a dictionary-based method. Given a dictionary of frequently used Chinese words, an input string is compared with words in the dictionary to find the one that matches the greatest number of characters of the input. This is called the maximum forward match heuristic. An alternative is to work backwards through the text, resulting in the maximum backward match heuristic. Both methods will fail in some situations.

The dictionary-based approaches are weak in dealing with the new words identification. New words, also called unknown words or OOV, are some words that are not listed in the dictionary. Most new words are nouns and we will discuss OOV problem in section 4.

2.2   Statistics-Based and Hybrid Approaches

The statistical-based approaches segment Chinese sentences using inherent statistical features of Chinese[2]. Training a large corpus of Chinese texts, the statistical-based approaches capture the statistical relationships between characters in the corpus. And then the texts are segmented according to the relationships of adjacent characters. Different statistical features are opted in statistical approaches, such as relative frequency, document frequency, local frequency, entropy, and mutual information.

The statistics-based approaches are weak in dealing with stop words. Stop words are words that appear frequently in the corpus but they do not convey any significant information to the document, for example, *of*, *the*, etc.

Any single approach will fail in some situations. So all the real Chinese word segmentation systems are hybrid approaches in practice. The dictionary-based approach and statistical-based approach will serve in different phases.

3   The Trigram Model Algorithm for Chinese Word Segmentation

3.1   Statistical Language Model

A language model is a fundamental component of many natural language applications like statistical machine translation, automatic speech recognition, spelling correction, handwriting recognition, augmentative communication, information retrieval, etc. The main object of language modeling is to help improve the performance of the natural language processing[3].

Basic language model is to compute the probability of a word sequence or usually a sentence by computing every conditional probability of each word given previous words and then multiplying hem together. A sequence of n words is denoted as $w_1, w_2, \ldots, w_n$ or w_1^n. For the joint probability of each word in a sentence, we use $P(w_1, w_2, \ldots, w_n)$ or $P(w_1^n)$. Then, using the chain rule of probability, $P(w_1^n)$ can be formulated as follows[4]:

$$P(w_1^n) = P(w_1)P(w_2 \mid w_1)P(w_3 \mid w_1 w_2) \cdots P(w_n \mid w_1 \ldots w_{n-1}) \qquad (1)$$

However, with limited corpus, it is too difficult to compute the exact probability of any word given a long sequence of preceding words since language is so creative and changeful that there are numerous possible n-grams given the last word.

According to the Markov assumption, this problem can be simplified to N-grams models. N-gram models are word prediction probabilistic models, which predict the next word of previous N-1 words instead of all the histories. Most common N-grams are bigram, trigram, 4-gram models. In trigram models, instead of computing $P(departure \mid Rhett\ Bulter\ is\ back\ from\ London\ after\ a\ long)$, we compute $P(departure \mid a\ long)$. Generally, using N-gram models, the conditional probability of the next word can be computed as follows:

$$P(w_n \mid w_1^{n-1}) \approx P(w_n \mid w_{n-N+1}^{n-1}) \qquad (2)$$

A simple way to estimate the above equation is called *maximum likelihood estimation* (MLE). We get the MLE estimate for the parameters of an N-gram model by taking the counts from corpus. If $C(w)$ is the number of times that w occurs in the corpus, then:

$$P(w_n \mid w_{n-N+1}^{n-1}) = \frac{C(w_{n-N+1}^{n-1} w_n)}{C(w_{n-N+1}^{n-1})} \qquad (3)$$

3.2   Chinese Word Segmentation Using N-Gram Language Model

N-gram models can be utilized to find the best segmentation of a sentence. With the aid of an n-gram model, the most probable segmentation $w_1, w_2, \ldots, w_k$ of a given Chinese sentence $S = c_1 c_2 \ldots c_m$ can be formulated as:

$$seg(S) = \underset{S = w_1 w_2 \cdots w_k}{arg\ max} \prod_{i}^{k} P(w_i | w_{i-N+1}^{i-1}) \tag{4}$$

Where $P(w_i | w_{i-N+1}^{i-1})$ denotes the language model probability of w_i in sentence S, denoted word sequence $w_{i-N+1} w_{i-N+2} \ldots w_{i-1}$ is the context (or history) of w_i, and N is the order of the N-gram model in use. For the sake of accuracy, trigram is chosen to calculate language model probability. And $P(w_i | w_{i-N+1}^{i-1})$ in (4) becomes $P(w_i w_{i-2} w_{i-1})$.

To segment a sentence $S = c_1 c_2 \ldots c_m$ into k words has C_{n-1}^{k-1} different segmentations. Considering k varies from 1 to m, the sentence S has total 2^{n-1} different segmentations. We can not try all 2^{n-1} segmentations and calculate there trigram language model probabilities to find out the best segmentation of S. Actually only possible words in S are considered.

First, a dictionary-based word segmentation system is opted to segment the unsegmented training corpus. An N-gram language models tool is implemented to train trigram language parameters from training corpus. The output of training corpus is a table of trigram language model parameters, together with a unigram table and a bigram one.

By extracting string from both unigram table and sentence S, we get possible words table T_p. Mathematically, we can define possible words table T_p as:

$$T_p = \{w | w \in unigram\_table \wedge w \in S\} \tag{5}$$

Where w in S denotes w is a consecutive substring of sentence S.

Only the words in table T_p are chosen to form S. To reduce the cost of searching, S is cut into strings by punctuations before segmentation. And all the separated strings will segment separately.

Regarding some very long sentences, we still encounter severe searching problem. The searching problem is mainly caused by word overlaps. As many overlaps among words exist in the vocabulary, words combined from possible characters are various. Here, for a given sentence, we define *none-overlapped* and $n(s_i)$.

Definition 1. *$s_1 s_2$ denotes the adjacent two parts in the sentence. $s(f)$ and $s(l)$ are the first character and the last one in a consecutive substring. If $s_1(f) s_2(l)$ is not found in any word of the vocabulary, s_1 and s_2 are none-overlapped.*

Definition 2. *For a consecutive substring s_i of a sentence, the number of different possible segmentations using the words in the given vocabulary to consist s_i is called $n(s_i)$.*

Given a sentence S which is composed of m *none-overlapped* parts, $S = s_1 s_2 s_m$. Therefore,

$$n(s) = n(s_1) n(s_2) \cdots n(s_m) \tag{6}$$

From Equation (6), we found that $n(s)$ is extremely large when m and every single $n(s_i)$ are relatively large numbers. Consequently we need search quite a lot of times for possible combinations of word segmentation.

The characteristic of N-gram language model leads a simply way to solve the problem. When the distance between two words in the substring is no less then N, the probability of one word does not affect that of another. Assume sentence S is made up of three *none-overlapped* parts, $S = s_1 s_2 s_3$, $n(s_2) = 1$ and s_2 has no less than N-1 words. Therefore, the number of possible segmentations of S is $n(s) = n(s_1)n(s_3)$. In practice, however, we only need to $n(s_1) + n(s_3)$ from candidate segmentation lists. In this way, the complexity of our algorithm has been greatly reduced.

3.3   The Algorithm of Training Corpus and Segmentation

Algorithm 1

Input	Training samples (corpus)
	sentence S (to be segmented)
Output	Final segmented sentence S

1. Segment training corpus with basic dictionary-based approach;
2. Train segmented samples by 3-gram language model tool and output 1-gram_table, 2-gram_table, 3-gram_table;
3. Resegment the training samples: for every clause delimitated by punctuations in the samples, segment it into m none-overlapped parts;
4. Segment every single part into a consecutive word string according to the dictionary and find all possible word strings;
5. For every possible segmentation, compute the probability of N-gram language model of the whole clause;
6. Search for the best segmentation with the maximum probability;
7. Output a newly constructed training corpus by collecting all the best segmentation of all the sentences in the training samples;
8. If no improvement has been obtained, go to 9, else go to 2;
9. Redo step 3 and 6 by substituting the candidate sentence for the training samples to segment it into a word string delimited by spaces.

4   Identification of OOV Word

OOV Word Identification, also called new word discovery, is usually considered as a separate issue in Chinese NLP research. Also there are many pragmatic methods have been proposed, some of which have achieved state-of-the-art performance in their experiments. OOV words can be categorized into four types: morphologically derived words, factoids, named entities, and other unlisted words [5].

4.1 Morphologically Derived Words, Factoids, and Name Entities

For morphologically derived words, there are five main categories each of which has several subcategories.

- Affixation(Prefix and Suffix): 同学→同学们
- Reduplication: 看→看一看
- Splitting: 睡觉→睡了觉
- Merging: 国内+国外→国内外
- Directional and Resultative Compounding: 想+得出→想得出

Most factoids are number related, others are foreign language related. Number related factoids include mathematical numbers, data, time, duration, money, phone number, measure, etc. Foreign language related factoids include English words, E-mail, website, etc.

- Number related: 三百六十五, 2007年2月18日, 12:20, 700公斤
- Foreign language related: feeling, IBM, admin@hotmail.com, www.yahoo.co.jp

Name entities refer to Chinese person names, Foreign person names in Chinese, location names, organization names, commodity names usually called brands, etc.

- Chinese person names: 张学友, 王小兵, 成龙
- Foreign person names: 乔治·布什, 大江健三郎, 莎士比亚
- Location names: 美利坚合众国(Country), 香港(City), 平安大道(Street)
- Organization names: 求是中学(Educational Orgnization), 发改委(Political Orgnization), 招商银行(Company)
- Brands: 索尼, 娃哈哈, 海飞丝, 迪斯尼

4.2 Unlisted Words

Unlisted words are partly due to the limitation of the dictionary. Most unlisted words are odd and infrequently used. However, there are several major causes of unlisted words. Old Chinese, sometimes known as *Archaic Chinese*, are quite different from modern Chinese in vocabulary and grammar. Some words can only be found in historical literatures. A lot of Chinese words also come from these Asian languages since Chinese characters are logograms used in writing Japanese, sometimes Korean, and formerly Vietnamese. Some of these words deriving from Japanese and Korean have been adopted by Chinese and widely used. Besides Asian languages which Chinese has had a great influence on, Chinese has a large quantity of subdivisions languages or dialects. There are seven main groups: Mandarin(National standard Chinese),Wu, Cantonese, Min, Hakka, Xiang and Gan. The majority of these dialects may not be intelligible for Mandarin users.

Terminologies are technical words used in a particular field, subject, science, or art. A dictionary can hardly contains all the terms in all scientific, technical or artistic fields. As language is so creative, there are many newly coined words every day, particularly from internet.

- Old Chinese words: 雅言 礼乐 娈童 交路
- Japanese words written in Chinese: 型录 会社 相手
- Korean words written in Chinese: 修道 受渡 手刀
- Cantonese: 唔该 屋企 泊车
- Terms: 盐酸左旋甲基苯丙胺 信噪 厚画法
- Internet words:版主 博客 贴图

Other than above reasons, abbreviations are neglected by most scholars while in most occasions abbreviations are regarded as unknown words, leading to segmentation errors or translation errors as a consequence. Google online translation system fail to recognize 北大(PKU) as 北京大学 (Peking University) and translated it into English as the North.

- Abbreviations: 奥运会(奥林匹克运动会) 港澳台 (香港 澳门 台湾) 计生办 (计 划生育办公室)

5 Experiments and Results

5.1 Preparations

We are undertaking a research programme of machine translation. For our Chinese-English machine translation system, the Chinese word segmentation is one of the key components. The accuracy of Chinese word segmentation greatly influences the quality of Chinese-English machine translation. Incorrect segmented sentences inevitably result in poor translation. Our experiment data including training data and test data is partly from our newly constructed Chinese-English parallel corpus for statistical Chinese-English machine translation. The parallel corpus contains 49,3632 Chinese-English sentence pairs. Randomly we take out 6,1368 Chinese sentences as our training set, 5193 sentences as test set. A research student major in Chinese language constructs a reference to the test set. The definition of Chinese words and the standard of segmentation vary in terms of specific purposes. In order to adapt to our machine translation system, both the principle of Chinese language and the requirement of Chinese-English translation are considered and merged into our standard. According to PRC Guidelines, 中华人民共和国 is a single word, while we divide it into three words 中华 人民 共和国 for it corresponds to three English words China, People's, Republic. We use SRLM as our N-gram language model tool for training corpus. SRILM is the most widely used language model tool, with a collection of C++ libraries, executable programs, and helper scripts designed to allow both production of and experimentation with statistical language models for speech recognition, machine translation and other NLP applications[6]. There are also other language model tools for substitution, the CMU-Cambridge toolkit and the HTK Lattice Toolkit, and more. We have also designed a language model tool for our machine translation system. All these tools have successfully resolved the problem of data sparseness.

5.2 Results

The performance of our segmentation system is presented in Table 1 in terms of
$Precision(P)$, $Recall(R)$ and F score in percentages.

$$Precision(P) = \frac{number\ of\ correct\ segmentation\ points}{total\ number\ of\ segmentation\ points\ by\ the\ system} \quad (7)$$

$$Recall(R) = \frac{number\ of\ correct\ segmentation\ points}{total\ number\ of\ segmentation\ points\ in\ standard\ answer} \quad (8)$$

$$F = \frac{2 \times P \times R}{P + R} \quad (9)$$

Table 1. Comparison of some results among Dictionary Based Approach, N-gram Based Approach and Manual Standard

DIC_BASED	N-GRAM_BASED	STANDARD
阿乔叔叔	阿乔叔叔	阿乔叔叔
烹煮	烹煮	烹煮
蒸腾	蒸腾	蒸腾
何安迪	何安迪	何安迪
黑胡椒酱	黑胡椒酱	黑胡椒酱
撞球场	撞球场	撞球场
牛仔竞技会	牛仔竞技会	牛仔竞技会
一百六十五	一百六十五	一百六十五
白小姐	白小姐	白小姐
《古墓奇兵》	《古墓奇兵》	《古墓奇兵》
你想要什么样式呢	你想要什么样式呢	你想要什么样式呢
谁人跟它比呢	谁人跟它比呢	谁人跟它比呢

Table 2. Comparison of Performance between Dictionary Based Approach and N-gram Based Approach

	P	R	F
DIC_BASED	0.937	0.946	0.941
N-GRAM_BASED	0.958	0.956	0.957
Improvement	2.24%	1.06%	1.70%

5.3 Analysis

Many errors are caused by word overlap and other more errors caused by dictionary based approach are due to the lack of OOV identification. Many names cannot be well recognized like 阿乔, 何安迪, 成龙史密斯. Sometimes it fails to recognize numbers and split numbers into halves. In table 2, we can see 一百六十五 is segmented into two words 一百六十 and 五. According to the rules of discovering morphologically derived words, we take 乔叔叔 and 白小姐 as instances of affixation. However, N-gram based approach cannot indentify 阿乔 as a whole name, which is an instances of affixation as well. Besides this kind of errors, N-gram based approach seems oversensitive for new word discovery. Most new words are well recognized, but quite a few new words are created by merging words in a phrase into one single word. The main side-effect of N-gram based approach is that the component of OOV identification may found new words and add the new words into the uni-gram table every time the iteration going from step 7 back to step 7 in the algorithm of segmentation. Since some new words are coined incorrectly, these errors may lead to more so-called new words by merging adjacent characters or words based on linguistic rules.

6 Conclusion and Future Work

Combined with the rules for OOV identification, trigram language model based method can solve both the two major problems: the resolution of ambiguity and the identification of OOV words. By computing the probability of trigram language model, a sentence can be measured whether it is good or bad in terms of related linguistic grammars and vocabulary. Hence, trigram models are also able to measure whether a Chinese sentence is well segmented. In this paper, we have an in-depth study for the causes of OOV words and construct relevant rules for discovering new words. Names and numbers can be well recognized. But our system seems oversensitive for OOV word identification. In the future, we are going to improve the algorithm of language model training and the strategy for OOV word identification by overcoming the defects mentioned above. To compare our performance against other more systems in a more comprehensive way, we will participate in the international Chinese word segmentation bake-off and experiment with public database offered by the bake-off.

References

1. Cheng, K.-S., Young, G.H., Wong, K.-F.: A study on word-based and integral-bit Chinese text compression algorithms. Journal of the American Society for Information Science 50(4), 18 C228 (1999)
2. Zou, F.: The Identification of Stop Words and Keywords: A Study of Automatic Term Weighting in Natural Language Text Processing. MPhil Thesis (June 2006)
3. Mao, J., Cheng, G., He, Y.: Phrase-based Statistical Language Modeling from Bilingual Parallel Corpus. In: The International Symposium on Combinatorics, Algorithms, Probabilistic and Experimental methodologies (April 2007)

4. Jurafsky, D., Martin, J.H.: Speech and Language Processing: An introduction to speech recognition, computational linguistics and natural language processing. Prentice-Hall, Englewood Cliffs (2006)
5. Gao, J., Wu, A., Li, M., Huang, C.-N.: Chinese Word Segmentation and Named Entity Recognition: A Pragmatic Approach. Computational Linguistics 31(4), 531–574 (2005)
6. Stolcke, A.: SRILM - An Extensible Language Modeling Toolkit. In: Proceeding of International Conference of Spoken Language Processing (September 2002)

An $O(nm)$-Time Certifying Algorithm for Recognizing HHD-Free Graphs*

Stavros D. Nikolopoulos and Leonidas Palios

Department of Computer Science, University of Ioannina
GR-45110 Ioannina, Greece
{stavros, palios}@cs.uoi.gr

Abstract. In this paper, we consider the recognition problem on a class of perfectly orderable graphs, namely, the HHD-free graphs, i.e., graphs that do not contain any induced subgraph isomorphic to a house, a hole, or a domino. We prove properties of the HHD-free graphs which enable us to present an $O(n\,m)$-time and $O(n+m)$-space algorithm for determining whether a given graph G on n vertices and m edges is HHD-free. The algorithm can be augmented to provide a certificate (an induced house, hole, or domino) whenever it decides that the input graph is not HHD-free; the certificate computation requires $O(n+m)$ additional time and $O(n)$ space.

Keywords: HHD-free graphs, perfectly orderable graphs, certifying algorithms, recognition.

1   Introduction

A linear order $\prec$ on the vertices of a graph G is *perfect* if the ordered graph $(G, \prec)$ contains no induced P_4 $abcd$ with $a \prec b$ and $d \prec c$ (such a P_4 is called an *obstruction*). In the early 1980s, Chvátal [2] defined the class of graphs that admit a perfect order and called them *perfectly orderable* graphs. The interest in perfectly orderable graphs comes from the fact that several problems in graph theory, which are NP-complete in general graphs, have polynomial-time solutions in graphs that admit a perfect order [1,5]; unfortunately, it is NP-complete to decide whether a graph admits a perfect order [11]. Since the recognition of perfectly orderable graphs is NP-complete, we are interested in characterizing graphs which form polynomially recognizable subclasses of perfectly orderable graphs. Many such classes of graphs, with very interesting structural and algorithmic properties, have been defined so far and shown to admit polynomial-time recognitions (see [1,5]); note however that not all subclasses of perfectly orderable graphs admit polynomial-time recognition [7].

* This research was co-funded by the European Union in the framework of the program "Pythagoras II" of the "Operational Program for Education and Initial Vocational Training" of the 3rd Community Support Framework of the Hellenic Ministry of Education, funded by national sources and the European Social Fund (ESF).

F.P. Preparata and Q. Fang (Eds.): FAW 2007, LNCS 4613, pp. 281–292, 2007.

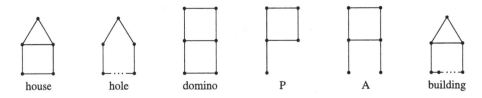

house   hole   domino   P   A   building

Fig. 1. Some simple graphs

In this paper, we consider the class of HHD-free graphs: a graph is *HHD-free* if it contains no induced subgraph isomorphic to a house, a hole (i.e., a chordless cycle on ≥ 5 vertices), or a domino (see Figure 1). The HHD-free graphs properly generalize the class of chordal (or triangulated) graphs [5]. In [8], Hoàng and Khouzam proved that the HHD-free graphs admit a perfect order, and thus are perfectly orderable. A superclass of the HHD-free graphs, which also properly generalizes the class of chordal graphs, is the class of HH-free graphs: a graph is *HH-free* if it contains no induced subgraph isomorphic to a house or a hole. Although an HH-free graph is not necessarily perfectly orderable, the complement of any HH-free graph is; this was conjectured by Chvátal and proved by Hayward [6].

Hoàng and Khouzam [8], while studying the class of brittle graphs (a well-known class of perfectly orderable graphs which contains the HHD-free graphs), showed that HHD-free graphs can be recognized in $O(n^4)$ time, where n denotes the number of vertices of the input graph. An improved result was obtained by Hoàng and Sritharan [9] who presented an $O(n^3)$-time algorithm for recognizing HH-free graphs and showed that HHD-free graphs can be recognized in $O(n^3)$ time as well; their algorithm processes each vertex v of the input graph by computing the chordal completion of the (ordered) non-neighbors of v, and by checking whether the resulting graph is chordal. A further improvement was achieved by Nikolopoulos and Palios [13]: based on properties characterizing the chordal completion of a graph, they were able to avoid performing the chordal completion step, which is the most time-consuming ingredient of the algorithm in [9], and described algorithms for recognizing HH-free and HHD-free graphs that require $O(n \min\{m\,\alpha(n,n), m + n \log n\})$ time and $O(n+m)$-space, where m is the number of edges of the input graph, and $\alpha(\,,\,)$ denotes the very slowly growing functional inverse of Ackerman's function.

On other related classes of perfectly orderable graphs, Eschen *et al.* [4] recently described recognition algorithms for several of them, among which a recognition algorithm for HHP-free graphs; a graph is HHP-free if it contains no hole, no house, and no "P" as induced subgraphs (see Figure 1). Their algorithm is based on the property that every HHP-free graph is HHDA-free graph (a graph with no induced hole, house, domino, or "A"), and thus a graph G is HHP-free graph if and only if G is HHDA-free and contains no "P" as an induced subgraph. The characterization of HHDA-free graphs due to Olariu [15] (a graph G is HHDA-free if and only if every induced subgraph of G either is chordal or contains a non-trivial module) and the use of modular decomposition [10] allowed Eschen *et al.* to present an $O(n\,m)$-time recognition algorithm for HHP-free graphs.

In this paper, we present a new, faster algorithm for recognizing HHD-free graphs. For each vertex v of a given graph G, our algorithm computes the partition of the non-neighbors of v into sets of vertices based on their common neighbors with v, and following that, the connected components of the subgraphs induced by these partition sets. We show that if G is HHD-free, the graph obtained from G by shrinking each of these connected components into a single vertex is "almost chordal." As a result, we obtain an $O(n\,m)$-time and $O(n+m)$-space algorithm for determining whether a graph on n vertices and m edges is HHD-free. We also describe how the algorithm can be augmented to provide a certificate (an induced house, hole, or domino) whenever it decides that the input graph is not HHD-free; the certificate computation requires $O(n+m)$ additional time and $O(n)$ space.

2 Terminology - Notation

We consider finite undirected graphs with no loops or multiple edges. Let G be such a graph; then, $V(G)$ and $E(G)$ denote the set of vertices and of edges of G respectively. The subgraph of G induced by a subset S of G's vertices is denoted by $G[S]$. The vertices adjacent to a vertex x of G form the *neighborhood* $N(x)$ of x; the cardinality of $N(x)$ is the *degree* of x. The *closed neighborhood* of x is defined as $N[x] := N(x) \cup \{x\}$. We extend the notion of the neighborhood to sets as follows: for a set $A \subseteq V(G)$, we define $N(A) := \left(\bigcup_{x \in A} N(x) \right) - A$ and $N[A] := N(A) \cup A$.

A *path* in a graph G is a sequence of vertices $v_0 v_1 \cdots v_k$ such that $v_{i-1} v_i \in E(G)$ for $i = 1, 2, \ldots, k$; we say that this is a path from v_0 to v_k and that its *length* is k. A path is called *simple* if none of its vertices occurs more than once; it is called *trivial* if its length is equal to 0. A path (simple path) $v_0 v_1 \cdots v_k$ is called a *cycle* (*simple cycle*) of length $k + 1$ if $v_0 v_k \in E(G)$. An edge connecting two non-consecutive vertices in a simple path (cycle) is called a *chord*; then, a simple path (cycle) $v_0 v_1 \cdots v_k$ of a graph G is *chordless* if G contains no chords of the path (cycle), i.e., $v_i v_j \notin E(G)$ for any two non-consecutive vertices v_i, v_j in the path (cycle). The chordless path (chordless cycle, respectively) on n vertices is commonly denoted by P_n (C_n, respectively).

A *connected component* of a graph G is a maximal set $A \subseteq V(G)$ such that the subgraph $G[A]$ is connected, i.e., there exists a path in G connecting any two vertices in A.

3 The Algorithm

In a fashion similar to the algorithms in [9,13], our algorithm processes each vertex v of the input graph G and checks whether v participates in a hole, is the top vertex of a house or a building (see Figure 1), or is a corner vertex of a domino. Note that all these subgraphs include a path $y_1 uvwy_2$ where y_1, y_2 are non-neighbors of v; it is interesting to observe that the vertices y_1, y_2 have different common neighbors with v. This suggests that it may be a good idea to

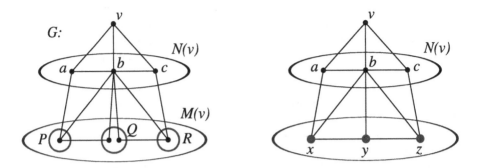

Fig. 2. Shrinking the partition sets into vertices may lead to error

partition the set of non-neighbors of v based on their common neighbors with v, to shrink each of these partition sets into a single super-vertex, and then to work with the resulting graph. This is reinforced by the following observation.

Observation 1. *Let G be a graph, v a vertex of G, and y_1, y_2 be two non-neighbors of v in G such that $|N(y_1) \cap N(v)| = |N(y_2) \cap N(v)|$ and $N(y_1) \cap N(v) \neq N(y_2) \cap N(v)$. If $y_1 y_2 \in E(G)$, then G contains an induced house or C_5.*

Proof. Since $N(y_1) \cap N(v) \neq N(y_2) \cap N(v)$ and $|N(y_1) \cap N(v)| = |N(y_2) \cap N(v)|$, there exist vertices $u, w \in N(v)$ such that $u \in N(y_1) - N(y_2)$ and $w \in N(y_2) - N(y_1)$. But then, the vertices v, u, y_1, y_2, w induce a house or a C_5 depending on whether u, w are adjacent or not. ∎

However, shrinking each of the different partition sets into a single vertex leads to error as the following example indicates: consider the graph G on the left of Figure 2 which contains no hole, house, or domino; the partition of the non-neighbors of v based on the common neighbors with v yields the sets P, Q, R; shrinking these sets into vertices x, y, z, respectively, yields the graph on the right of Figure 2, which contains the hole $vaxyzc$.

A closer look at the example reveals that the error is due to the fact that the two connected components of the subgraph $G[Q]$ induced by the partition set Q in Figure 2 were shrunk into the same vertex. This suggests that if we intend to apply a shrinking mechanism, we need to treat the connected components of the partition sets as separate entities. In detail, we do the following:

▷ consider the partition of the non-neighbors $M(v)$ of vertex v in G based on the common neighbors of the vertices in $M(v)$ with v and let $(S_1, S_2, \ldots, S_\ell)$ be an ordering of the partition sets by non-decreasing number of such common neighbors;

▷ for each set S_i, consider the connected components of the subgraph $G[S_i]$;

▷ we construct an auxiliary graph G_v by shrinking each of these connected components into a single vertex: namely, for each $i = 1, 2, \ldots, \ell$, we define

$$Z_i = \{ z_{C_1}, z_{C_2}, \ldots, z_{C_{t_i}} \} \tag{1}$$

where $C_1, C_2, \ldots, C_{t_i}$ are the conn. components of the subgraph $G[S_i]$; then

$$V(G_v) = \{v\} \cup N(v) \cup \left(\bigcup_{i=1}^{\ell} Z_i\right)$$

$$\begin{aligned} E(G_v) = {} & \{\, uw \mid u, w \in \{v\} \cup N(v) \text{ such that } uw \in E(G) \,\} \\ & \cup \{\, u\, z_C \mid u \in N(v) \text{ and } \exists\, ux \in E(G), \text{ where} \\ & \qquad x \in \text{a conn. component } C \text{ of a } G[S_i] \,\} \\ & \cup \{\, z_C\, z_{C'} \mid \exists\, xy \in E(G), \text{ where} \\ & \qquad x, y \in \text{conn. components } C \text{ and } C' \text{ of } G[S_i] \text{ and } G[S_j], \text{ resp.,} \\ & \qquad \text{and } i \neq j \,\}. \end{aligned}$$

Note that when a component C is shrunk into a vertex z_C, then (i) z_C is adjacent to a vertex $u \in N(v)$ iff there exists a vertex $x \in C$ such that $ux \in E(G)$, and (ii) z_C is adjacent to vertex $z_{C'}$ corresponding to a component $C' \neq C$ iff there exist vertices $x \in C$ and $y \in C'$ such that $xy \in E(G)$. The following result has important implications for the graph G_v.

Lemma 1. *Let G be an HHD-free graph, v a vertex of G, and A, B, C connected components of the subgraphs $G[S_i], G[S_j], G[S_k]$ induced by three distinct partition sets S_i, S_j, S_k, respectively, where $i < j < k$. Suppose further that there exist non-neighbors x, x', y, z of v in G, where $x, x' \in A$, $y \in B$, $z \in C$, such that $xz \in E(G)$ and $x'y \in E(G)$. Then, $yz \in E(G)$.*

In terms of the graph G_v, Lemma 1 implies the following corollary:

Corollary 1. *Let G be an HHD-free graph, v a vertex of G, G_v the auxiliary graph described earlier in terms of G and v, and z_A, z_B, z_C be vertices of G_v such that $z_A \in Z_i$, $z_B \in Z_j$, and $z_C \in Z_k$ where $i < j < k$. If $z_A z_B \in E(G_v)$ and $z_A z_C \in E(G_v)$, then $z_B z_C \in E(G_v)$.*

Lemma 1 and Corollary 1 prove very useful in the special case in which the graph G is such that for every edge $xy \in E(G)$, where x belongs to a connected component A of a subgraph $G[S_i]$ and y belongs to a connected component B of $G[S_j]$ with $j > i$, no vertex in A is adjacent to any vertex in $S_j - B$. In this case, in the auxiliary graph G_v, for all $1 \leq i < j \leq \ell$, any vertex in Z_i (i.e., corresponding to a connected component of the subgraph $G[S_i]$) is adjacent to at most one vertex in each Z_j (i.e., corresponding to a component of $G[S_j]$). Then, if G is HHD-free, Corollary 1 implies that the subgraph $G_v[\bigcup_{t=1}^{\ell} Z_t]$ of G_v induced by the non-neighbors of v is chordal, and hence no chordal completion is needed. In general, however, G may have edges xy and $x'z$, where x, x' belong to a connected component of a subgraph $G[S_i]$ and y, z belong to distinct connected components of $G[S_j]$ with $j > i$; then, we take advantage of the following lemma.

Lemma 2. *Let G be an HHD-free graph, A a connected component of the subgraph $G[S_i]$, and B, B' distinct connected components of $G[S_j]$, where $j > i$, such that G contains edges connecting a vertex in A to a vertex in B, and a vertex in A to a vertex in B'. Then:*

(i) In G, each vertex in A that is adjacent to at least one vertex in $B \cup B'$ is adjacent to all the vertices in $B \cup B'$.

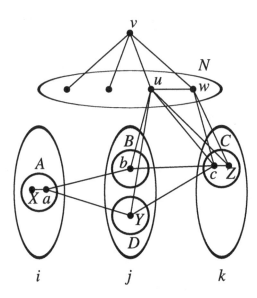

Fig. 3. The vertices in $B \cup B'$ are not necessarily adjacent to all the vertices in $A \cup C$

(ii) *If C is a connected component of the subgraph $G[S_k]$, where $k > j$, such that G contains an edge connecting a vertex in B to a vertex in C. Then, in G, each vertex in C that is adjacent to at least one vertex in $B \cup B'$ is adjacent to all the vertices in $B \cup B'$.*

It is worth noting that although in an HHD-free graph, each vertex in $A \cup C$ that is adjacent to at least one vertex in $B \cup B'$ is in fact adjacent to each vertex in $B \cup B'$, it is not necessarily true that each vertex in $B \cup B'$ is adjacent to each vertex in $A \cup C$; consider the HHD-free graph shown in Figure 3.

Lemma 2, statement (ii) implies that if in an HHD-free graph G there exist edges connecting vertices of a connected component A of a subgraph $G[S_i]$ to vertices in at least 2 connected components $B_1, B_2, \ldots, B_k$ of a subgraph $G[S_j]$ where $j > i$, then the vertices in $B_1 \cup B_2 \cup \ldots \cup B_k$ are adjacent to the exact same neighbors in $\left(\bigcup_{r=j+1}^{\ell} S_r\right)$; additionally, as the components $B_1, B_2, \ldots, B_k$ are all subsets of S_j, the vertices in $B_1 \cup B_2 \cup \ldots \cup B_k$ are adjacent to the same vertices in $N(v)$, and thus they are adjacent in G to the exact same neighbors in $\left(\bigcup_{r=j+1}^{\ell} S_r\right) \cup N(v)$. In terms of the graph G_v, this implies that the vertices $z_{B_1}, z_{B_2}, \ldots, z_{B_k}$ corresponding to the connected components $B_1, B_2, \ldots, B_k$ of $G[S_j]$ are adjacent to the exact same neighbors in $\left(\bigcup_{r=j+1}^{\ell} Z_r\right) \cup N(v)$. Moreover, the transitivity of equality leads to the following corollary.

Corollary 2. *Let G be an HHD-free graph, and $A_1, A_2, \ldots, A_h$ be connected components of subgraphs $G[S_r]$, where $r < j$, such that for each $i = 1, 2, \ldots, h$,*

- *the graph G contains edges connecting vertices in at least 2 distinct components of $G[S_j]$ to vertices in A_i, and*

– *there exists a component B of $G[S_j]$ such that at least one vertex in A_i and at least one vertex in $\bigcup_{t=1}^{i-1} N(A_t)$ are adjacent to a vertex in B.*

(in terms of the graph G_v, for each i, we have that $|N_{G_v}(z_{A_i}) \cap Z_j| \geq 2$ and $\left(N_{G_v}(z_{A_i}) \cap \left(\bigcup_{t=1}^{i-1} N_{G_v}(z_{A_t}) \right) \right) \cap Z_j \neq \emptyset$). Then,

(i) *in G, all the vertices in $\bigcup_{i=1}^{h} \left(N(A_i) \cap S_j \right)$ are adjacent to the exact same neighbors in $\left(\bigcup_{r=j+1}^{\ell} S_r \right) \cup N(v)$;*

(ii) *in G_v, all the vertices in $\bigcup_{i=1}^{h} \left(N_{G_v}(z_{A_i}) \cap Z_j \right)$ are adjacent to the exact same neighbors in $\left(\bigcup_{r=j+1}^{\ell} Z_r \right) \cup N(v)$.*

In order to be able to check that case (ii) of Corollary 2 holds for the graph G_v, we maintain a partition $\mathcal{P}_{Z_j}$ of the vertices in Z_j that correspond to the connected components of the subgraph $G[S_j]$: initially, each partition set contains a single vertex of Z_j; every time we process a vertex z_{A_i} which is adjacent to vertices $z_{B_1}, z_{B_2}, \ldots, z_{B_t} \in Z_j$, where $t \geq 2$, we union the sets of the partition $\mathcal{P}_{Z_j}$ that contain the vertices $z_{B_1}, z_{B_2}, \ldots, z_{B_t}$. Then, when we process the vertices in Z_j, we check that the vertices in each partition set are adjacent to the exact same neighbors in $\left(\bigcup_{r=j+1}^{\ell} Z_r \right) \cup N(v)$.

Based on these results, in Figure 4 we give Algorithm Recognize-HHD-free, in which, for each vertex v of the given graph G, we construct the auxiliary graph G_v by shrinking each of the connected components of each of the subgraphs $G[S_i]$, $i = 1, 2, \ldots, \ell$, into a single vertex, and then, for each $i = 1, 2, \ldots, \ell$, we process the set Z_i of vertices corresponding to the connected components of $G[S_i]$ together (note that, by construction, in the graph G_v there are no edges between any two vertices $z_C, z_{C'} \in Z_i$). We also mention that Step 1.4 of our algorithm, which is applied on G_v, is an extension of the linear-time algorithm for testing whether an ordering of the vertices of a given graph is a perfect elimination ordering [5,16].

We note that when we check that all the vertices in a set P_j are adjacent to the same neighbors, we check only the neighbors in $\bigcup_{r=i+1}^{\ell} Z_r$, and not in $\left(\bigcup_{r=i+1}^{\ell} Z_r \right) \cup N(v)$ as suggested by Corollary 2, statement (ii); this is so, because all the vertices in P_j correspond to connected components of the subgraph $G[S_i]$ for the same set S_i, and so, by construction, they have the same neighbors in $N(v)$. Additionally, when this check is successful, we take advantage of the fact that all the vertices in P_j are adjacent to the same neighbors in $\bigcup_{r=i+1}^{\ell} Z_r$, when we union the sets of the partition $\mathcal{P}_{Z_k}$ that contain the vertices in W; we need only do the unioning once for the entire set P_j rather than once for each vertex in the set.

The correctness of the Algorithm Recognize-HHD-free is established in Theorem 1 with the help of Lemma 3.

Lemma 3. *Let G be a graph, v a vertex of G, G_v the auxiliary graph described earlier in terms of G and v, and Z_i $(i = 1, 2, \ldots, \ell)$ the sets of vertices of G_v defined in Eqn. (1). Suppose that there exist vertices $z_A \in Z_i$ and $z_{A'} \in Z_j$ of*

Algorithm Recognize-HHD-free

Input: an undirected graph G
Output: a message stating that G is HHD-free or not

1. **for** each vertex v of G **do**
   1.1. Compute the neighbors $N(v)$ and non-neighbors $M(v) = V(G) - (N(v) \cup \{v\})$ of v in G;
      compute the partition $\mathcal{S}_v$ of the set $M(v)$ based on the common neighbors of the vertices in $M(v)$ with v in G, and order the partition sets by non-decreasing number of such common neighbors; let $\mathcal{S}_v = (S_1, S_2, \ldots, S_\ell)$ be the resulting ordering;
   1.2. **for** each edge xy of G, where $x, y \in M(v)$, **do**
      if x, y belong to different partition sets of $\mathcal{S}_v$
         and $|N(x) \cap N(v)| = |N(y) \cap N(v)|$
         then print("The graph G is not HHD-free"); **exit**;
   1.3. Construct the auxiliary graph G_v by shrinking each connected component C of the subgraphs $G[S_i]$, $i = 1, 2, \ldots, \ell$, into a single vertex z_C;
   1.4. **for** $i \leftarrow 1$ to ℓ **do**
      form a partition $\mathcal{P}_{Z_i}$ of the set Z_i of vertices of G_v that correspond to the connected components of $G[S_i]$, by placing each of these vertices in a separate partition set;
      for each vertex $z_C \in Z_i$ **do**
         associate with z_C an initially empty set $A(z_C)$;
      for $i \leftarrow 1$ to ℓ **do**
      let the partition $\mathcal{P}_{Z_i}$ of Z_i be $\mathcal{P}_{Z_i} = \{P_1, P_2, \ldots, P_t\}$;
      for $j \leftarrow 1$ to t **do**
         let z_C be any vertex contained in the set P_j;
         $X' \leftarrow N_{G_v}(z_C) \cap \left(\bigcup_{r=i+1}^{\ell} Z_r \right)$;
         if P_j is not a singleton set
         then if there exists a vertex in P_j that is not adjacent in G_v to a vertex in X' or is adjacent to a vertex in $\left(\bigcup_{r=i+1}^{\ell} Z_r \right) - X'$
               then print("The graph G is not HHD-free"); **exit**;
         if $X' \neq \emptyset$
         then let Z_k be the minimum-index set such that $X' \cap Z_k \neq \emptyset$;
               $W \leftarrow X' \cap Z_k$;
               union the sets of the partition $\mathcal{P}_{Z_k}$ (of Z_k) that contain the vertices in W;
               $X \leftarrow X' \cup (N_{G_v}(z_C) \cap N(v))$;
               choose any $z_B \in W$ and concatenate the set $X - W$ to $A(z_B)$;
         if $\left(\bigcup_{z \in P_j} A(z) \right) - N_{G_v}(z_C) \neq \emptyset$
         then print("The graph G is not HHD-free"); **exit**;
2. **print**("The graph G is HHD-free");

Fig. 4. Algorithm Recognize-HHD-free

G_v corresponding to connected components A, A' of subgraphs $G[S_i]$ and $G[S_j]$, respectively, where $i < j$, such that $z_A z_{A'} \in E(G_v)$ and there exists a vertex $x \in$

$(\bigcup_{r=j+1}^{\ell} Z_r) \cup N(v)$ such that $x \in N_{G_v}(z_A) - N_{G_v}(z_{A'})$. Then if Algorithm Recognize-HHD-free is run on G, it reports that G is not HHD-free and stops.

Theorem 1. When Algorithm Recognize-HHD-free is run on a graph G, it reports that G is not HHD-free if and only if G is indeed not HHD-free.

Remark

We note that in the course of Algorithm Recognize-HHD-free we do not check whether the conditions of Lemma 2, statement (i), hold, although we know that if they do not hold then the input graph G is not HHD-free. In fact, we do not check the conditions of Lemma 2, statement (ii), either; instead, we check the weaker conditions stated in Corollary 2. Nevertheless, the conditions that we check suffice to enable us to recognize HHD-free graphs.

3.1   Time and Space Complexity

Let n be the number of vertices and m be the number of edges of the graph G. Since each of the forbidden subgraphs that we are looking for (a house, a hole, or a domino) is connected, we may assume that G is connected, otherwise we work on G's connected components which we can compute in $O(n+m)$ time [3]; thus, $n = O(m)$. Below, we give the time and space complexity of each step of Algorithm Recognize-HHD-free.

For a vertex v, its neighbors and non-neighbors in the graph G can be stored in $O(n)$-size arrays for constant-time access; this takes $O(n)$ time. The partition $\mathcal{S}_v$ can be computed in $O(m + n\, deg(v))$ time time and $O(n)$ space, where $deg(v)$ denotes the degree of v in G; see [12][1]. After having computed for a vertex of each of the partition sets of $\mathcal{S}_v$, the number of its common neighbors with v, which can be done in $O(n + m)$ time, we can form the ordered sequence $(S_1, S_2, \ldots, S_\ell)$ in $O(\ell + deg(v)) = O(n)$ time and $O(n)$ space using bucket sorting. Thus, Step 1.1 takes $O(m + n\, deg(v))$ time and $O(n)$ space in total.

Step 1.2 takes $O(m)$ time assuming that each non-neighbor of vertex v stores the index of the set of the partition $\mathcal{S}_v$ to which it belongs; storing this information on each such vertex can be done in $O(n)$ time by traversing the sets $S_1, S_2, \ldots, S_\ell$. Thus, Step 1.2 takes $O(n + m)$ time and $O(n)$ space.

Adjacency-list representations of the subgraphs $G[S_i]$, $i = 1, 2, \ldots, \ell$, can be obtained in $O(n+m)$ time and space by appropriate partitioning of a copy of an adjacency-list representation of the graph G and removal of unneeded records; then, computing the connected components of all these subgraphs takes a total of $O(n + m)$ time and space, from which the graph G_v can be constructed in $O(n + m)$ additional time and space. Thus, Step 1.3 takes a total of $O(n + m)$ time and space.

---

[1] An algorithm to construct a partition of a set L_2 in terms of adjacency to elements of a set L_1 is given in Section 3.2 of [12] with a stated time complexity of $O(m+n\,|L_2|)$; yet, it can be easily seen that the algorithm has a time complexity of $O(m+|L_1|\cdot|L_2|)$, which in our case gives $O(m + n\, deg(v))$ since $|L_1| = deg(v)$ and $|L_2| = O(n)$.

Crucial for Step 1.4 is the construction and processing of the partitions $\mathcal{P}_{Z_i}$, $i = 1, 2, \ldots, \ell$. These are maintained by means of an auxiliary (multi-)graph H_v: members of the same partition set belong to the same connected component of H_v. The graph H_v has one vertex for each connected component of each subgraph $G[S_i]$, $i = 1, 2, \ldots, \ell$; hence, with a slight abuse of notation, we can write that $V(H_v) = \bigcup_{i=1}^{\ell} Z_i$. Initially, the graph H_v has no edges.

▷ In order to compute the partition $\mathcal{P}_{Z_i}$ of the set Z_i, we compute the connected components of the subgraph $H_v[Z_i]$; graph traversing algorithms, such as, depth-first search and breadth-first search, can be used on $H_v[Z_i]$ to yield the connected components in time linear in the number of vertices and edges of $H_v[Z_i]$.

▷ In order to union the sets of a partition $\mathcal{P}_{Z_k}$ that contain the vertices in a set W, we pick a vertex, say, $z \in W$, and add edges in H_v connecting z to all the other vertices in W; this takes $O(|W|)$ time and space.

The above description implies that forming the initial partitions $\mathcal{P}_{Z_i}$ for $i = 1, 2, \ldots, \ell$, where each vertex in Z_i is placed in a separate partition set, corresponds to constructing the graph H_v without any edges; this can be done in $O(n)$ time and space. Initializing the sets $A(\ )$ for all the vertices in $\bigcup_{i=1}^{\ell} Z_i$ also takes $O(n)$ time and space. Now, for each $i = 1, 2, \ldots, \ell$, computing the partition $\mathcal{P}_{Z_i}$ takes time linear in the number of vertices and edges of $H_v[Z_i]$. Let us consider the processing of a set P_j of the partition $\mathcal{P}_{Z_i}$. Computing the set X' takes $O(deg_{G_v}(z_C))$ time and space, where $deg_{G_v}(z_C)$ denotes the degree of vertex z_C in the graph G_v. Checking if P_j is a singleton set takes $O(1)$ time, and checking if all the vertices in P_j are adjacent to exactly the vertices in X' among the vertices in $\bigcup_{i=i+1}^{\ell} Z_i$ takes $O(\sum_{z \in P_j} deg_{G_v}(z))$ time. Next, checking whether X' is non-empty takes $O(1)$ time while doing all the processing if $X' \neq \emptyset$ takes $O(deg_{G_v}(z_C))$ time and space; note that $|W| \leq |X'| \leq deg_{G_v}(z_c)$ and recall that unioning the sets of the partition $\mathcal{P}_{Z_k}$ that contain the vertices in W involves adding $|W| - 1$ edges in H_v. Finally, checking whether $(\bigcup_{z \in P_j} A(z)) - N_{G_v}(z_C) \neq \emptyset$ takes $O(\bigcup_{z \in P_j} |A(z)|)$ time. In summary, processing the set P_j takes $O(\sum_{z \in P_j} (deg_{G_v}(z) + |A(z)|))$ time and $O(deg_{G_v}(z_C))$ space. Since the sets of each partition $\mathcal{P}_{Z_i}$ of the set Z_i are disjoint and the sets Z_i are disjoint, we have that $O(\sum_{i=1}^{\ell} \sum_{P_j \in \mathcal{P}_{Z_i}} \sum_{z \in P_j} deg_{G_v}(z)) = O(|V(G_v)| + |E(G_v)|) = O(n + m)$. Additionally, since the sets $A(\ )$ are formed by concatenating some of the neighbors in G_v of one vertex z_C from each set P_j, we have that $O(\sum_{i=1}^{\ell} \sum_{P_j \in \mathcal{P}_{Z_i}} \sum_{z \in P_j} |A(z)|) = O(\sum_{i=1}^{\ell} \sum_{P_j \in \mathcal{P}_{Z_i}} \sum_{z \in P_j} deg_{G_v}(z)) = O(|V(G_v)| + |E(G_v)|) = O(n + m)$ as well. Thus, in total, Step 1.4 takes $O(n + m)$ time and space.

Since Steps 1.1-1.4 are executed for each vertex v of the input graph G and Step 2 takes $O(1)$ time, we have that the overall time complexity of Algorithm Recognize-HHD-free is:

$$\left[\sum_{v \in V(G)} \left(O(m + n \, deg(v)) \right) + O(n + m) \right] + O(1) = O(nm).$$

Therefore, we have:

Theorem 2. *Let G be an undirected graph on n vertices and m edges. Then, Algorithm Recognize-HHD-free determines whether G is an HHD-free graph in $O(nm)$ time and $O(n+m)$ space.*

3.2 Providing a Certificate

Algorithm Recognize-HHD-free can be made to provide a certificate (a house, a hole, or a domino) whenever it decides that the input graph G is not HHD-free. The algorithm reports that the graph G is not HHD-free in three occasions, once in Step 1.2 and twice in Step 1.4. Next, we give some of the details of how we handle these cases.

Step 1.2: In this case, we have two non-neighbors x, y of v which have the same number of common neighbors with v and belong to different partition sets. Then, there exist vertices $u \in (N(x) \cap N(v)) - N(y)$ and $w \in (N(y) \cap N(v)) - N(x)$; we can find these vertices in $O(n)$ time using $O(n)$ space by traversing the adjacency lists of x and of y and marking the neighbors of x and y in two $O(n)$-size arrays, and then by traversing these two arrays. The vertices v, u, w, x, y induce a house or a C_5 depending on whether u, w are adjacent in G or not.

Step 1.4: In order to be able to efficiently produce a certificate when Algorithm Recognize-HHD-free reports that the input graph G is not HHD-free in Step 1.4, we do the following additional work:

W_1: Whenever, during the processing of a set P_j of a partition $\mathcal{P}_{Z_i}$, we need to union the sets of a partition $\mathcal{P}_{Z_k}$ containing the vertices in the set W, which is done by adding edges in the auxiliary multi-graph H_v (as explained in Section 3.1), we associate with each such edge the selected vertex z_C of P_j.

W_2: When processing a set P_j, we store with each element of the set $X - W$, which is added to $A(z_B)$ for $z_B \in W$, a reference to the selected vertex z_C of P_j; in this way, for each vertex z, each element of the set $A(z)$ carries a reference to a vertex of the set during whose processing this element was added to $A(z)$.

Note that this additional work does not increase asymptotically the time and space complexity of the algorithm. Moreover, thanks to this work, we can produce a certificate whenever Algorithm Recognize-HHD-free reports in Step 1.4 that the graph G is not HHD-free in $O(n+m)$ additional time using $O(n)$ additional space (due to space limitation, the details are omitted but can be found in [14]). Thus, we have:

Theorem 3. *Let G be an undirected graph on n vertices and m edges. Then, Algorithm Recognize-HHD-free can be augmented to produce a house, a hole, or a domino whenever it decides that G is not an HHD-free graph in $O(n+m)$ additional time and $O(n)$ additional space.*

4   Concluding Remarks

We have presented a recognition algorithm for the class of HHD-free graphs that runs in $O(n\,m)$ time and requires $O(n+m)$ space, where n is the number of vertices and m is the number of edges of the input graph. The algorithm can be augmented to yield, in $O(n+m)$ time and $O(n)$ space, a certificate (a house, a hole, or a domino) whenever it decides that the input graph is not HHD-free.

Despite the close relation between HHD-free and HH-free graphs, our results do not lead to an improvement in the recognition time complexity for HH-free graphs; therefore, we leave as an open problem the design of an $O(n\,m)$-time algorithm for recognizing HH-free graphs. Additionally, it would be interesting to obtain faster recognition algorithms for other related classes of graphs, such as, the brittle and the semi-simplicial graphs.

References

1. Brandstädt, A., Le, V.B., Spinrad, J.P.: Graph classes: A survey. SIAM Monographs on Discrete Mathematics and Applications (1999)
2. Chvátal, V.: Perfectly ordered graphs. Annals of Discrete Math. 21, 63–65 (1984)
3. Cormen, T.H., Leiserson, C.E., Rivest, R.L., Stein, C.: Introduction to Algorithms, 2nd edn. MIT Press, Cambridge (2001)
4. Eschen, E.M., Johnson, J.L., Spinrad, J.P., Sritharan, R.: Recognition of some perfectly orderable graph classes. Discrete Appl. Math. 128, 355–373 (2003)
5. Golumbic, M.C.: Algorithmic Graph Theory and Perfect Graphs. Academic Press, London (1980)
6. Hayward, R.: Meyniel weakly triangulated graphs I: co-perfect orderability. Discrete Appl. Math. 73, 199–210 (1997)
7. Hoàng, C.T.: On the complexity of recognizing a class of perfectly orderable graphs. Discrete Appl. Math. 66, 219–226 (1996)
8. Hoàng, C.T., Khouzam, N.: On brittle graphs. J. Graph Theory 12, 391–404 (1988)
9. Hoàng, C.T., Sritharan, R.: Finding houses and holes in graphs. Theoret. Comput. Sci. 259, 233–244 (2001)
10. McConnell, R.M., Spinrad, J.: Linear-time modular decomposition and efficient transitive orientation. In: Proc. 5th Annual ACM-SIAM Symp. on Discrete Algorithms (SODA'94), pp. 536–545 (1994)
11. Middendorf, M., Pfeiffer, F.: On the complexity of recognizing perfectly orderable graphs. Discrete Math. 80, 327–333 (1990)
12. Nikolopoulos, S.D., Palios, L.: Algorithms for P_4-comparability graph recognition and acyclic P_4-transitive orientation. Algorithmica 39, 95–126 (2004)
13. Nikolopoulos, S.D., Palios, L.: Recognizing HH-free, HHD-free, and Welsh-Powell opposition graphs. Discrete Math. and Theoret. Comput. Sci. 8, 65–82 (2006)
14. Nikolopoulos, S.D., Palios, L.: An O(n m)-time Certifying Algorithm for Recognizing HHD-free Graphs, Technical Report TR-05-2007, Department of Computer Science, University of Ioannina (2007)
15. Olariu, S.: Weak bipolarizable graphs. Discrete Math. 74, 159–171 (1989)
16. Rose, D.J., Tarjan, R.E., Lueker, G.S.: Algorithmic aspects of vertex elimination on graphs. SIAM J. Comput. 5, 266–283 (1976)

Easy Problems for Grid-Structured Graphs

Detlef Seese

Institute AIFB, University Karlsruhe (TH)
D-76128 Karlsruhe, Germany
seese@aifb.uni-karlsruhe.de

Abstract. This article concentrates on classes of graphs containing large grids and having a very regular structure. Grid-structured hierarchical graphs are defined in [19] by giving a static graph defining the content of a cell of a d-dimensional grid, repeating this static graph in each cell and by connecting the vertices in cells of a local neighborhood corresponding to a finite transit function in a uniform way. It is shown that for each finitely represented class K of dynamic graphs all monotone graph properties and all first order (FO) problems can be solved in constant time O(1). This result improves the linear time computability of FO problems for graphs of bounded degree from [25].

Keywords: grid structured hierarchical graphs, first order problems, monotone properties, constant time.

1 Introduction

Many algorithmic problems of practical or theoretical interest are NP hard and permit at least till now no efficient algorithmic solution. Usually one tries to get solutions in polynomial time by restricting the structure of the input objects. While on one side almost all problems for arbitrary graphs are NP-hard, they become usually solvable in polynomial or even linear time when the input structures are restricted to structures of bounded width (band-width, tree-width, VF tree-width, strong tree-width, branch-width, path-width, clique-width, rank-width, cut-width, see [2], [14] for surveys on recent developments). For all these width-concepts large grids imply high width. More precisely let K be a class of graphs having arbitrary large grids (see below) as minors then the X-width of K is not bounded, where X is an arbitrary width-concept mentioned above. So it can not be expected that the ideas and concepts to prove that a large class of algorithmic problems can be solved in polynomial or even in linear time for graphs of bounded X-width work also for classes of graphs containing arbitrary large grids. In the contrary the containment of large grids often indicates high complexity of many problems, e.g. by coding tiling problems in a canonical way (see [23] or [14]).

But grids appear very often in many areas of application, e.g. in VLSI circuits (see e.g. [21]), in the process of adaptive grid refinement required for given accuracy of the numerical solution of partial differential equations (see [22]), and picture processing. Though grids can not be avoided in many applications, the input structures in applications are often not arbitrary – they result by using special repeated

Easy Problems for Grid-Structured Graphs, pp. 293–304, 2007.

operations of an engineering process or they posses many similar parts. Key examples are memory units, gate arrays, customized circuits and reusable software components.

We will concentrate here on graphs containing large grids but having a very regular, hierarchical structure. This regularity will be used to show that each problem definable in first order logic (FO logic) for such graphs is decidable in constant time (O(1)). This improves for these structures the results that FO problems for arbitrary graphs are in P and can be solved in logarithmic space ([16]) and that FO problems for graphs of bounded degree or locally bounded tree-width are computable in linear time (see [25], [9]). Moreover we will show that each monotone property, i.e. a property which is closed under taking subgraphs can be solved in constant time for such classes of structures.

The article is composed as follows. In section 2 grids and grid-structured hierarchical graphs are introduced and it is proved that all graph properties closed under taking substructures can be decided in constant time for each class of finitely represented grid structured graphs. Some tools from logic are introduced in section 3 to show that each FO problem can be solved in constant time for each class of finitely represented grid structured graphs. The article ends with some conclusions and open problems in section 4.

2 Grid Structured Graphs

Throughout the paper the following notations and conventions will be used. N denotes the set of non negative integers (natural numbers) and i, j, k, l, m, n are used to denote elements of N. For each natural number n we use n also as an abbreviation for the set $\{0,1, \ldots ,n-1\}$. The cardinality of a set X, i.e. the number of its elements is denoted as $|X|$. In the article we use standard terminology of graph theory, logic and computational complexity as it can be found in [4], [23], [6] or [5].

A *simple graph* $G := (V, E)$ with n *vertices* and m *edges* consists of a vertex set $V := V(G) = \{v_1, \ldots ,v_n\}$ and an edge set $E := E(G) = \{e_1, \ldots ,e_m\}$, where each edge is an unordered pair of (different) vertices. We write uv for the edge $\{u,v\}$ and say that u and v are adjacent, when $uv \in E(G)$. Remember, that in this interpretation uv = vu. In this case the vertices u and v are denoted the *endpoints* of the edge e = uv. All graphs are assumed to be finite. Exceptions of this rule are given explicitly. Let n, m > 0 be arbitrary natural numbers. The *n×m-grid* $Q_{n,m}$ (sometimes denoted also as *solid n×m-grid*) is the graph defined by $V(Q_{n,m}) := \{(i, j) : 0 \leq i < n \text{ and } 0 \leq j < m\}$ and $E(Q_{n,m}) := \{\{(i,j), (k,l)\}: |i - k| + |j - l| = 1, 0 \leq i,k < n \ 0 \leq j, l < m \}$. For n=m we write simply Q_n. Define the *infinite grid* $Q_{\infty,\infty} := (V(Q_{\infty,\infty}), E(Q_{\infty,\infty}))$ by $V(Q_{\infty,\infty}) = \{(i, j) : i \in N \text{ and } j \in N\}$ and $E(Q_{\infty,\infty}) := \{\{(i,j), (k,l)\}: |i - k| + |j - l| = 1 \text{ and } i, j, k, l \in N \}$. We will use also the natural generalization of this concept from 2 to an arbitrary finite number of dimensions. For integers $n_1, n_2, \ldots, n_d$, we will speak then of the *solid $n_1 \times n_2 \times \ldots \times n_d$-grid*. The solid $n_1 \times n_2 \times \ldots \times n_d$-grid is isomorphic to $P_{n_1} \times P_{n_2} \times \ldots \times P_{n_d}$, where P_{n_i} is the path of n_i vertices. The *toroidal $n_1 \times n_2 \times \ldots \times n_d$-grid* is the product $C_{n_1} \times C_{n_2} \times \ldots \times C_{n_d}$, where C_{n_i} is the cycle of n_i vertices. A *grid graph* is a finite node-induced subgraph of the infinite two-dimensional integer grid $Q_{\infty,\infty}$. A *solid grid graph* is a grid graph all of whose bounded faces have area one. Here it is assumed that this infinite grid is embedded in a canonical way in the Euclidean plane.

For such classes of structures it is easy to see that many algorithmic problems can be solved by using the regular structure of these graphs, e. g. CHROMATIC NUMBER, MONOCHROMATIC TRIANGLE, PARTITION INTO TRIANGLES, COVERING BY CLIQUES, CLIQUE, INDEPENDENT SET, BIPARTITE SUBGRAPH, PLANAR SUBGRAPH are trivial or follow from known results (see [10] for notation). This list is not complete and it can be assumed that there are many other problems, also other NP-complete problems from [10], which can be solved easily for d-dimensional solid grids just by considering the regular structure of these grids. Some more interesting results exist for the HAMILTONIAN CYCLE problem (see [17]), for the DOMINATION NUMBER (see [13]) and for PARITY DOMINATION problems (see e.g. [11]). Every graph contains an odd dominating set. These and related results motivate the question, whether there could be general principles to deduce efficient algorithms for graphs with a very regular grid structure.

We will consider here a more general class of graphs, the grid-structured hierarchical graphs. Such graphs have interesting applications in VLSI-design, systolic arrays, in the parallelization of algorithms and especially in building multi-dimensional schedules for system of uniform recurrence equations (see e.g. [19], [18], [21], [20]). Such uniform recurrence equations typically appear when finite difference approximations to systems of partial differential equations are constructed.

Grid-structured hierarchical graphs are defined by giving a static graph defining the content of a cell of a d-dimensional grid, repeating this static graph in each cell and by connecting the vertices in cells of a local neighborhood corresponding to a finite transit function in a uniform way (see [19], [15] for details). More precisely, a *d-dimensional static graph* is a system $S = (V, E, f)$, where $S\backslash f := (V, E)$ is a directed graph and $f: E \to Z^d$ is a transit function mapping a d-dimensional integer vector to each edge E. The *dynamic graph* S^x, for $x \in N^d$, is a (directed) graph (V', E') with:

$$V' := \{u^z : u \in V \land 0^\to \leq z < x\} \text{ and}$$
$$E' := \{(u^z, v^{z+f((u,v))}) : (u,v) \in E \land 0^\to \leq z, z+f((u,v)) < x\}.$$

Here we $0^\to$ denotes the vector of 0's, $\leq$ and $+$ are defined component-wise and $<$ is defined to be equivalent to "$\leq$ and at least one component has to be less". More precisely the relation $\leq$ on N^d is defined as follows. For all $x=(x_1,\ldots,x_d)$, $y=(y_1,\ldots,y_d) \in N^d$: $x \leq y$ if and only if $x_i \leq y_i$ for all i with $1 \leq i \leq d$. An *infinite dynamic graph* results by setting some of the components to infinity. The resulting grid structures play an important role in parallel programming where they are a natural model of flow graphs of uniform iterative or recursive programs. Graphs appearing in mesh-like chip architectures can be easily translated into such grid-structured graphs.

Each complete, solid $n_1 \times n_2 \times \ldots \times n_d$ grid can be represented as a dynamic graph $S^{(n_1,n_2,\ldots,n_d)}$ of a corresponding static graph S. A class K of d-dimensional dynamic graphs is *finitely represented* if there is a finite set D of static graphs, such that K is the set $K = \{S^{(n_1,n_2,\ldots,n_d)} : n_1, n_2, \ldots ,n_d \in N, S \in D\}$. It is an interesting question to characterize those problems which can be solved efficiently (in polynomial or linear time) for the classes if d-dimensional finitely represented dynamic graphs. When one considers the class of NP-complete problems as the basis to select problems, then with the exception of some isolated problems (see above) this problem is widely open. Of course two known results can be applied here. The first is to restrict dynamic graphs to more restricted subclasses. If only one of the dimensions is allowed to grow to infinity then the corresponding dynamic graphs have bounded path-width.

Observation 1. Assume that K is defined as $K = \{S^{(n_1, n_2, \ldots, n_d)}: n_1 \in N_1, n_2 \in N_2, \ldots , n_d \in N_d, S \in D\}$, where D is a finite set of static graphs, $N_1, N_2, \ldots , N_d, \subseteq N$ and only one N_i is allowed to be infinite, while all the other have to be finite. Then K has bounded path-with. Hence each existential locally verifiable (ELV, see [24]) problem, each monadic second order (MSO, see [3]) problem and each extended monadic second order problem (EMS, see [1]) can be solved in linear time for such classes.

It is easy to see that the path-width depends only on the maximum of the dimensions in the sets N_j, which are finite, and of the "reach" of the functions f in the corresponding underlying static graphs $S = (V, E, f) \in D$. Moreover we can observe that each finitely represented class of d-dimensional dynamic graphs has bounded degree. The bound depends only on the dimension d and on the maximum of the degrees of the members of the corresponding set of static graphs. Hence each first order problem can be solved for such classes in linear time, using a corresponding result from [25] (see also [8], [9]).

Observation 2. Let K be an arbitrary finitely represented class of dynamic graphs. Then each first order problem (FO problem, see section 3 for exact notation) can be solved for K in linear time.

In the following we will improve this observation by showing that each FO-problem can be solved for such classes in constant time $O(1)$.

Theorem 1. Let K be an arbitrary finitely represented class of dynamic graphs. Then each first order problem (FO problem, see section 3 for exact notation) can be solved for K in constant time $O(1)$, when the representation, i.e. the static graphs S and the vector x of dimensions, is given together with the input structure S^x.

The basic idea of the proof is to show that for each FO-formula φ there is a size of vectors x and y, such that dynamic graphs S^x and S^y can not be distinguished by φ if x and y are sufficiently large. This idea can not be generalized to arbitrary MSO-formulas, since one can distinguish for some structures an odd and an even number of vertices by such formulas, what is not possible for FO-formulas. The proof follows in the next section after presenting the logical machinery. At first the following observation follows directly from the corresponding definitions.

Observation 3. Let S be a static graph and assume that x, y $\in N^d$ are d-tuples of natural numbers, then x≤y implies the subgraph inclusion $S^x \subseteq S^y$ for the dynamic graphs.

A decision problem P is *monotone* if for all graphs H and G with H $\subseteq$ G, P(G) implies P(H). Examples of monotone properties of a graph G are "chromatic number (G) $\leq$ k", "genus (G) $\leq$ k", "crossing number of G $\leq$ k " and "H is not a minor of G" for fixed k$\in$ N and fixed graph H.

Theorem 2. Let K be a class of finitely represented dynamic graphs. Each monotone decision problem P can be solved in constant time for K when the representation, i.e. the static graph S and the vector x of dimensions, is given together with the input S^x.

As a corollary follows that CHROMATIC NUMBER can be solved in constant time, since it can be bounded by the maximum degree + 1.

Proof. Let P be the monotone property and assume, that K is given as K = $\{S^{(n_1,n_2,\ldots,n_d)}: n_1 \in N_1, n_2 \in N_2, \ldots, n_d \in N_d, S \in D\}$, for d and D as above. Define T to be the set of vectors x $\in N^d$ for which S^x has the property P. For technical simplicity we present the technical details of the proof only in case |D| = 1 and dimension d=2. The arbitrary case is then a straightforward generalization. We have to distinguish several cases. The basic idea is to subdivide N^d in a finite number of rectangular areas, where the answer to P is either "yes" or "no" in the whole area. This subdivision procedure starts with the whole area N^d and subdivides an area (as described below) if the solution is not constant in it.

In case T = N^d we are done since then the answer to P for each instance is always "yes", so we have nothing to compute. So let us assume T $\neq N^d$. Hence there exists an $i_0 \in N$ with $(i_0,i_0) \in T$ and $(i_0+1,i_0+1) \notin T$. Otherwise we would have T=N^d by observation 3, since P is closed under substructures. By the same argument we know that there is no y$\in$ T with $(i_0+1,i_0+1) \leq y$. So we have distinguished two areas where the decision problem P can be solved in a trivial way. The first is a finite area containing all x$\in N^d$ with x$\leq(i_0,i_0)$ and the answer is for each such x "yes", since the problem is monotone. The second is the infinite area of all y$\in N^d$ with $(i_0+1,i_0+1)\leq$y. As stated above no such y is in T and hence no such y has property P. Now there remain two infinite areas R = $\{x=(i,j): i_0<i$ and $0\leq j\leq i_0\}$ and L = $\{x=(i,j): 0\leq i\leq$ i0 and $i_0<j\}$. We start with R. When R$\subseteq$T is satisfied, then P can be answered trivially "yes" on R. In case R$\cap$T=$\varnothing$ we are also done, since then the answer for the whole area R is trivially "no". So we can assume R$\cap$T $\neq\varnothing$. Let j_1 be defined as $j_1 :=$ max $\{j \in N:$ there is a i with $(i,j) \in$R$\cap$T$\}$ and let i_1 be defined as $i_1 :=$ max$\{i\in N: (i,j_1)\in$R$\cap$T$\}$. j_1 does exist in our case and is bounded by i_0, by the definition of R. For $i_1 \neq i_0$ it can happen that the second maximum i_1 does not exist. In this case we set $i_1=\infty$. The subdivision procedure can be finished at this branch, since then the whole area R is subdivided into $\{(i,j)\in N^d: i_0\leq i$ and $j_1 < j \leq i_0\}$ and $\{(i,j)\in N^d: i_0\leq i$ and $0 \leq j\leq j_1\}$. In the first area P has the constant answer "yes" and in the second are P has the constant answer "no". If i_1 is finite then the answer to P in the area $\{(i,j)\in N^d: i_0\leq i\leq i_1$ and $0\leq j\leq j_1\}$ is constant "yes". In this case we proceed in the same way for the area $\{(i,j)\in N^d: i^1<i$ and $0\leq j\leq j_1\}$ as we did for area R. The area L can be handled in the same way, only the role of i and j has to be interchanged. This subdivision procedure ends after a finite number of steps, since in each subdivision step we either reach an infinite area which has not to be divided any more or one of the dimensions of the subdivision points (i in the left

area L, or j in the right area R) becomes smaller, hence the process has to end since there are no infinite descending sequences of natural numbers.

The subdivision process in d≠2 dimensions is handled in the same way. In the general case we have to handle only some more cases. This procedure gives us at the end a finite set $C \subseteq (N \cup \{\infty\})^d$, such that:

$$T = \bigcup_{a=(a_1,\dots,a_d) \in C} \left\{ x = (x_1,\dots,x_d) \in N^d : x_i < a_i, \ 1 \le i \le d \right\} \qquad (1)$$

Here the relation < is defined in the canonical way when $a_i = \infty$. So the algorithm to compute P for an input S^x has simply to compare the vector x of dimensions of S^x with the vectors a in C. If all the components of x are less than all the components of a vector $a \in C$, then the answer is "yes", otherwise it is "no". To see that this is O(1) one has to remember only, that we do not have to read the whole structure S^x. It is sufficient to check the size of the dimensions. If they exceed the size of the corresponding component of the vectors in C it can stop. This ends the proof of theorem 2.

Remark 1. For many classes of finitely represented dynamic graphs it is possible to prove a stronger form of theorem 2, in which the assumption that the representation is given together with the input is omitted. This holds e.g. for solid rectangular grids of dimension d and for all dynamic graphs, where d-dimensional "grid-structures" can be found.

But in general a lot of case analysis has to be done here. So we leave it as a conjecture to the reader to find an elegant proof of such a generalization. While theorem 2 has a pure combinatorial proof some terminology and machinery from logic is used to prove theorem 1. This is presented in the next section.

3 First Order Logic and n-Equivalence

The formulas from *first-order logic* are build in the usual way (see e.g. [6]) from *atomic* formulas (formulas of the form x=y or $R(x_1,\dots,x_n)$ for variables $x,y,x_1,\dots,x_n$ and n-ary relation symbols R) using the *connectives* ¬, ∧, ∨, →, ↔, and the *quantifiers* ∀, ∃. We will restrict our attention here only to relational structures, especially graphs, hence we do not allow functions in the atomic formulas. A decision problem P over a class K of input structures is *first-order definable* (FO), when there does exist a first order formula φ such that for all structures $G \in K$:

G has property P ⟺ G ⊨ φ.

Examples of FO-properties are k-DOMINATING-SET, k-INDEPENDENT-SET and k-CLIQUE for fixed k, H-SUBGRAPH-ISOMORPHISM, H-INDUCED-SUBGRAPH and H-HOMOMORPHISM for fixed H (see [10] for definition). Examples of properties which are provably not FO properties are "G is connected",

"G has an even number of vertices" and "H is a minor of G" for many graphs H, e.g. $H=K_3$.

FO problems for arbitrary finite structures are in P and can be solved in logarithmic space ([16]). For relational structures of bounded degree FO problems can be solved in linear time.

Theorem 3 ([25]). Let d > 0 be an arbitrary natural number and let P be a first order property for a class of graphs of degree bounded by d. Then P can be solved in linear time over K.

The basic idea of the proof of theorem 3, which is also the basis of the proof of theorem 1, is that first order properties can be reduced to properties of local neighborhoods for the vertices of the regarded graphs. For graphs of bounded degree these properties degenerate to a check whether the isomorphism type of each local neighborhood is in a finite set of possible types (and we count the number of the occurring types up to a certain size, depending an the size of the formula), what can be checked by an algorithm visiting each vertex once and testing its local isomorphism type. Hence each FO problem reduces to a check of the local structure of the input objects. A very interesting generalization of this result for locally tree-decomposable structures can be found in [8]. Such structures are roughly those structures whose local neighborhoods have bounded tree width. Examples of classes, which are locally *tree-decomposable* are structures of bounded tree width, structures of bounded degree, planar graphs, graphs of bounded genus and all graphs with an arbitrary forbidden minor.

Theorem 4 ([8]). Assume that K is a class of relational structures that is locally tree-decomposable and let P be a first-order property. Then there is a linear time algorithm deciding whether a given structure G ∈ K has property P.

Theorem 1 can be viewed as a refinement of theorem 3, since finitely represented grid structured graphs have bounded degree. The basic idea of the proof is to show that for each static graph S, for each dimension d∈N and for each first order formula φ there is a natural number s such that the dynamic graphs S^x and S^y can not be distinguished, when x,y∈N^d and all components of x and y are larger than s.

To prove this theorem we will use a localization technique based on a method that was originally developed by Hanf in [12] to show that elementary theories of two structures are equal under certain conditions, i.e., that two structures agree on all first-order sentences. Fagin, Stockmeyer and Vardi [7] developed a variant of this technique, which is applicable in descriptive complexity theory to classes of finite relational structures of uniformly bounded degree. The essential content of this result, which is that two relational structures of bounded degree satisfy the same first-order sentences of a certain quantifier-rank if both contain, up to a certain number m, the same number of isomorphism types of substructures of bounded radius r.

The *quantifier-rank* of a formula φ, denoted as qr(φ), is defined by induction on the structure of formulas: qr(φ) := 0 if φ is atomic, qr(φ∧ψ) := max(qr(φ),qr(ψ)), qr(φ∨ψ) := max(qr(φ),qr(ψ)), qr(φ→ψ) := max(qr(φ),qr(ψ)), qr(φ↔ψ) := max(qr(φ),qr(ψ)),

$qr(\neg\varphi) := qr(\varphi)$, $qr(\forall x\varphi) := qr(\varphi)+1$, $qr(\exists x\varphi) := qr(\varphi)+1$, Let G and H be two structures for the same vocabulary and assume $n \in N$. We define a relation $\equiv_n$ between these structures by: $G \equiv_n H$ if and only if for each FO-formula φ of the corresponding language with $qr(\varphi) \leq n$, $G \models \varphi \Leftrightarrow H \models \varphi$. The two structures G and H are called *n-equivalent*, what means in other words that they both satisfy the same sentences of quantifier-rank $\leq$ n. For the class of relational structures of finite signature the relation $\equiv_n$ is an equivalence relation with finitely many equivalence classes and for each equivalence class Γ there is a formula γ of quantifier-rank n such that $\Gamma = \{H: H$ is a structure for the corresponding vocabulary and $H \models \gamma\}$ (see [6] for details).

For a graph G and two of its vertices, define $d^G(a,b)$ to be the distance of a and b in G, i.e. the length of the shortest path between a and b. We write $d^G(a,b)=\infty$, in case that a and b can not be connected by a path. For $r \geq 1$ we define $N_r^G(a) := \{b \in V(G): d^G(a,b) \leq r\}$, the *r-neighbourhood* of $a \in V(G)$ and $N_r^G(X) := \cup\{N_r^G(a): a \in X\}$ for some $X \subseteq V(G)$. The 1-neighborhood corresponds to the closed neighborhood. When this notion is used for tuples $(a_1,..,a_n)$ then it is defined as $N_r^G((a_1,..,a_n)) := N_r^G(\{a_1,..,a_n\})$. For a graph G Hanf defined the *r-type* of a vertex of G, to be the isomorphism type of $(G\downarrow N_r^G(a),a)$ where $(G\downarrow N_r^G(a),a)$ is the restriction of G to the r-neighbourhood of a in G, where a is designated as the value of a new constant, which is called the *centre*. More precisely, individuals a and b in the domain of two structures G and H, respectively, have the same r-type if $(G\downarrow N_r^G(a),a) \cong (H\downarrow N_r^H(b),b)$ under an isomorphism mapping a to b. r will be called the radius of the r-type. We will denote these structures as r-neighborhoods of vertex a in G. Let $r \in N$ and $m \geq 1$ be given. Following [7] we define two structures G and H to be *(r,m)-equivalent* if and only if for every r-type τ, either G and H have the same number of individuals with r-type τ, or both have at least m individuals with r-type τ. Hence, G and H are (r,m)-equivalent if they have both the same number of individuals with r-type τ, where we can count only as high as m, for arbitrary r-types τ. The following result is due to Fagin, Stockmeyer and Vardi.

Theorem 5 ([7]). Let n and t be positive integers. There are positive integers r, m, where r depends only on n, such that whenever G and H are (r,m)-equivalent structures of degree at most t, then $G \equiv_n H$.

This result was used in [25] to prove theorem 3, since it enabled us to reduce an arbitrary FO-problem to a local problem, i.e. a problem that can be decided by visiting once each vertex of the structure and looking only at its neighborhood to a certain fixed radius. Using this result we are now able to prove theorem 1.

To prove theorem 1 assume that K is a finitely represented class of dynamic graphs and that P is a first-order property defined by a FO sentence φ. Let n be the quantifier rank of φ. It is sufficient to prove the theorem for each of the static graphs S, representing K, since K is finitely represented. Since S is finite all dynamic graphs S^x, for $x \in N^d$, where d is the dimension corresponding to S, have a universally bounded degree, say by t. For this n and t we choose now r and m corresponding to theorem 5. It is now a technical, but easy exercise to compute from r and m a size k_0, such that S^x

$\equiv_n S^y$, for all x and y with $k_0 < x_i$, y_i (for all i: $1 \le i \le d$). The basic idea is that the r-types occurring in the dynamic graphs S^x can be subdivided into different classes:

i.   those occurring in the "corners" of S^x (i.e. those related to components, where all of the coordinates are either maximal or minimal, up to a certain distance),

ii. those occurring at the "sides/faces" of S^x (i.e. those related to components, where at least one of the coordinates is up to a certain distance either maximal or minimal – of course the dimension of the "side/face" has to be considered here), and

iii. those occurring in the "center" of S^x (i.e. those related to components, where none of the coordinates is maximal or minimal, up to a certain distance).

The key observation here is that when the structure S^x is enlarged or changed to S^y then the types occurring close to the corners do not change their number. All the other types become more or less depending on the factor of enlargement, but they never disappear. So the same types occur in S^x and S^y, only the number of the "central" types differs. When now k_0 is chosen in such a way that in the central and side area the number of occurrences of each type is larger than m then S^x and S^y are n-equivalent via theorem 5. To choose k_0 let s_0 be the size of the largest component of a vector $f(e)$ for all edges of the static graph (where f is the transit function from the definition of a dynamic graph). Now we define $k_0 := 2*(r+1)*s_0 + m*s_0$, where r and m are chosen as in theorem 5 for t and n. Now let $x=(x_1,...,x_d)$, $y=(y_1,...,y_d) \in N^d$ be arbitrary vectors with $k_0 > x_i$, y_i for all i with $1 \le i \le d$. Then by the definition of the dynamic graph S^x all r-types of the neighborhoods $(S^x \downarrow N_r^{S^x}(a),a)$ for vertices $a=u^z$ (as in the definition of a dynamic graph) with $z=(z_1,...z_d)$, for which there is an i_0 with $(r+1)*s_0 < z_{i_0} < x_{i_0} - (r+1)*s_0$ occur at least m times. These types are r-types represented at the "center" or at the "center of the sides / faces". The "center" area are just the vertices $a=u^z$ (as in the definition of a dynamic graph) with $z=(z_1,...z_d)$, for which $(r+1)*s_0 < z_i < x_i - (r+1)*s_0$ holds for all i with $1 \le i \le id$. Also here all r-types occur at least m times. All the other r-types are represented near the "corner", i.e. the area of vertices $a=u^z$ (as in the definition of a dynamic graph) with $z=(z_1,...z_d)$, for which for all i either $0 \le z_i \le (r+1)*s_0$, or $x_i-(r+1)*s_0 \le z_i \le x_i$ holds. But the "corner"-types occur as often in S^x as in S^y, when they do not occur at least m times, since the "corner"-areas are isomorphic by the definition of these dynamic graphs. Thus we get $S^x \equiv_n S^y$, for all x and y with $k_0 < x_i$, y_i (for all i: $1 \le i \le d$), via theorem 5. But there are only a finite number of vectors $x=(x_1,...,x_d) \in N^d$ with $\max(x_1,...,x_d) \le k_0$. With the exception of these vectors all other structures S^x are n-equivalent. Hence property P, i.e. $S^x \models \varphi$, has to be checked only for one of them. e.g. for $x_{S,0} = (k_0+1,k_0+1,...,k_0+1) \in N^d$.

The idea of the O(1) algorithm is now as follows. For each of the finite number of static graphs S representing K we compute $x_{S,0}$ as above. Then we compute the values of "$S^x \models \varphi$" for all $x \in N^d$ with $x \le x_{S,0}$. Assume that these values are stored in a table (S_i,x,t), where t is the corresponding truth value of "$S_i^x \models \varphi$" and i is the index of the corresponding representations. Moreover assume that the vectors $x_{S_i,0}$ are also stored together with their corresponding static graphs S_i in a table $(S_i,x_{S_i,0})$, where i is the index of the corresponding representation. This is the pre-processing of the algorithm, which can be done in constant time. Now assume that $T^y \in K$ is given. The only things we have to do is to find the index of T, i.e. to find the i with $T \cong S_i$ and then to check $y < x_{S_i,0}$. In case the latter is true the entry t of (S_i,y,t) gives the truth value of "$T^y \models \phi$", i.e. the truth value of property P for T^y. Otherwise the entry t of $(S_i, x_{S_i,0}, t)$ gives the corresponding output. But these checks can be done in constant time. Remember that it is not necessary to know the complete y to perform the check $y < x_{S_i,0}$, it is sufficient just to know a small part of the bits of the corresponding components. This holds for a uniform, logarithmic and also for a unary encoding of the components of y. The other checks can be done obviously in constant time, since we assumed that the corresponding representations are given together with the input. This proves theorem 1.

Remark 2. As in remark 1 it can be shown that the assumption, that the representation has to be given together with the input S^x, can be omitted for many classes of finitely represented dynamic graphs. As there we conjecture that the result holds in general without this assumption. As in theorem 1 this conjecture can be proved when complete grid-structures can be found in S^x, hence especially for the complete rectangular solid grids.

4 Concluding Remarks

The results of this article can be generalized to more general classes of structures. Theorem 1 holds also for toroidal grids and for a corresponding generalization of grid structured hierarchical graphs to toroidal grid structured hierarchical graphs. Just substitute in the definition of dynamic graphs the rectangular d-dimensional grids by d-dimensional toroidal grids. Moreover the definition of dynamic graphs and grid structured hierarchical graphs can be generalized to triangular, hexagonal and other regular grids and also to structures with more than one relation. This concept generalizes for arbitrary relational structures using the underlying graph or the concept of Gaifman-graphs. Moreover it can be shown that the degree bound, which was needed in theorem 5 is not necessary by using variants of Ehrenfeucht-Fraissé game theoretic equivalences instead of n-equivalence. Using linear time interpretability as introduced in [1] and the theorem of Feferman and Vaught one can show that each class of finitely represented grid-structured hierarchical graphs has a decidable first order theory. Moreover these ideas can lead to a different proof of theorem 1 by reducing grid-structured hierarchical graphs via interpretability to a variant of a Cartesian product of paths P_i of a sufficient length i. For further details we refer the reader to the final version of the full paper. For all the above and related classes of graphs with a highly regular structure (e.g. generated by graph grammars or special operations as e.g. in [15], [26]) it is an interesting question to find a general characterization of those algorithmic problems with efficient solutions. Moreover it could be of interest to find suitable parameters (in the sense of parameterized complexity [5]) to characterize the regularity or homogeneity of these and related structures which can be exploited for the design of efficient algorithms for (almost) regular or homogeneous structures or huge networks (see possibility B to reduce complexity in the introductory section 1.2 of [14]).

Acknowledgments. The work at this article was done during a visit of the City University Hong Kong, the Chinese University Hong Kong, the University of New South Wales Sydney and the University of Newcastle during my research semester 2006/2007. I am thanking Xiaotie Deng, Cai Leizhen, Pradeep Ray, Nandan Parameshwaran, Fethi Rabhi, Regina Betz and Stephan Chalup for their interest and their support. Special thanks go to Mike Fellows for stimulating discussions on the very first ideas of this subject during a walk in the Barrington Tops in April 2002.

References

1. Arnborg, S., Lagergren, J., Seese, D.: Easy Problems for Tree-Decomposable Graphs. Journal of Algorithms 12, 308–340 (1991)
2. Bodlaender, H.L., Koster, A.M.C.A.: Combinatorial Optimization on Graphs of bounded treewidth. The Computer Journal, special Volume on FPT, pp. 1–26, December 2006 (to appear)
3. Courcelle, B.: The monadic second order theory of graphs III: Tree decompositions, minors, and complexity issues. Theoret. Inform. Applic. 26(2), 257–286 (1992)
4. Diestel, R.: Graph Theory. Springer, Heidelberg (2000)
5. Downey, R.G., Fellows, M.R.: Parameterized Complexity. Springer, Heidelberg (1997)
6. Ebbinghaus, H.-D., Flum, J., Thomas, W.: Mathematical Logic, 2nd edn. Springer, Heidelberg (1994)
7. Fagin, R., Stockmeyer, L., Vardi, M.: On monadic NP vs. monadic co-NP. In: The Proceedings of the 8th Annual IEEE Conference on Structure in Complexity Theory, pp. 19–30 (1993) (Full paper in Information and Computation 120(1), 78–92 (1995))
8. Frick, M., Grohe, M.: Deciding first-order properties of locally tree-decomposable structures. In: Wiedermann, J., van Emde Boas, P., Nielsen, M. (eds.) ICALP 1999. LNCS, vol. 1644, Springer, Heidelberg (1999)
9. Frick, M., Grohe, M.: The complexity of first-order and monadic second-order logic revisited. In: Proceedings of the 17th IEEE Symposium on Logic in Computer Science (LICS'02), pp. 215–224 (2002)
10. Garey, M.R., Johnson, D.S.: Computers and Intractability. Bell Tel. Laboratories (1979)
11. Goldwasser, J., Klostermeyer, W., Trapp, G.: Characterizing Switch-Setting Problems. Linear and Multilinear Algebra 43, 121–135 (1997)
12. Hanf, W.: Model-theoretic methods in the study of elementary logic. In: Addison, J.W., Henkin, L.A., Tarski, A. (eds.) Symposium on the Theory of Models, pp. 132–145. North-Holland Publ. Co., Amsterdam (1965)
13. Hare, E.O., Hedetniemi, S.T., Hare, W.R.: Algorithms for Computing the domination number of KxN complete grid graphs. Congr. Numerantium 55, 81–92 (1986)
14. Hlineny, P., Oum, S., Seese, D., Gottlob, G.: Width Parameters Beyond Tree-width and Their Applications. The Computer Journal, spec. Vol. on FPT, pp. 1–71(to appear)
15. Hoefting, F., Lengauer, T., Wanke, E.: Processing of hierarchically defined graphs and graph families. In: Monien, B., Ottmann, T. (eds.) Data Structures and Efficient Algorithms. LNCS, vol. 594, pp. 44–69. Springer, Heidelberg (1992)
16. Immerman, N.: Descriptive complexity. Springer, New York (1999)

17. Itai, A., Papadimitriou, C.H., Szwarcfiter, J.L.: Hamiltonian paths in grid graphs. SIAM Journal of Comp. 11(4), 676–686 (1982)

18. Iwano, K., Steiglitz, K.: Planarity testing of double periodic infinite graphs. Networks 18(3), 205–222 (1988)

19. Karp, R.M., Miller, R.E., Winograd, A.: The organization of computations for uniform recurrence equations. Journal of the ACM 14(3), 563–590 (1967)

20. Kaufmann, M., Mehlhorn, K.: Routing problems in grid graphs. In: Korte, B., Lovasz, L., Promel, H.J., Schrijver, A. (eds.) Paths, Flows and VLSI-Layout, pp. 165–184. Springer, Heidelberg (1990)

21. Lengauer, T.: Combinatorial Algorithms for Integrated Circuit Layout. B.G.Teubner and John Wiley & Sons, Chichester (1990)

22. Mitchell, W.F.: Hamiltonian paths through two- and three-dimensional grids. Journal of Research of the National Institute of Standards and Technology 110(2), 127–136 (2005)

23. Papadimitriou, C.H.: Computational Complexity. Addison-Wesley, Reading (1994)

24. Seese, D.: Tree-partite graphs and the complexity of algorithms (extended abstract). In: Budach, L. (ed.) FCT 1985. LNCS, vol. 199, pp. 412–421. Springer, Heidelberg (1985)

25. Seese, D.: Linear time computable problems and first-order descriptions. Math. Struct. in Comp. Science, 505–526 (1996)

26. Wagner, K.: The complexity of problems concerning graphs with regularities. In: Chytil, M.P., Koubek, V. (eds.) Friedrich-Schiller-Universitaet Jena 1984 (extended abstract appeared in Mathematical Foundations of Computer Science 1984. LNCS, vol. 176, pp. 544–552. Springer, Heidelberg (1984))

Long Alternating Cycles in Edge-Colored Complete Graphs

Hao Li[1,3], Guanghui Wang[1,2], and Shan Zhou[1,3]

[1] Laboratoire de Recherche en Informatique, UMR 8623, C.N.R.S-Université de Paris-sud, 91405-Orsay cedex, France
[2] School of Mathematics and System Science, Shandong University, 250100 Jinan, Shandong, China
[3] School of Mathematics and Statistics, Lanzhou University, 730000 Lanzhou, China

Abstract. Let K_n^c denote a complete graph on n vertices whose edges are colored in an arbitrary way. And let $\Delta(K_n^c)$ denote the maximum number of edges of the same color incident with a vertex of K_n^c. A properly colored cycle (path) in K_n^c, that is, a cycle (path) in which adjacent edges have distinct colors is called an alternating cycle (path). Our note is inspired by the following conjecture by B. bollobás and P. Erdős(1976): If $\Delta(K_n^c) < \lfloor n/2 \rfloor$, then K_n^c contains an alternating Hamiltonian cycle. We prove that if $\Delta(K_n^c) < \lfloor n/2 \rfloor$, then K_n^c contains an alternating cycle with length at least $\lceil \frac{n+2}{3} \rceil + 1$.

Keywords: alternating cycle, color degree, edge-colored graph.

1 Introduction and Notation

We use [2] for terminology and notations not defined here. Let $G = (V, E)$ be a graph. An *edge-coloring* of G is a function $C : E \to N$(N is the set of nonnegative integers). If G is assigned such a coloring C, then we say that G is an *edge-colored graph*, or simply *colored graph*. Denote by (G, C) the graph G together with the coloring C and by $C(e)$ the *color* of the edge $e \in E$. For a subgraph H of G, let $C(H) = \{C(e) : e \in E(H)\}$ and $c(H) = |C(H)|$. For a color $i \in C(H)$, let $i_H = |\{e : C(e) = i \text{ and } e \in E(H)\}|$ and say that *color i appears i_H times in H*. For an edge-colored graph G, if $c(G) = c$, we call it a *c-edge colored* graph.

A properly colored cycle (path) in an edge-colored graph, that is, a cycle(path) in which adjacent edges have distinct colors is called an *alternating* cycle (path). In particular, an *alternating Hamiltonian* cycle (path) is a properly colored Hamiltonian cycle (path). For $l \geq 3$, let AC_l denote an alternating cycle with length l. Besides a number of applications in graph theory and algorithms, the concept of alternating paths and cycles, appears in various other fields: genetics (cf. [9,10,11]), social sciences (cf. [8]). A good resource on alternating paths and cycles is the survey paper [2] by Bang-Jensen and Gutin.

F.P. Preparata and Q. Fang (Eds.): FAW 2007, LNCS 4613, pp. 305–309, 2007.
© Springer-Verlag Berlin Heidelberg 2007

Grossman and Häggkvist [12] were the first to study the problem of the existence of the alternating cycles in c-edge colored graphs. They proved Theorem 1.1 below in the case $c = 2$. The case $c \geq 3$ was proved by Yeo [17]. Let v be a cut vertex in an edge-colored graph G. We say that v *separates colors* if no component of $G - v$ is joined to v by at least two edges of different colors.

Theorem 1.1 (Grossman and Häggkvist [12], and Yeo [17]). *Let G be a c-edge colored graph, $c \geq 2$, such that every vertex of G is incident with at least two edges of different colors. Then either G has a cut vertex separating colors, or G has an alternating cycle.*

Given an edge-colored graph G, let $d^c(v)$, named the color degree of a vertex v, be defined as the maximum number of edges adjacent to v, that have distinct colors. In [16], some color degree conditions for the existence of alternating cycles are obtained as follows.

Theorem 1.2 (Li and Wang [16]). *Let G be an edge-colored graph with order $n \geq 3$. If $d^c(v) > \frac{n+1}{3}$ for every $v \in V(G)$, then G has an alternating cycle AC such that each color in $C(AC)$ appears at most two times in AC.*

Theorem 1.3 (Li and Wang [16]). *Let G be an edge-colored graph with order $n \geq 3$. If $d^c(v) \geq \frac{37n-17}{75}$ for every $v \in V(G)$, then G contains at least one alternating triangle or one alternating quadrilateral.*

Theorem 1.4 (Li and Wang [16]). *Let G be an edge-colored graph with order n. If $d^c(v) \geq d \geq 2$, for every vertex $v \in V(G)$, then either G has an alternating path with length at least $2d$, or G has an alternating cycle with length at least $\lceil \frac{2d}{3} \rceil + 1$.*

Consider the edge-colored complete graph, we use the notation K_n^c to denote a complete graph on n vertices, each edge of which is colored by a color from the set $\{1, 2, \cdots, c\}$. And $\Delta(K_n^c)$ is the maximum number of edges of the same color adjacent to a vertex of K_n^c. And we have the following conjecture due to Bollobás and Erdős [4].

Conjecture 1.5 (Bollobás and Erdős [4]). *If $\Delta(K_n^c) < \lfloor \frac{n}{2} \rfloor$, then K_n^c contains an alternating Hamiltonian cycle.*

Bollobás and Erdős managed to prove that $\Delta(K_n^c) < \frac{n}{69}$ implies the existence of an alternating Hamiltonian cycle in K_n^c. This result was improved by Chen and Daykin [7] to $\Delta(K_n^c) < \frac{n}{17}$ and by Shearer [15] to $\Delta(K_n^c) < \frac{n}{7}$. So far the best asymptotic estimate was obtained by Alon and Gutin [1].

Theorem 1.6 (Alon and Gutin [1]). *For every $\epsilon > 0$ there exists an $n_o = n_0(\epsilon)$ so that for every $n > n_o$, K_n^c satisfying $\Delta(K_n^c) \leq (1 - \frac{1}{\sqrt{2}} - \epsilon)n$ has an alternating Hamiltonian cycle.*

In the present paper, we study the long alternating cycles of edge-colored complete graphs and gain the following result.

Theorem 1.7 *If* $\Delta(K_n^c) < \lfloor \frac{n}{2} \rfloor$, *then* K_n^c *contains an alternating cycle with length at least* $\lceil \frac{n+2}{3} \rceil + 1$.

2 Proof of Theorem 1.7

If $P = v_1 v_2 \cdots v_p$ is a path, let $P[v_i, v_j]$ denote the subpath $v_i v_{i+1} \cdots v_j$, and $P^-[v_i, v_j] = v_j v_{j-1} \cdots v_i$.

Lemma 2.1 (Bang-Jensen, Gutin and Yeo[3]). *If* K_n^c *contains a properly colored 2-factor, then it has a properly colored Hamiltonian path.*

Häggkvist [13] announced a non-trivial proof of the fact that every edge-colored complete graph graph satisfying above Bollobás-Erdős condition contains a properly colored 2-factor. Lemma 2.1 and Häggkvist's result imply that every edge-colored complete graph satisfying Bollobás-Erdős condition has an alternating Hamiltonian path.

Proof of Theorem 1.7. If $n = 3$, the conclusion holds clearly. So we assume that $n \geq 4$. By contradiction. Suppose that our conclusion does not hold. Then let $P = v_1 v_2 \cdots v_n$ be an alternating Hamiltonian path of K_n^c. Choose s satisfying the followings:

R_1. $C(v_1 v_s) \neq C(v_1 v_2)$.
R_2. $s \geq \lceil \frac{n+2}{3} \rceil + 1$.
R_3. Subject to R_1, R_2, s is minimum.

Lemma 2.2
(1.1) $s \leq \lfloor \frac{n}{2} \rfloor + \lceil \frac{n+2}{3} \rceil - 1$.
(1.2) For $i \geq s$, if $C(v_1 v_i) \neq C(v_1 v_2)$, then $C(v_1 v_i) \neq C(v_i v_{i+1})$.

Proof. By R_3, for $\lceil \frac{n+2}{3} \rceil + 1 \leq j \leq s - 1$, we have that $C(v_1 v_j) = C(v_1 v_2)$. If $s \geq \lfloor \frac{n}{2} \rfloor + \lceil \frac{n+2}{3} \rceil$, then there exist at least $\lfloor \frac{n}{2} \rfloor + \lceil \frac{n+2}{3} \rceil - (1 + \lceil \frac{n+2}{3} \rceil) + 1 \geq \lfloor \frac{n}{2} \rfloor$ edges with the color $C(v_1 v_2)$ incident with v_1, a contradiction with $\Delta(K_n^c) < \lfloor \frac{n}{2} \rfloor$.

Since P is an alternating Hamiltonian path, then $C(v_{i-1} v_i) \neq C(v_i v_{i+1})$. If there exists $i \geq s$ such that $C(v_1 v_i) \neq C(v_1 v_2)$ and $C(v_1 v_i) = C(v_i v_{i+1})$, then $P[v_1, v_i] v_i v_1$ is an alternating cycle with length $i \geq s \geq \lceil \frac{n+2}{3} \rceil + 1$, a contradiction.

Then choose t satisfying the followings:

R_1'. $C(v_t v_n) \neq C(v_{n-1} v_n)$.

R'_2. $t \leq n - \lceil \frac{n+2}{3} \rceil$.

R'_3. Subject to R'_1, R'_2, t is maximum.

Similarly, we have the following lemma, here we omit the details.

Lemma 2.3

(3.1) $t \geq \lceil \frac{n}{2} \rceil - \lceil \frac{n+2}{3} \rceil + 2$.

(3.2) For $i \leq t$, if $C(v_i v_n) \neq C(v_{n-1} v_n)$, then $C(v_i v_n) \neq C(v_{i-1} v_i)$.

Lemma 2.4. $s < t$.

Proof. Otherwise, we have that $s \geq t$. If $s > t$, then $AC^0 = v_1 v_s P[v_s, v_n] v_l v_t$ $P^-[v_t, v_1]$ is an alternating cycle. And $|AC^0| = |P[v_s, v_n]| + |P[v_1, v_t]| \geq (n - \lfloor \frac{n}{2} \rfloor - \lceil \frac{n+2}{3} \rceil + 2) + (\lceil \frac{n}{2} \rceil - \lceil \frac{n+2}{3} \rceil + 2) = 2(\lceil \frac{n}{2} \rceil - \lceil \frac{n+2}{3} \rceil) + 4 \geq \lceil \frac{n+2}{3} \rceil + 1$, a contradiction.

So we assume that $s = t$. For $s + 1 \leq j \leq n - 1$, we conclude that $C(v_1 v_j) = C(v_1 v_2)$. Otherwise, there is an alternating cycle $AC^1 = v_1 v_j P[v_j, v_n] v_n v_s P^-[v_s, v_n]$ with length $|AC^1| \geq 2 + |V(P[v_1, v_s])| \geq 3 + \lceil \frac{n+2}{3} \rceil$, which gives a contradiction. Similarly, for $2 \leq j \leq s - 1$, it holds that $C(v_j v_n) = C(v_{n-1} v_n)$. Then by $\Delta(K_n^c) < \lfloor \frac{n}{2} \rfloor$, consider vertex v_1 and the color $C(v_1 v_2)$, it holds that $n - s < \lfloor \frac{n}{2} \rfloor$, then $s > \lceil \frac{n}{2} \rceil$. Similarly, consider vertex v_n and the color $C(v_{n-1} v_n)$, we have that $s - 1 < \lfloor \frac{n}{2} \rfloor$, then $s < \lfloor \frac{n}{2} \rfloor + 1$, a contradiction.

Lemma 2.5. For $2 \leq j \leq s-1$, $C(v_n v_j) = C(v_{n-1} v_n)$; And for $t+1 \leq j \leq n-1$, $C(v_1 v_j) = C(v_1 v_2)$.

Proof. By symmetry, we only prove the first part. Otherwise, there exists $2 \leq j \leq s - 1$ such that $C(v_j v_n) \neq C(v_{n-1} v_n)$. Clearly, $j \leq t$, thus by Lemma 2.3 we have that $C(v_{j-1} v_j) \neq C(v_j v_n)$. Then we get an alternating cycle $AC^2 = v_1 v_s P[v_s, v_n] v_n v_j P^-[v_j, v_1]$. And it holds that $|AC^2| \geq |V(P[v_s, v_n])| + 2 \geq |V(P[v_t, v_n])| + 3 \geq \lceil \frac{n+2}{3} \rceil + 3$, a contradiction.

Denote $A = \{v : C(v_1 v) \neq C(v_1 v_2)\}$ and $B = \{v : C(v_n v) \neq C(v_{n-1} v_n)\}$.

Lemma 2.6. $|A \cap V(P[v_s, v_t])| + |B \cap V(P[v_s, v_t])| \geq 2(\lceil \frac{n}{2} \rceil - \lceil \frac{n+2}{3} \rceil + 1)$.

Proof. By R_1, $|A \cap V(P[v_s, v_n])| \geq n - (\lfloor \frac{n}{2} \rfloor - 1) - (\lceil \frac{n+2}{3} \rceil - 1) \geq \lceil \frac{n}{2} \rceil - \lceil \frac{n+2}{3} \rceil + 2$. Then by Lemma 2.5, we obtain that $A \cap V(P[v_s, v_n]) = A \cap (V(P[v_s, v_t]) \cup \{v_n\}) = (A \cap V(P[v_s, v_t])) \cup (A \cap \{v_n\})$. It follows that $|A \cap V(P[v_s, v_t])| \geq \lceil \frac{n}{2} \rceil - \lceil \frac{n+2}{3} \rceil + 1$. Similarly, we can obtain that $|B \cap V(P[v_s, v_t])| \geq \lceil \frac{n}{2} \rceil - \lceil \frac{n+2}{3} \rceil + 1$. Then $|A \cap V(P[v_s, v_t])| + |B \cap V(P[v_s, v_t])| \geq 2(\lceil \frac{n}{2} \rceil - \lceil \frac{n+2}{3} \rceil + 1)$.

Now we completes the proof of Theorem 1.7 as follows. We have that $|V(P[v_s, v_t])| \leq n - |V(P[v_1, v_{s-1}])| - |V(P[v_{t+1}, v_l])| \leq n - 2\lceil \frac{n+2}{3} \rceil$. And by Lemma 2.6, $|A \cap V(P[v_s, v_t])| + |B \cap V(P[v_s, v_t])| = |A| + |B| \geq 2(\lceil \frac{n}{2} \rceil - \lceil \frac{n+2}{3} \rceil + 1) > n - 2\lceil \frac{n+2}{3} \rceil + 1 > |V(P[v_s, v_t])|$, then it follows that there exists v_j $(s+1 \leq j \leq t)$ such that $v_j \in A$ and $v_{j-1} \in B$. So we get an alternating Hamiltonian cycle $v_1 v_j P[v_j, v_n] v_n v_{j-1} P^-[v_{j-1}, v_1]$, a contradiction. This completes the proof.

References

1. Alon, N., Gutin, G.: Properly colored Hamiltonian cycles in edge colored complete graphs. Random Structures and Algorithms 11, 179–186 (1997)
2. Bang-Jensen, J., Gutin, G.: Alternating cycles and paths in edge-colored multigraphs: a survey. Discrete Math., pp. 165–166, pp. 39–60 (1997)
3. Bang-Jensen, J., Gutin, G., Yeo, A.: Properly coloured Hamiltonian paths in edge-colored complete graphs. Discrete Appl. Math. 82, 247–250 (1998)
4. Bollobás, B., Erdös, P.: Alternating Hamiltonian cycles, Israel. J. Math. 23, 126–131 (1976)
5. Bondy, J.A., Murty, U.S.R.: Graph Theory with Applications. Macmillan Press, New York (1976)
6. Caccetta, L., Häggkvist, R.: On minimal digraphs with given girth. In: Proceedings, Ninth S-E Conference on Combinatorics, Graph Theory and Computing, pp. 181–187 (1978)
7. Chen, C.C., Daykin, D.E.: Graphs with Hamiltonian cycles having adjacent lines different colors. J. Combin. Theory Ser. B21, 135–139 (1976)
8. Chow, W.S., Manoussakis, Y., Megalakaki, O., Spyratos, M., Tuza, Z.: Paths through fixed vertices in edge-colored graphs. J. des Mathematqiues, Informatique et Science, Humaines 32, 49–58 (1994)
9. Dorninger, D.: On permutations of chromosomes. In: Contributions to General Algebra, vol. 5, Verlag Hölder-Pichler-Tempsky, Wien, Teubner, Stuttgart, pp. 95–103 (1987)
10. Dorninger, D.: Hamiltonian circuits determing the order of chromosomes. Discrete Appl. Math. 50, 159–168 (1994)
11. Dorninger, D., Timischl, W.: Geometrical constraints on Bennett's predictions of chromosome order. Heredity 58, 321–325 (1987)
12. Grossman, J.W., Häggkvist, R.: Alternating cycles in edge-partitioned graphs. J. Combin. Theory Ser B 34, 77–81 (1983)
13. Häggkvist, R.: A talk at Intern. Colloquium on Combinatorics and Graph Theory at Balatonlelle, Hungary (July 15-19, 1996)
14. Hahn, G., Thomassen, C.: Path and cycle sub-Ramsey numbers and edge-coloring conjecture. Discrete Math. 62(1), 29–33 (1986)
15. Shearer, J.: A property of the colored complete graph. Discrete Math. 25, 175–178 (1979)
16. Li, H., Wang, G.: Color degree and alternating cycles in edge-colored graphs, RR L.R.I NO1461(2006)
17. Yeo, A.: Alternating cycles in edge-coloured graphs. J. Combin. Theory Ser. B 69, 222–225 (1997)

Notes on Fractional $(1, f)$-Odd Factors of Graphs[*]

Jiguo Yu[1],[**] and Guizhen Liu[2]

[1] School of Computer Science, Qufu Normal University,
Ri-zhao, Shandong, 276826, P.R. China
jiguoyu@sina.com
[2] School of Mathematics and System Science, Shandong University,
Ji-nan, Shandong, 250100, P.R. China
gzliu@sdu.edu.cn

Abstract. Let G be a simple graph and f an odd integer-valued function defined on $V(G)$. A spanning subgraph F of G is called a fractional $(1, f)$-odd factor if $d_F(v) \in \{1, 3, \cdots, f(v)\}$ for all $v \in V(G)$, where $d_F(v)$ is the fractional degree of v in F. In this paper, we discuss the existence for a graph to have a fractional $(1, f)$-odd-factor. A necessary and sufficient condition for a tree to have a fractional $(1, f)$-odd factor is given.

1 Introduction

Fractional factor theory has extensive applications in some areas such as network design, combinatorial topology, decision lists and so on. For example, in the communication networks, if we permit that large date package can be partitioned into some parts to send to some different destinations by different channels, then the running efficiency of networks will be greatly improved. Feasible and efficient assignment for date package can be viewed as problem to find a fractional factor satisfying some special conditions.

All graphs considered are finite simple graphs. Let G be a graph with vertex set $V(G)$ and edge set $E(G)$. For a vertex $x \in V(G)$, the degree of x in G is denoted by $d_G(x)$. We write $N_G(x)$ for the set of vertices adjacent to x in G. If S is a subset of $V(G)$, $o(G \backslash S)$ denotes the number of odd components of $G \backslash S$. The set of isolated vertices of $G \backslash S$ is denoted by $I(G \backslash S)$ and $|I(G \backslash S)| = i(G \backslash S)$. For two disjoint subsets S, T of $V(G)$, $E_G(S, T)$ denotes the set of edges in G whose one vertex in S and another in T and $|E_G(S, T)| = e_G(S, T)$.

Let g and f be two integer-valued functions defined on $V(G)$ such that $0 \le g(x) \le f(x)$ for all $x \in V(G)$. A (g, f)-factor F of G is a spanning subgraph of G satisfying $g(x) \le d_F(x) \le f(x)$ for all $x \in V(G)$. If $g(x) = f(x)$ for every $x \in V(G)$, then a (g, f)-factor F is called an f-factor. If $f(x) = k$ for all

[*] The work is supported by NNSF (10471078) of China and RFDP (20040422004), Promotional Foundation (2005BS01016) for Middle-aged or Young Scientists of Shandong Province, DRF and UF(XJ0609)of QFNU.
[**] The corresponding author.

F.P. Preparata and Q. Fang (Eds.): FAW 2007, LNCS 4613, pp. 310–316, 2007.
© Springer-Verlag Berlin Heidelberg 2007

$x \in V(G)$, then an f-factor is called a k-factor. In particular, if f is a function defined on $V(G)$ such that $f(x)$ is a positive odd integer for every $x \in V(G)$, then we denote such a function by $f : V(G) \rightarrow \{1, 3, 5, \cdots\}$. Then a spanning subgraph F of G is called an $(1, f)$-odd factor if $d_F(x) \in \{1, 3, 5, \cdots, f(x)\}$ for all $x \in V(G)$. Let n be a positive integer and $f(x) = 2n - 1$ for all $x \in V(G)$. Then a $(1, f)$-odd factor is called a $\{1, 3, 5, \cdots, 2n - 1\}$-factor of G.

A *fractional (g, f)-indicator function* is a function h that assigns to each edge of graph G a number in $[0, 1]$ so that for each vertex $x \in V(G)$ we have $g(x) \leq d_{G_h}(x) \leq f(x)$, where $d_{G_h}(x) = \sum_{e \in E_x} h(e)$ is the fractional degree of $x \in V(G)$ with $E_x = \{e : x \text{ is incident to } e\}$. Let h be a fractional (g, f)-indicator function of a graph G. Set $E_h = \{e : e \in E(G) \text{ and } h(e) \neq 0\}$. If G_h is a spanning subgraph of G such that $E(G_h) = E_h$, then G_h is called a *fractional (g, f)-factor* of G. In particular, if f is a positive odd integer-valued function defined on $V(G)$, and set $g(x) = 1$ for all $x \in V(G)$, G_h is a fractional $(1, f)$-factor of G and $d_{G_h}(x) \in \{1, 3, 5, \cdots, f(x)\}$, then G_h is called a *fractional $(1, f)$-odd factor* of G. Set $f(x) = 2n - 1$ for all $x \in V(G)$, a fractional $(1, f)$-odd factor changes into a *fractional $\{1, 3, 5, \cdots, 2n - 1\}$-factor*. Similarly, we can define *fractional f-factors, fractional k-factors* etc.. Some other notations can be founded in [3].

Various authors studied $(1, f)$-odd factors ([1], [5], [6], [9], [10]) of graphs.

For a vertex subset S of G and a integer-valued function f defined on $V(G)$, we write $f(S)$ and $d_G(S)$ for $\sum_{x \in S} f(x)$ and $\sum_{x \in S} d_G(x)$.

Anstee obtained a necessary and sufficient condition for a graph to have a fractional (g, f)-factor by algorithmic methods. Liu and Zhang gave a new proof by graph-theoretic methods.

Theorem A ([2], [7]). *Let G be a graph. Then G has a fractional (g, f)-factor if and only if for every subset S of $V(G)$*

$$g(T) - d_{G \setminus S}(T) \leq f(S),$$

where $T = \{x : x \in V(G) \setminus S \text{ and } d_{G \setminus S}(x) \leq g(x)\}$.

Let $g(x) = f(x) = 1$ for all $x \in V(G)$, we have the following

Theorem A' ([8]). *Let G be a graph. Then G has a fractional perfect matching if and only if for every subset S of $V(G)$*

$$i(G \setminus S) \leq |S|.$$

The following necessary and sufficient condition for a graph to have a $(1, f)$-odd factor was obtained by Cui and Kano [5].

Theorem B ([5]). *Let G be a tree and $f : V(G) \rightarrow \{1, 3, 5, \cdots\}$. Then G has a $(1, f)$-odd factor if and only if*

$$o(G - x) \leq f(x) \quad for \quad all \quad x \in V(G).$$

Theorem C ([5]). *Let G be a graph and $f : V(G) \to \{1, 3, 5, \cdots\}$. Then G has a $(1, f)$-odd factor if and only if*

$$o(G \setminus S) \le f(S) \quad for \ all \quad S \subseteq V(G).$$

Now, it is natural to imagine that the conditions for a graph G to have a fractional $(1, f)$-odd factor have a similar form to fractional 1-factor. In the following section, we discuss the existence of fractional $(1, f)$-odd factor of graphs. We always assume that $f(x) \le d_G(x)$ for all $x \in V(G)$.

2   Main Results

Lemma 1. *Let G be a tree and $f : V(G) \to \{1, 3, 5, \cdots\}$ be an odd integer-valued function. If G has a fractional $(1, f)$-odd factor, then $i(G - x) \le f(x)$ for all $x \in V(G)$.*

Proof. If G has a fractional $(1, f)$-odd factor $F = G_h$, where h is the indicator function. Then $I(G - x) = \{y : x$ is adjacent with y in G and $d_G(y) = 1\}$. Obviously, for any $y \in I(G - x)$, we have $d_G(y) = 1$ and $h(e) = 1$ $(e \in E_y)$ since G has a fractional $(1, f)$-odd factor. Let $E_x = \{e : x$ is incident to e in $G\}$. Thus we have

$$
\begin{aligned}
i(G - x) &= \sum_{y \in I(G-x)} \sum_{e \in E_y} h(e) \\
&= \sum_{e \in E_G(x, I(G-x))} h(e) \\
&\le \sum_{e \in E_G(x, I(G-x))} h(e) + \sum_{e \in E_x \setminus E_G(x, I(G-x))} h(e) \\
&= \sum_{e \in E_x} h(e) \\
&= d_{G_h}(x) \\
&\le f(x).
\end{aligned}
$$
□

Remark 1. The condition $i(G - x) \le f(x)$ is not sufficient for a tree to have a fractional $(1, f)$-odd-factor. For example, let $f(u) = f(w) = 3$ and $f(x) = 1$ for all $x \in V(G_1) \setminus \{u, w\}$ in the following graph G_1. Then $i(G - x) \le f(x)$ for $x \in V(G_1)$, but G_1 has no fractional $(1, f)$-odd-factor.

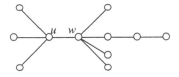

Fig. 1. Tree G_1 has no fractional $(1, f)$-odd-factor

As we have known, a graph G has a (g, f)-factor, it must have a fractional (g, f)-factor. But in general, graph G having a fractional (g, f)-factor may not have a (g, f)-factor. In fact, if a tree G has a fractional $(1, f)$-odd factor, we have the following

Lemma 2. *Let G be a tree and $f : V(G) \rightarrow \{1, 3, 5, \cdots\}$ be an odd integer-valued function. Then G has a fractional $(1, f)$-odd factor if and only if G has a $(1, f)$-odd factor. Furthermore, if G_h is a fractional $(1, f)$-odd factor of G, then G_h is also a $(1, f)$-odd factor of G.*

Proof. Suppose that $F = G_h$ is a fractional $(1, f)$-odd factor of G and h is the indicator function. Then $1 \leq d_{G_h}(x) \leq f(x)$ and $d_{G_h}(x)$ is an odd integer. $d_{G_h}(x) = h(E_x)$. Let $E'_h = \{e : e \in E_h$ and $0 < h(e) < 1\}$ and $E''_h = \{e : e \in E_h$ and $h(e) = 1\}$, then obviously $E_h = E'_h \cup E''_h$ since $E_h = \{e : e \in E(G)$ and $h(e) \neq 0\}$. We have the following

Claim. $E'_h = \phi$.

For any edge $e_0 = uv \in E'_h$, then e_0 is not a pendent edge. If e_0 is a pendent edge of G, then $h(e_0) = 1$ since G has a fractional $(1, f)$-odd factor. Hence the end-vertices u and v of e_0 are not pendent vertices. Moreover, u and v can not be incident only with pendent edges besides e_0 since every pendent edge e_0 in G has weight 1. Without loss of generality, for any vertex x, we suppose that the number of pendent edges incident with x is 1 or 2. We obtain two components after deleting e_0. Denote that the component containing u by G_u, and the component containing v by G_v, respectively. In G_u, we can find a longest path $p_u = u_k u_{k-1} \cdots u_1 u_0$ starting from u and $u_k = u$, and then u_0 is a pendant vertex since G is a tree. Thus, $h(u_0 u_1) = 1$ since G_h is a fractional $(1, f)$-odd factor. Furthermore, we have $h(u_1 u_2) = 0$ if $|N_G(u_1) \setminus \{u_2\}|$ is odd and $h(u_1 u_2) = 1$ if $|N_G(u_1) \setminus \{u_2\}|$ is even. If u_2 has another branch besides the branch containing u_1, we suppose that $u_2 u'_1 \in E(G)$ and obtain a longest path $p_{u_2} = u_2 u'_1 u'_0$ starting from u_2 containing no u_1. Repeat the above procedure to p_{u_2}, we can obtain $h(u_2 u'_1) = 1$ or 0. Moreover we can obtain $h(u_2 u_3) = 1$ or 0. If u_3 has another branch besides the branch containing u_2, then we obtain $h(u_3 u'_2) = 1$ or 0 by the same procedure. Thus we obtain $h(u_3 u_4) = 1$ or 0. Repeating the above procedure, we have $h(e) = 1$ or 0 for any $e \in \{e : e = ux \in E(G)$ and $x \neq v\}$. Consider G_v by the same way, we obtain $h(e) = 1$ or 0 for all $e \in \{e : e = vx \in E(G)$ and $x \neq u\}$ in the end. Thus $h(e_0) = 1$ or 0, which contradicts to $e_0 \in E'_h$. Hence $E'_h = \phi$ and the claim is proved.

Thus, if G has a fractional $(1, f)$-odd factor G_h, then G_h itself is a $(1, f)$-odd factor of G by Claim. On the other hand, a $(1, f)$-odd factor F, obviously, is a fractional $(1, f)$-odd factor. The theorem is proved. □

By Lemma 2 and Theorem B, we have the following

Theorem 1. *A tree G has a fractional $(1, f)$-odd factor if and only if for all $x \in V(G)$,*

$$o(G - x) \leq f(x).$$

Now we consider the existence of fractional $(1, f)$-odd-factors in a graph which is not a tree. Without loss of generality, assume that G is connected, we have the following

Theorem 2. *Let G be a connected graph and $f : V(G) \rightarrow \{1, 3, 5, \cdots\}$ be a positive odd integer-valued function with $f(v) \leq d_G(v)$ for any $v \in V(G)$. Then the following statements hold.*

(1) If G has a fractional $(1, f)$-odd factor, then $i(G \setminus S) \leq f(S)$ for every subset S of $V(G)$.

(2) If G is of odd order and has a fractional $(1, f)$-odd-factor, then G has at least one odd cycle. In other words, bipartite graph G of odd order has no fractional $(1, f)$-odd-factor.

(3) All k-regular graphs ($k \geq 1$ is an integer) have fractional $(1, f)$-odd-factors. In particular, each odd cycles has a fractional $(1, f)$-odd-factor.

(4) There exist infinite many graphs of even order which have fractional $(1, f)$-odd-factor but have no $(1, f)$-odd-factor.

Proof. (1) Set $g(x) = 1$. By Theorem A, a graph has a fractional $(1, f)$-factor if and only if for every subset S of $V(G)$, $|T| - d_{G \setminus S}(T) \leq f(S)$ holds, where $T = \{x : x \in V(G) \setminus S \text{ and } d_{G \setminus S}(x) \leq 1\}$. Note that $d_{G \setminus S}(T) = |T| - i(G \setminus S)$, thus $|T| - d_{G \setminus S}(T) \leq f(S)$ changes into $|T| - (|T| - i(G \setminus S)) \leq f(S)$, that is, $i(G \setminus S) \leq f(S)$. Since a fractional $(1, f)$-odd-factor is a fractional $(1, f)$-factor, thus (1) holds.

(2) We prove that if $G = (X, Y, E)$ is a bipartite graph of odd order, then G has no fractional $(1, f)$-odd-factor. Without loss of generality, assume that $|X|$ is odd and $|Y|$ is even. If G has a fractional $(1, f)$-odd-factor G_h with indicator function h, then $d_{G_h}(X) = \sum_{x \in X} h(E_x) = h(E_G(X, Y))$ is an odd number, and $d_{G_h}(Y) = \sum_{y \in Y} h(E_y) = h(E_G(X, Y))$ is an even number. A contradiction.

(3) For any k-regular graph G ($k \geq 1$ is an integer), let $f(x) = 1$ for all $x \in V(G)$, and $h(e) = \frac{1}{k}$ for all $e \in E(G)$. Then we obtain a fractional 1-factor G_h of G. Clearly, G_h is also a fractional $(1, f)$-odd-factor of G. For any odd cycle C_{2n+1} ($n \geq 1$ is an integer), let $f(x) = 1$ for all $x \in V(G)$ and $h(e) = \frac{1}{2}$ for all $e \in E(G)$, we can obtain a fractional 1-factor.

(4) Consider the following graph G_2. Let $f(u) = f(w) = 3$ and $f(x) = 1$ for all $x \in V(G_2) \setminus \{u, w\}$. Choose $S = \{u, w\}$, then $o(G_2 \setminus S) = 8 > 6 = f(u) + f(w) = f(S)$. Hence G_2 has no $(1, f)$-odd-factor by Theorem C. Replace C_5 in G_2 by any other odd cycle C_{2n+1}, we can obtain a new graph G which has no $(1, f)$-odd-factor.

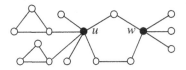

Fig. 2. Graph G_2 has no $(1, f)$-odd-factor

We can find a fractional $(1, f)$-odd-factor by assigning values to edges in the following way.

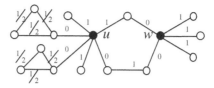

Fig. 3. Graph G_2 has a fractional $(1, f)$-odd-factor   □

Remark 2. In Theorem 2 (1), the condition $i(G \setminus S) \le f(S)$ is not sufficient for a bipartite graph to have a fractional $(1, f)$-odd-factor. In the following example, let $f(u) = f(v) = f(w) = 3$, and $f(x) = 1$ for all $x \in V(G) \setminus \{u, v, w\}$. Clearly, $i(G \setminus S) \le f(S)$ for all $S \subseteq V(G)$, but G has no fractional $(1, f)$-odd-factor.

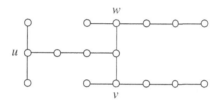

Fig. 4. Graph G_3 has no fractional $(1, f)$-odd-factor

By above discussion, we are faced with the following

Problem 1. Whether there exists a non-bipartite graph G such that $i(G \setminus S) \le f(S)$ for all $S \subseteq V(G)$, but G has no fractional $(1, f)$-odd-factor?

If there exists no such a graph, combine with Theorem 2, whether we can prove that "Let G be a connected non-bipartite graph, then G has fractional $(1, f)$-odd-factor if and only if $i(G \setminus S) \le f(S)$ for all $S \subseteq V(G)$."If this holds, in fact, we obtain the following equivalent result, i.e., "A connected non-bipartite graph G has a fractional $(1, f)$-odd-factor if and only if G has a fractional $(1, f)$-factor."

Acknowledgement. The authors would like to thank Professor Qinglin Yu for his valuable remarks.

References

1. Amahashi, A.: On factors with all degrees odd. Graphs and Comb. 1, 111–114 (1985)
2. Ansteen, R.P.: An algorithmic proof Tutte's f-factors of graphs. J. Algor., 112–131 (1985)
3. Bondy, J.A., Murty, U.S.R.: Graph Theory with Applications. MacMillan Press, London (1976)

4. Chen, C.P., Wang, J.F.: Factors in graphs with odd-cycle property. Discrete Math. 112, 29–40 (1993)
5. Cui, Y., Kano, M.: Some Results on Odd factors of graph. J. Graph Theory 12(3), 327–323 (1988)
6. Kano, M., Katona, G.Y.: Odd subgraphs and matchings. Discrete Math. 250, 265–272 (2002)
7. Liu, G.Z., Zhang, L.J.: Fractional (g, f)-factors of graphs. Acta Math. Scientia 21B4, 541–545 (2001)
8. Scheinerman, E.R., Ullman, D.H.: Fractional Graph Theory. John Wiley and Sons, Inc., New York (1997)
9. Topp, J., Vestergaard, P.D.: Odd Factors of a graph. Graphs and Comb. 9, 371–381 (1993)
10. Yu, Q.L., Zhang, Z.: Extremal properties of (1, f)-odd factors in graphs. Ars Comb. (to appear)

Some New Structural Properties of Shortest 2-Connected Steiner Networks

Shuying Peng[1], Meili Li[2], Shenggui Zhang[3,4], and T.C. Edwin Cheng[4]

[1] College of Science, Tianjin Polytechnic University, Tianjin 300160, P.R. China
[2] Department of Mathematics, Xi'an Shiyou University,
Xi'an, Shaanxi 710065, P.R. China
[3] Department of Applied Mathematics, Northwestern Polytechnical University,
Xi'an, Shaanxi 710072, P.R. China
[4] Department of Logistics, The Hong Kong Polytechnic University,
Hung Hom, Kowloon, Hong Kong, P.R. China

Abstract. In this paper we give a number of structural results for the problem of constructing minimum-weight 2-connected Steiner networks for a set of terminals in a graph and in the plane. A sufficient condition for a minimum-weight 2-connected Steiner network on a set of points in the plane to be basic is also obtained. Using the structural results, we show that the minimum-weight 2-connected Steiner network on a set of terminals Z is either a minimum-weight 2-connected spanning network on Z or isomorphic to one of several specific networks when $|Z| = 6$ or 7.

1 Introduction

Let $G = (V, E)$ be a complete undirected graph with a weight function $w : E \to R$, where w is assumed to fulfil the requirements of a metric, i.e., it is nonnegative, symmetric and satisfies the triangle inequity. For a subgraph H of G, its *weight* is defined as the sum of the weights of its edges. Let $Z \subseteq V$ be a set of *terminals*. A subgraph $G' = (V', E')$ of G is called a *Steiner network* on Z if $Z \subseteq V'$, and particularly a *spanning network* on Z if $Z = V$. The vertices in $V \setminus Z$ are called *Steiner vertices*. The *2-connected Steiner network problem* (2SNPG) is to find a minimum-weight 2-edge-connected Steiner network on Z. Since a minimum-weight 2-edge-connected network is necessarily 2-vertex-connected when the weight function is a metric [5], we use the shorthand 2-connected in the following.

The 2SNPG has important applications in the design of low-cost survival networks. It has been studied in the literature [1,5,6]. Monma *et al.* [5] mainly focused on the special case $Z = V$, where they proved that there always exists an optimal solution in which all the terminals have degree 2 or 3, and all Steiner vertices have degree 3. Grötschel *et al.* [1] proved that 2SNPG is NP-hard. Winter and Zachariasen [6] gave some structural results and presented an algorithmic framework for this problem.

The *Euclidean 2-connected Steiner network problem in the plane* (2SNPP) is a special case of 2SNPG. In this problem, the terminals are points in the plane, and

F.P. Preparata and Q. Fang (Eds.): FAW 2007, LNCS 4613, pp. 317–324, 2007.
© Springer-Verlag Berlin Heidelberg 2007

the task is to find a shortest 2-connected network that interconnects Z, where the length (weight) of an edge pq is the Euclidean distance between the two points p and q. In this case Steiner vertices are referred as *Steiner points*. Some structural properties of the optimal solutions for this problem were obtained in [2,3,4]. Luebke and Provan [4] proved that the problem is NP-hard. Bounds for the length of a shortest 2-connected Steiner network divided by the length of a shortest 2-connected spanning network were given in [2], and approximation algorithms for solving this problem were discussed in [2,4].

As in [6], we use SMN to denote an arbitrary minimum-weight 2-connected Steiner network such that the total degree of all the vertices has been minimized for 2SNPG, and use SMNP to denote an arbitrary minimum-weight 2-connected Steiner network for 2SNPP. Luebke [3] proved that all the vertices have degree 2 or 3, and all Steiner vertices has degree 3 in an SMN. For SMNPs, Hsu and Hu [2] proved that exactly three edges meet at 120^0 angles for every Steiner point.

In this paper we will give some further structural properties for both SMNs and SMNPs, and present a sufficient condition for an SMNP to be basic. Using the structural results, we show that, for a set of terminals Z, the optimal solution of an SMNG is either a shortest 2-connected spanning network on Z (which is a Hamilton cycle) or isomorphic to one of several specific networks when $|Z| = 6$ or 7.

In order to avoid the trivial case, throughout the paper, we assume that there does not exist a straight line containing all the terminals in SMNPs. For two points p and q in the plane, we use pq to denote the edge connecting them. Mathematically, pq also refers to the length of pq, i.e., the Euclidean distance between p and q. The total length of a graph H is denoted by $l(H)$.

2   Structural Properties of SMNs

Given a cycle C in a network N and two distinct vertices u and v on C, a *chord-path* between u and v is a (u, v)-path in N that, except from u and v, shares neither vertices nor edges with C. Note that the interior vertices of a chord-path are not required to be of degree 2 in N. If a chord-path has only one edge, then it is called a *chord-edge*. Winter and Zachariasen [6] gave the following property of chord-paths.

Theorem 1 ([6]). *Any chord-path in an SMN must have a pair of consecutive terminals in its interior.*

The following consequences are obvious.

Corollary 1 ([6]). *Any chord-path in an SMN must have at least three edges.*

Corollary 2. *Let N be an SMN and C be a cycle in N. Then every component of $N - V(C)$ contains at least two terminals.*

Theorem 1 and Corollary 1 can be viewed as properties of SMNs concerning forbidden subnetworks. The following result shows that SMNs cannot contain another kind of subnetworks.

Theorem 2. *Let N be an SMN. Then N contains no subnetwork isomorphic to the network shown in Figure 1, where u, v, w and x are four different vertices, $uw, vx \in E(N)$, and P_1 and P_2 are two internally-disjoint (u,v)-paths with $|V(P_1)| \leq 3$ and $|V(P_2)| \leq 6$.*

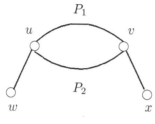

Fig. 1.

Proof. Let $C = P_1 \cup P_2$. Then C is a cycle with at most 7 edges. Since N is 2-connected, there exist two internally-disjoint (w,x)-paths Q_1 and Q_2 in N. From Corollary 1 we know that $V(C) \cap V(Q_1) \neq \emptyset$ and $V(C) \cap V(Q_2) \neq \emptyset$.

By the fact that each vertex of N is of degree 2 or 3, we have $|V(C)| \geq 4$, $E(Q_1) \cap E(C) \neq \emptyset$ and $E(Q_2) \cap E(C) \neq \emptyset$. Since C has at most seven edges, there must exist a path R on C with $|V(R) \cap V(Q_1)| = |V(R) \cap V(Q_2)| = 1$ and $|E(R)| = 1$ or 2. This implies that R is a chord-path of some cycle of N, contradicting Corollary 1.

From Corollary 1 and Theorem 2, we have

Corollary 3. *Let N be an SMN. Then the neighbors of a vertex of degree 3 are pairwise nonadjacent.*

3   Structural Properties of SMNPs

Let N be an SMNP. Then clearly N consists of a cycle, denoted by $\mathcal{C}(N)$, and the union of some connected subnetworks inside $\mathcal{C}(N)$. If $N - V(\mathcal{C}(N))$ contains no cycle, then N is called *basic*, otherwise, *nonbasic*. These notions were introduced by Hsu and Hu [2], where some properties for basic SMNPs were proved.

In this section we will give some further structural properties for both general SMNPs and basic SMNPs. At the same time, we present a sufficient condition for an SMNP to be basic.

Theorem 3. *Let N be an SMNP on a set of points P in the plane. Then*
(1) All the points in P on the sides of its convex hull lie on $\mathcal{C}(N)$;
(2) All the Steiner points in N lie inside the sides of the convex hull of P.

Proof. (1) By contradiction. Suppose that there exists a point p in P that is on the sides of its convex hull and lies inside $\mathcal{C}(N)$. Then by the definition of convex hulls, there exists a straight line l passing through p such that

(i) l intersects $C(N)$ at two points s_0 and s_{k+1};
(ii) all the vertices of N lying on a segment of $C(N)$ from s_0 to s_{k+1} (not including s_0 and s_{k+1}) are Steiner points.

We denote these Steiner points by $s_1, s_2, \ldots, s_k$ $(k \geq 1)$, and assume that these points appear on the corresponding (s_0, s_{k+1})-segment of $C(N)$ successively. We claim that in the polygon $s_0 s_1 \cdots s_k s_{k+1}$, at least one of the inner angles $\angle s_{i-1} s_i s_{i+1}$ $(1 \leq i \leq k)$ is $120°$. Otherwise, from the 120^0 angle conditions at the Steiner points,

$$\angle s_{k+1} s_0 s_1 + \sum_{i=1}^{k} \angle s_{i-1} s_i s_{i+1} + \angle s_k s_{k+1} s_0 = \angle s_{k+1} s_0 s_1 + \angle s_k s_{k+1} s_0 + k * 240°.$$

On the other hand, the sum of the inner angles of the polygon $s_0 s_1 \cdots s_k s_{k+1}$ is

$$\angle s_{k+1} s_0 s_1 + \sum_{i=1}^{k} \angle s_{i-1} s_i s_{i+1} + \angle s_k s_{k+1} s_0 = (k + 2 - 2) * 180° = k * 180°,$$

a contradiction. Without loss of generality, we assume $\angle s_0 s_1 s_2 = 120°$. Then, again from the 120^0 angle conditions at the Steiner points, there exists a vertex of N that is adjacent to s_1 and outside of $C(N)$, contradicting the definition of $C(N)$.

(2) First, it can be proved that every vertex of N on $C(N)$ cannot be outside of the sides of the convex hull of P by an analysis similar to that in the proof of (1). This means that every Steiner point cannot be outside of the sides of the convex hull of P. Furthermore, if there is a Steiner point s lying on the sides of the convex hull of P, then from the 120^0 angle conditions at the Steiner points, there exists a vertex of N that is adjacent to s and outside of $C(N)$, a contradiction. The result follows.

Theorem 4. *Let N be an SMNP and C a cycle of N. Then the subnetwork of N induced by the set of vertices of N on and inside (outside) C is 2-connected.*

Proof. We only prove that the subnetwork of N induced by the set of vertices of N on and inside C is 2-connected. The other assertion can be proved similarly.

Denote the subnetwork of N induced by the vertices on and inside C by H. Then it suffices to prove that any pair of vertices u and v in H are contained in a cycle of H. If both u and v are on C, then there is nothing to prove. So we assume that at least one of u and v is inside C.

If exactly one of u and v (say u) is inside C, then by the 2-connectedness of N, H contains a path P passing through u such that $|V(P) \cap V(C)| = 2$. It is easy to see that u and v are contained in a cycle in H.

If both of u and v are inside C, then by the 2-connectedness of N, H contains a path Q_1 passing through u and a path Q_2 passing through v such that $|V(Q_i) \cap V(C)| = 2$ for $i = 1, 2$. If $Q_1 \cup Q_2$ contains a cycle passing through both u and v, then we are done. Otherwise, it is not difficult to see that $Q_1 \cup Q_2$ contains a path Q passing through u and v such that $|V(Q) \cap V(C)| = 2$. Thus, we can find a cycle in H passing through both u and v. This completes the proof.

For basic SMNPs, we have

Theorem 5. *Let N be an SMNP, and u and v be two adjacent vertices of degree 3 in N. If N is basic, then $u, v \in V(\mathcal{C}(N))$ and $uv \in E(\mathcal{C}(N))$.*

Proof. First, we prove that $u, v \in V(\mathcal{C}(N))$. If it is not true, we consider the following two cases:

Case 1. Both u and v are not contained in $V(\mathcal{C}(N))$.

Since N is basic, the component of $N - V(\mathcal{C}(N))$ containing u and v is a tree. Therefore, there exist two internally-disjoint paths Q_1 and Q_2 from u to two distinct vertices on $\mathcal{C}(N)$ and two internally-disjoint paths R_1 and R_2 from v to two distinct vertices on $\mathcal{C}(N)$, such that $V(Q_i) \cap V(R_j) = \emptyset$ for $i, j = 1, 2$. Then uv is a chord-edge, contradicting Corollary 1.

Case 2. Exactly one of u and v is not contained in $V(\mathcal{C}(N))$.

Suppose that $u \in V(\mathcal{C}(N))$ but $v \notin V(\mathcal{C}(N))$. Then apart from uv, there are two other disjoint paths from v to $\mathcal{C}(N)$. Again, uv is a chord-edge, contradicting Corollary 1.

Finally, $uv \in E(\mathcal{C}(N))$. Otherwise, uv is a chord-edge in N, contradicting Corollary 1.

Let C be a cycle, and u and v be two vertices on C. Then C is divided into two paths. In the following by $C[u, v]$ we denote the path from u to v in the clockwise orientation.

Theorem 6. *Let N be an SMNP on a set of points P in the plane. If at most four points in P lie inside the sides of its convex hull, then N is basic.*

Proof. Let u, v, w, x be the four points of P that are not on the sides of its convex hull. Then, from Theorem 3 (1), they are the only vertices of N that could not lie on $\mathcal{C}(N)$. If all these points lie on $\mathcal{C}(N)$, then we are done. Otherwise, there must be a chord-path Q connecting two distinct vertices of $\mathcal{C}(N)$. By Theorem 1, Q contains at least two of u, v, w, x, say u and v, and these two vertices are adjacent. If w and x also lie on Q, then N is basic. Otherwise, there is another chord-path R with two end-vertices on $V(\mathcal{C}(N)) \cup V(Q)$ such that both w and x lie on R. Clearly we need only to consider the case that both of the end-vertices of R lie on Q.

Denote the end-vertices of Q by v_1 and v_2, the end-vertices of R by u_1 and u_2, and the cycle formed by Q and R by C_0. We assume the vertices v_1, u_1, u, v, u_2, v_2 appear on Q successively and the vertices u_1, u, v, u_2, w, x appear on C_0 along the clockwise orientation. From Corollary 1, $|V(\mathcal{C}(N)[v_1, v_2])| \geq 4$. Hence, there exist two different vertices v_3 and v_4 in $V(\mathcal{C}(N)[v_1, v_2]) \setminus \{v_1, v_2\}$ such that $v_1 v_3, v_2 v_4 \in E(\mathcal{C}(N))$.

By the triangle inequality, we have

$$uv_3 \le l(Q[u, u_1]) + l(Q[u_1, v_1]) + v_1v_3 \tag{1}$$

and

$$vv_4 \le l(Q[v, u_2]) + l(Q[u_2, v_2]) + v_2v_4. \tag{2}$$

Therefore, at least one of the following two inequalities

$$uv_3 \le l(Q[u, u_1]) + l(Q[u_2, v_2]) + v_1v_3$$

and

$$vv_4 \le l(Q[v, u_2]) + l(Q[u_1, v_1]) + v_2v_4$$

holds. Without loss of generality, we assume that $uv_3 \le l(Q[u, u_1]) + l(Q[u_2, v_2]) + v_1v_3$.

Case 1. $uv_3 < l(Q[u, u_1]) + l(Q[u_2, v_2]) + v_1v_3$.

Let $N' = \mathcal{C}(N)[v_3, v_1] + Q[u_1, v_1] + uv_3 + C_0[u, u_1]$. Then N' is a 2-connected Steiner network on P and $l(N') < l(N)$, a contradiction.

Case 2. $uv_3 = l(Q[u, u_1]) + l(Q[u_2, v_2]) + v_1v_3$.

It follows that $vv_4 \le l(Q[v, u_2]) + l(Q[u_1, v_1]) + v_2v_4$. If $vv_4 < l(Q[v, u_2]) + l(Q[u_1, v_1]) + v_2v_4$, then similar to the proof in Case 1, we can construct a new 2-connected Steiner network on P that is shorter than N, a contradiction. So we assume that $vv_4 = l(Q[v, u_2]) + l(Q[u_1, v_1]) + v_2v_4$. Then, from (1) and (2), we have $l(Q[u_1, v_1]) = l(Q[u_2, v_2])$. Thus, we obtain

$$uv_3 = l(Q[u, u_1]) + l(Q[u_1, v_1]) + v_1v_3$$

and

$$vv_4 = l(Q[v, u_2]) + l(Q[u_2, v_2]) + v_2v_4.$$

This implies that u, u_1, v_1 and v_3 lie on a straight line, and v, u_2, v_2 and v_4 lie on a straight line.

From Corollary 1, $|V(\mathcal{C}(N)[v_2, v_1])| \ge 4$. Hence, there exists a vertex $v_5 \in V(\mathcal{C}(N)[v_2, v_1]) \setminus \{v_1, v_2\}$ such that $v_1v_5 \in E(\mathcal{C}(N))$.

Case 2.1. u, u_1, v_1 and v_5 do not lie on a common straight line.

By the triangle inequality, $uv_5 < l(Q[u, u_1]) + l(Q[u_1, v_1]) + v_1v_5 = l(Q[u, u_1]) + l(Q[u_2, v_2]) + v_1v_5$. Let $N' = \mathcal{C}(N)[v_1, v_5] + Q[u_1, v_1] + uv_5 + C_0[u, u_1]$. Then N' is a new 2-connected Steiner network on P with $l(N') < l(N)$, a contradiction.

Case 2.2. u, u_1, v_1 and v_5 lie on a common straight line.

Let $N' = \mathcal{C}(N)[v_1, v_5] + v_3v_5 + Q[u_1, v_1] + Q[u_2, v_2] + C_0$ if $v_1v_5 > v_1v_3$, and $N' = \mathcal{C}(N)[v_3, v_1] + v_3v_5 + Q[u_1, v_1] + Q[u_2, v_2] + C_0$ if $v_1v_5 < v_1v_3$. Then N' is a new 2-connected Steiner network on P with $l(N') < l(N)$, a contradiction.

The proof is complete.

Corollary 4. *Let N be an SMNP on a set of points P in the plane. If at most three points in P lie inside the sides of its convex hull, then every two vertices of degree 3 in N are nonadjacent.*

Proof. By contradiction. Suppose that there exist two vertices u and v of degree 3 such that $uv \in E(N)$. By Theorem 6, N is basic. From Theorem 5, we have $u, v \in V(\mathcal{C}(N))$ and $uv \in E(\mathcal{C}(N))$. Then each of u and v has a neighbor inside $\mathcal{C}(N)$. Denote the neighbors of u and v inside $\mathcal{C}(N)$ by u_1 and v_1, respectively. By Theorem 5, $d(u_1) = d(v_1) = 2$. This implies that u_1 and v_1 are terminals of N.

By Theorem 3 (1), there are at most three terminals in $\mathcal{C}(N)$. Then by Corollary 2, we know that u_1 and v_1 must lie in a same component of $N - V(\mathcal{C}(N))$. So, there exists a (u_1, v_1)-path in $N - V(\mathcal{C}(N))$, which implies that uv is a chord-edge, contradicting Corollary 1.

4   Structure of SMNs on 6 and 7 Terminals

For 2SNPP, Hsu and Hu [2] proved that a shortest 2-connected Steiner network on a set of points P in the plane is a shortest 2-connected spanning network on P if all the points except at most one in P lie on the sides of its convex hull or if $|P| \leq 5$. It can readily be seen that these results are immediate consequences of Corollary 1 and Theorem 3.

As a generalization of Hsu and Hu's results, Winter and Zachariasen [6] proved that an SMN cannot have vertices of degree 3 unless it contains at least six terminals and showed that this bound is tight for 2SNPG. At the same time, they conjectured that the smallest number of terminals needed for an SMNP to have vertices of degree 3 is eight and gave a problem instance with eight terminals where the SMNP in fact has 2 Steiner vertices.

We have the following result concerning the conjecture of Winter and Zachariasen.

Theorem 7. *Let N be an SMN on a set of terminals Z with $|Z| = 6$ or 7.*
(1) If $|Z| = 6$, then N is either a shortest 2-connected spanning network on Z, which is a shortest Hamilton cycle on Z, or isomorphic to the network shown in Figure 2, where $Z = \{a, b, c, d, e, f\}$, and s_1 and s_2 are two Steiner vertices.

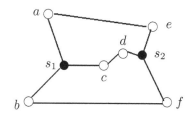

Fig. 2.

(2) *If $|Z| = 7$, then N is either a shortest 2-connected spanning network on Z, which is a shortest Hamilton cycle on Z, or isomorphic to the networks shown in Figure 3, where $Z = \{a, b, c, d, e, f, g\}$, and s_0, s_1 and s_2 are Steiner vertices.*

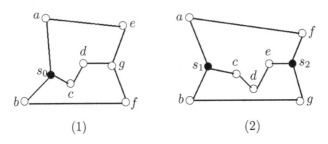

(1)   (2)

Fig. 3.

Proof. Suppose that N is not a cycle. Choose a cycle C in N. Then N has a chord-path P connecting two distinct vertices of C. By Theorem 1 it is clear that P contains two consecutive terminals. At the same time, C is divided into two paths and each of these two paths is a chord-path in N. Again by Theorem 1, each of these paths contains two consecutive terminals. The result follows from $|Z| = 6, 7$ immediately.

Acknowledgements

This work was supported by NSFC (No. 10101021) and SRF for ROCS of SEM. The third and fourth authors were supported in part by The Hong Kong Polytechnic University under grant number G-YX42.

References

1. Grötschel, M., Monma, C.L., Stoer, M.: Design of survivable networks. In: Ball, M.O., Magnanti, T.L., Monma, C.L., Nemhauser, G.L. (eds.) Network Models, Handbook in Operations Research and Management Sciences Series, pp. 617–672. North-Holland, New York (1995)
2. Hsu, D.F., Hu, X.-D.: On shortest two-connected Steiner networks with Euclidean distance. Networks 32(2), 133–140 (1998)
3. Luebke, E.L.: K-connected Steiner network problems, PhD Thesis, University of North Carolina, USA (2002)
4. Luebke, E.L., Provan, J.S.: On the structure and complexity of the 2-connected Steiner network problem in the plane. Oper. Res. Lett. 26, 111–116 (2000)
5. Monma, C.L., Munson, B.S., Pulleyblank, W.R.: Minimum-weight two-connected spanning networks. Math. Prog. 46, 153–171 (1990)
6. Winter, P., Zachariasen, M.: Two-connected Steiner networks: structural properties. Oper. Res. Lett. 33, 395–402 (2005)

The Parameterized Complexity of the Induced Matching Problem in Planar Graphs[*]

Hannes Moser[1,**] and Somnath Sikdar[2]

[1] Institut für Informatik, Friedrich-Schiller-Universität Jena,
Ernst-Abbe-Platz 2, D-07743 Jena, Germany
moser@minet.uni-jena.de
[2] The Institute of Mathematical Sciences,
C.I.T Campus, Taramani, Chennai 600113, India
somnath@imsc.res.in

Abstract. Given a graph G and an integer $k \geq 0$, the NP-complete INDUCED MATCHING problem asks for an edge subset M such that M is a matching and no two edges of M are joined by an edge of G. The complexity of this problem on general graphs as well as on many restricted graph classes has been studied intensively. However, little is known about the parameterized complexity of this problem. Our main contribution is to show that INDUCED MATCHING, which is W[1]-hard in general, admits a linear problem kernel on planar graphs. Additionally, we generalize a known algorithm for INDUCED MATCHING on trees to graphs of bounded treewidth using an improved dynamic programming approach.

1 Introduction

A *matching* in a graph is a set of edges no two of which have a common endpoint. An *induced matching* M of a graph $G = (V, E)$ is an edge-subset $M \subseteq E$ such that (1) M is a matching and (2) no two edges of M are joined by an edge of G. In other words, the subgraph induced by $V(M)$ is precisely the graph consisting of the edges in M. Let $\mathrm{im}(G)$ denote the size of a largest induced matching in G. The decision version of the INDUCED MATCHING problem asks, given a graph G and an integer k, whether G has an induced matching of size at least k. The optimization version asks for an induced matching of maximum size. The IN-DUCED MATCHING problem was introduced as a variant of the maximum matching problem and motivated by Stockmeyer and Vazirani [22] as the "risk-free" marriage problem[1]. This problem has been intensively studied in recent years. It is known to be NP-complete for planar graphs of maximum degree 4 [17], bipartite graphs of maximum degree 3, r-regular graphs for $r \geq 5$, line-graphs and

---

[*] Supported by a DAAD-DST exchange program, D/05/57666.
[**] Supported by the Deutsche Forschungsgemeinschaft, project ITKO (iterative compression for solving hard network problems), NI 369/5.
[1] Find the maximum number of pairs such that each married person is compatible with no married person except the one he or she is married to.

F.P. Preparata and Q. Fang (Eds.): FAW 2007, LNCS 4613, pp. 325–336, 2007.
© Springer-Verlag Berlin Heidelberg 2007

Hamiltonian graphs [18]. The problem is polynomial time solvable for trees [23] and weakly chordal graphs [6]. There exist many other results on special graph classes (see, e.g., [5,12,19]). Regarding polynomial-time approximability, it is known that the INDUCED MATCHING problem is APX-hard on $4r$-regular graphs, for all $r \geq 1$ [23], and bipartite graphs with maximum degree 3 [10]. Moreover, for r-regular graphs it is NP-hard to approximate INDUCED MATCHING within a factor of $r/2^{O(\sqrt{\ln r})}$ [8]. There exists an approximation algorithm for the problem on d-regular graphs ($d \geq 3$) with performance ratio $d - 1$ [10], which has subsequently been improved to $0.75d + 0.15$ [13].

In contrast to these results, little is known about the parameterized complexity of INDUCED MATCHING. To the best of our knowledge, the only known result is that the problem is $W[1]$-hard (with respect to the matching size as parameter) in the general case [20], and hence unlikely to be fixed-parameter tractable. Therefore, it is of interest to study the parameterized complexity of the problem in those restricted graph classes where it remains NP-complete. In this paper, we focus on planar graphs. The parameterized complexity of various NP-complete problems in planar graphs has already been studied. An interesting aspect of such studies are *linear* problem kernels. The intuitive idea behind kernelization is that a polynomial-time preprocessing step removes the "easy" parts of a problem instance such that only the "hard" core of the problem remains, which can then be solved by other methods. We call such a core a linear kernel if we can prove that its size is a linear function of the parameter. For a recent survey about problem kernelization, see [14].

Using a newly introduced technique, the question of whether DOMINATING SET has a linear kernel in planar graphs was answered positively by Alber et al. [2]. The kernel size has subsequently been improved by Chen et al. [7]. Moreover, they show lower bounds on the kernel size for DOMINATING SET, VERTEX COVER, and INDEPENDENT SET in planar graphs. The technique developed by Alber et al. [2] has been exploited by Guo et al. [16] in developing a linear kernel for FULL-DEGREE SPANNING TREE, a maximization problem. Moreover, Fomin and Thilikos [11] extended the technique to graphs of bounded genus. Very recently, Guo and Niedermeier [15] gave a generic kernelization framework for NP-hard problems in planar graphs based on that technique. Thus far, the technique has been applied to problems whose solutions are vertex subsets. We give the first application of this technique for a maximization problem whose solutions are edge subsets.

As our main result, we show that INDUCED MATCHING in planar graphs admits a linear problem kernel. We adapt and extend the known kernelization technique [2,16,15]. The corresponding data reduction rules can be carried out in linear time. Moreover, we generalize an algorithm for INDUCED MATCHING on trees by Zito [23] to graphs of bounded treewidth using an improved dynamic programming approach, which runs in $O(4^\omega \cdot n)$ time, where ω is the width of the given tree decomposition.

In Section 2, we start out with some basic definitions and notation. In Section 3, we present the kernelization proof, which is the main technical contribution of this paper. Finally, in Section 4, we outline the algorithm on graphs of bounded treewidth and give an outlook on possible future research.

2   Preliminaries

In this paper, we deal with fixed-parameter algorithms that emerge from the field of parameterized complexity [9,21]. A parameterized problem is *fixed-parameter tractable* if it can be solved in $f(k) \cdot n^{O(1)}$ time, where f is a computable function depending only on the parameter k. A common method to prove that a problem is fixed-parameter tractable is to provide data reduction rules that lead to a problem kernel. Given a problem instance (I, k), a *data reduction rule* replaces that instance by another instance (I', k') in polynomial time, such that (I, k) is a yes-instance iff (I', k') is a yes-instance. An instance to which none of a given set of data reduction rules applies is called *reduced* with respect to this set of rules. A parameterized problem is said to have a *problem kernel* if the resulting reduced instance has size $f(k)$ for a function f depending only on k. If $f(k) = c \cdot k$ for some constant c, then we call the kernel *linear*. The basic complexity class for fixed-parameter intractability is $W[1]$ [9].

In this paper we assume that all graphs are simple and undirected. For a graph G, let $V(G)$ denote its vertex set and $E(G)$ denote its edge set. For a subset $V' \subseteq V$, let $G[V']$ denote the subgraph of G induced by V'. Let $G \setminus V' := G[V \setminus V']$, and for $v \in V$, let $G - v := G \setminus \{v\}$. Let $N(v) := \{u \in V : \{u, v\} \in E\}$ be the *(open) neighborhood* of v. We assume that paths are *simple*, that is, each vertex appears at most once in a path. A path P from a to b is denoted as a vector $P = (a, \dots, b)$, and a and b are called the *endpoints* of P. The *length* of a path $(a_1, a_2, \dots, a_q)$ is $q - 1$, that is, the number of edges on it. For an edge set M we define $V(M) := \bigcup_{e \in M} e$. The *distance* $d(u, v)$ between two vertices u, v is the length of a shortest path between them. The *distance* between two edges e_1, e_2 is the minimum distance between two vertices $v_1 \in e_1$ and $v_2 \in e_2$.

If a graph can be drawn in the plane without edge crossings then it is *planar*. A *plane* graph is a planar graph with a fixed embedding in the plane. Given a plane graph, a cycle $C = (a, \dots, a)$ of length at least three encloses an *area* A of the plane. The cycle C is called the *boundary* of A, all vertices in the area A are *inside* A. A vertex is *strictly inside* A if it is inside A and not part of C.

3   A Linear Kernel on Planar Graphs

In order to show our kernel, we employ the following data reduction rules. Compared to the data reduction rules applied in other proofs of planar kernels [2,7,16], these data reduction rules are quite simple and can be carried out in $O(n + m)$ time on general graphs and thus in $O(n)$ time in planar graphs.

(R0) *Degree Zero Rule:* Delete vertices of degree zero.
(R1) *Degree One Rule:* If a vertex u has two distinct neighbors x, y of degree 1, then delete x.

(R2) *Degree Two Rule:* If u and v are two vertices such that $|N(u) \cap N(v)| \geq 2$ and if there exist two vertices $x, y \in N(u) \cap N(v)$ with $\deg(x) = \deg(y) = 2$, then delete x.

Note that these data reduction rules are parameter-independent. The following is our main theorem whose proof spans the remainder of this section.

Theorem 1. *Let $G = (V, E)$ be a planar graph reduced with respect to the rules R0, R1, and R2. Then $|V| \leq c \cdot \mathrm{im}(G)$ for some constant c. That is, the MAXIMUM INDUCED MATCHING problem in planar graphs admits a linear problem kernel.*

The basic observation is that if M is a maximum induced matching of a graph $G = (V, E)$ then for each vertex $v \in V$ there exists a $u \in V(M)$ such that $d(u, v) \leq 2$. Otherwise, we could add edges to M and obtain a larger induced matching. Since every vertex in the graph is within distance at most two to some vertex in $V(M)$, we know, roughly speaking, that the edges in M have distance at most four to other edges in M. This leads to the idea of regions "in between" matching edges that are close to each other. We will see that these regions cannot be too large if the graph is reduced with respect to the above data reduction rules. Moreover, we show that there cannot be many vertices that are not contained within such regions.

This idea of a region decomposition was introduced in [2], but the definition of a region as it appears there is much simpler since the regions are defined between vertices, and they are smaller. The remaining part of this section is dedicated to the proof of Theorem 1. First, in Section 3.1 we show how to find a "maximal region decomposition" of a reduced graph that contains only $O(|M|)$ regions, where M is the size of a maximum induced matching of the graph. Then, in Section 3.2 we show that a region in such a maximal region decomposition contains only a constant number of vertices. Finally, in Section 3.3 we show that in any reduced graph there are only $O(|M|)$ vertices which lie outside of regions.

3.1   Finding a Maximal Region Decomposition

Definition 1. *Let G be a plane graph and M a maximum induced matching of G. For edges $e_1, e_2 \in M$, a region $R(e_1, e_2)$ is a closed subset of the plane such that*

1. *the boundary of $R(e_1, e_2)$ is formed by two length-at-most-four paths*
   - *$(a_1, \ldots, a_2)$, $a_1 \neq a_2$, between $a_1 \in e_1$ and $a_2 \in e_2$,*
   - *$(b_1, \ldots, b_2)$, $b_1 \neq b_2$, between $b_1 \in e_1$ and $b_2 \in e_2$, and*
   by e_1 if $a_1 \neq b_1$ and e_2 if $a_2 \neq b_2$;
2. *for each vertex x in the region $R(e_1, e_2)$, there exists a $y \in V(\{e_1, e_2\})$ such that $d(x, y) \leq 2$;*
3. *no vertices inside the region other than endpoints of e_1 and e_2 are from M.*

The set of boundary vertices of R is denoted by δR. We write $V(R(e_1, e_2))$ to denote the set of vertices of a region $R(e_1, e_2)$, that is, all vertices strictly inside $R(e_1, e_2)$ together with the boundary vertices δR. A vertex in $V(R(e_1, e_2))$ is inside R.

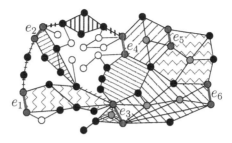

Fig. 1. An example of an M-region decomposition: black vertices denote boundary vertices; gray vertices lie strictly inside a region and white vertices lie outside of regions. Each region is hatched with a different pattern. Note the special cases, as for instance regions that consist of a path like the region between e_1 and e_2, or regions that are created by only one matching edge (the region on the left side of e_3).

Note that the two enclosing paths may be identical; the corresponding region then consists solely of a simple path of length at most four. Note also that e_1 and e_2 may be identical.

Definition 2. *An M-region decomposition of $G = (V, E)$ is a set $\mathcal{R}$ of regions such that no vertex in V lies strictly inside more than one region from $\mathcal{R}$. For an M-region decomposition $\mathcal{R}$, we define $V(\mathcal{R}) := \bigcup_{R \in \mathcal{R}} V(R)$. An M-region decomposition $\mathcal{R}$ is maximal if there is no $R \notin \mathcal{R}$ such that $\mathcal{R} \cup \{R\}$ is an M-region decomposition with $V(\mathcal{R}) \subsetneq V(\mathcal{R}) \cup V(R)$.*

For an example of an M-region decomposition, see Fig. 1.

Lemma 1. *Given a plane reduced graph $G = (V, E)$ and a maximum induced matching M of G, there exists an algorithm that constructs a maximal M-region decomposition with $O(|M|)$ regions.*

Lemma 1 can be proved by exhibiting a greedy algorithm that builds a maximal M-region decomposition in a stepwise manner by searching a region of maximal size that is not yet in the region decomposition at the actual step of the algorithm. Since this approach is similar to the algorithms by Alber et al. [2] and Guo et al. [16], we omit the details here.

3.2   Bounding the Size of a Region

To upper-bound the size of a region R we make use of the fact that any vertex strictly inside R has distance at most two from some vertex in δR. For this reason, the vertices strictly inside R can be arranged in two layers. The first layer consists of the neighbors of boundary vertices, and the second of all the remaining vertices, that is, all vertices at distance at least two from every boundary vertex. The proof strategy is to show that if any of these layers contains too many vertices then there exists an induced matching M' with $|M'| > |M|$. An important structure for our proof are areas enclosed by 4-cycles, called *diamonds*.

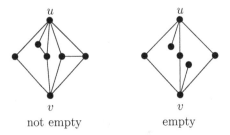

not empty   empty

Fig. 2. A diamond (left) and an empty diamond (right) in a reduced plane graph

Definition 3. *Let u and v be two vertices in a plane graph. A* diamond[2] *is a closed area of the plane with two length-2 paths between u and v as boundary. A diamond $D(u,v)$ is* empty, *if every edge e in the diamond is incident to either u or v.*

Fig. 2 shows an empty and a non-empty diamond. In a reduced plane graph empty diamonds have a restricted size. We are especially interested in the maximum number of vertices strictly inside an empty diamond $D(u,v)$ that have both u and v as neighbors. The following lemma is easy to show.

Lemma 2. *Let $D(u,v)$ be an empty diamond in a reduced plane graph. Then there exists at most one vertex strictly inside $D(u,v)$ that has both u and v as neighbors.*

Lemma 2 shows that if there are more than three edge-disjoint length-two paths between two vertices u,v, then there must be an edge e in an area enclosed by two of these paths such that e is not incident to u or v. This fact is used in the following lemma to show that the number of length-two paths between two vertices of a reduced plane graph is bounded.

Lemma 3. *Let u and v be two vertices of a reduced plane graph G such that there exists two distinct length-2 paths (u,x,v) and (u,y,v) enclosing an area A of the plane. Let M be a maximum induced matching of G. If neither x nor y is an endpoint of an edge in M and no vertex strictly inside A is contained in $V(M)$, then the following holds:*

If neither u nor v is an endpoint of an edge in M, then there are at most 5 edge-disjoint length-2 paths between u and v inside A. If exactly one of u or v is an endpoint of an edge in M, then there are at most 10 such paths, and if both u and v are endpoints of edges in M, then there are at most 15 such paths.

Proof. The idea is to show that if there are more than the claimed number of length-2 paths between u and v, then we can exhibit an induced matching M' with $|M'| > |M|$, which would then contradict the optimality of M.

---

[2] In standard graph theory, a diamond denotes a 4-cycle with exactly one chord. We abuse this term here. Note that diamonds also play an important role in proving linear problem kernels in planar graphs for other problems [2,15].

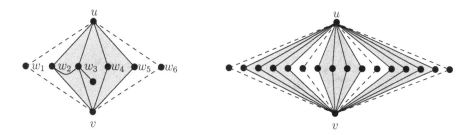

Fig. 3. Left: An embedding of the vertices $w_1, \ldots, w_6$ for the first case in the proof of Lemma 3. Right: An embedding of 16 neighbors of u and v for the last case of the proof. The diamonds are shaded and the "isolation paths" are drawn with dashed lines.

First, we consider the case when neither u nor v is contained in $V(M)$. Suppose for the purpose of contradiction that there are 6 common neighbors $w_1, \ldots, w_6$ of u and v that lie inside A (that is, strictly inside and on the enclosing paths). Without loss of generality, suppose that these vertices are embedded as shown in Fig. 3 (left-hand side), with w_1 and w_6 lying on the enclosing paths. Consider the diamond D with the boundary induced by the vertices u, v, w_2, w_5. Since w_3 and w_4 are strictly inside D and are incident to both u and v, by Lemma 3, we know that D is not empty. That is, there exists an edge e in D which is not incident to u or v. Clearly e is incident to neither w_1 nor w_6 and the endpoints of e are at distance at least 2 from every vertex in $V(M)$. Therefore, we can add e to M and obtain a larger induced matching, which contradicts the optimality of M.

Next, consider the case when exactly one of u or v is an endpoint of an edge e in M. Using the same idea as above, it is easy to see that if there exist 11 length-2 paths between u and v, then there are at least two non-empty diamonds (using (u, w_1, v), (u, w_6, v) and (u, w_{11}, v) as "isolation paths") whose boundaries share only u and v. We can then replace e in M by edges e_1 and e_2, one from each nonempty diamond, and obtain a larger induced matching.

The last case, when both u and v are endpoints of edges in M, can be handled in the same way using three non-empty diamonds (see Fig. 3).   □

Lemma 3 is needed to upper-bound the number of vertices inside and outside of regions that are connected to at least two boundary vertices.

The next two lemmas are needed to upper-bound the number of vertices that are connected to exactly one boundary vertex. First, Lemma 4 upper-bounds the number of such vertices under the condition that they are contained in an area which is enclosed by a short cycle. Lemma 4 is then used in Lemma 5 to upper-bound the total number of such vertices for a given boundary vertex. The proofs of both lemmas are omitted due to space restrictions.

Lemma 4. *Let u be a vertex in a reduced plane graph G and let $v, w \in N(u)$ be two distinct vertices that have distance at most three in $G - u$. Let P denote a shortest path between v and w in $G - u$ and let A denote the area of the plane*

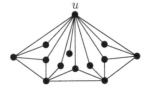

Fig. 4. Worst-case embeddings to illustrate Lemma 4

enclosed by P and the path (v, u, w). If there are at least 9 neighbors of u strictly inside A, then there is at least one edge strictly inside A.

Fig. 4 shows, for two different situations, the maximum number of neighbors of u that can be strictly inside A such that no edge lies strictly inside A.

Lemma 5. *Let u be a boundary vertex of a region $R(e_1, e_2)$ in a reduced plane graph G, and let M be a maximum induced matching of G. If u has at least 41 neighbors strictly inside R that are not adjacent to any other boundary vertex, then we can find an induced matching M' with $|M'| > |M|$.*

Using Lemma 3 and Lemma 5, we can now upper-bound the number of vertices inside a region.

Lemma 6. *A region $R(e_1, e_2)$ of an M-region decomposition of a reduced plane graph contains $O(1)$ vertices.*

Proof. We prove the lemma by partitioning the vertices strictly inside $R(e_1, e_2)$ into two sets A and B, where A consists of all vertices at distance exactly one from some boundary vertex, and B consists of all vertices at distance at least two from every boundary vertex, and then showing that $|A|$ and $|B|$ are upper-bounded by a constant.

To this end, partition A into A_1 and A_2, where A_1 contains all vertices in A that have exactly one neighbor on the boundary, and A_2 all vertices that have at least two neighbors on the boundary. To upper-bound the size of A_1, observe that due to Lemma 5, a vertex $v \in \delta R$ on the boundary can have at most 41 neighbors in A_1. Since a region has at most ten boundary vertices, we conclude that A_1 contains at most 410 vertices.

Next we upper-bound the size of A_2. Consider the planar graph G' induced by $\delta R \cup A_2$. Every vertex in A_2 is adjacent to at least two boundary vertices in G'. Replace every vertex $v \in A_2$ with an edge connecting two arbitrary neighbors of v on the boundary. Merge multiple edges between two boundary vertices into a single edge. Since G' is planar, the resulting graph must also be planar. As $|\delta R| \le 10$, using the Euler formula we conclude that the resulting graph has at most $3 \cdot 10 - 6 = 24$ newly added edges. By Lemma 3, each such edge represents at most 15 length-two paths, and thus $|A_2| \le 24 \cdot 15 = 360$.

To upper-bound the size of B, observe that $G[B]$ must be a graph without edges (that is, B is an independent set). By the Degree One Rule, each vertex in A has at most one neighbor in B of degree one. Therefore, there are $O(1)$ degree-one vertices in B. To bound the number of degree-at-least-two vertices in B, we use the same argument as the one used to bound the size of A_2. Since $|A| = O(1)$, there is a constant number of degree-at-least-two vertices in B. Therefore $|B| = O(1)$. This completes the proof.   □

Proposition 1. *Let G be a reduced plane graph and let M be a maximum induced matching of G. There exists an M-region decomposition such that the total number of vertices inside all regions is $O(|M|)$.*

Proof. Using Lemma 1, there exists a maximal M-region decomposition for G with at most $O(|M|)$ regions. By Lemma 6, each region has a constant number of vertices. Thus there are $O(M)$ vertices inside regions.   □

3.3   Bounding the Number of Vertices Lying Outside of Regions

In this section, we upper-bound the number of vertices that lie outside of regions of a maximal M-region decomposition. The strategy to prove this bound is similar to that used in the last section. We subdivide the vertices lying outside of regions into several disjoint subsets and upper-bound their sizes separately.

Note again that the distance from any vertex of the graph to a vertex in $V(M)$ is at most two. We partition the vertices lying outside of regions into two sets A and B, where A is the set of vertices at distance exactly one from some vertex in $V(M)$, and B is the set of vertices at distance at least two from every vertex in $V(M)$. We bound the sizes of these two sets separately.

Partition A into two subsets A_1 and A_2, where A_1 is the set of vertices that have exactly one boundary vertex as neighbor, and A_2 is the set of vertices that have at least two boundary vertices as neighbors. Note that each vertex v in A can be adjacent to exactly one vertex $u \in V(M)$. For if it is adjacent to distinct vertices $u, w \in V(M)$, then the path (u, v, w) can be added to the region decomposition, contradicting its maximality (recall that regions can consist of simple paths between two vertices in $V(M)$). To bound the number of vertices in A_1 we need the following lemma, which is easy to prove.

Lemma 7. *Let v be a vertex in A_1 and let u be its neighbor in $V(M)$. Then for all $w \in V(M) \setminus \{u\}$, the distance between v and w in $G - u$ is at least three.*

Lemma 8. *Given a maximal M-region decomposition consisting of $O(|M|)$ regions, the set A contains $O(|M|)$ vertices.*

Proof. To bound the size of A_1, we claim that each vertex $u \in V(M)$ has at most 20 neighbors in A_1. Suppose, for the purpose of contradiction, that 21 vertices $v_1, \ldots, v_{21}$ in A_1 are adjacent to $u \in V(M)$. Also assume that they are embedded in a clockwise fashion around u in that order. Let e be the edge in M incident to u. First, suppose that v_1 and v_{11} have distance at least four in $G - u$. Then there exist edges e_a, e_b in $G - u$ incident to v_1 and v_{11}, respectively, that form an induced matching of size 2. Moreover by Lemma 7, the endpoints of e_a

and e_b are not adjacent to any vertex of $V(M)$ in $G - u$. Therefore, $M' = (M \setminus \{e\}) \cup \{e_a, e_b\}$ is an induced matching of size larger than that of M, a contradiction to the maximum cardinality of M. The same holds if the distance between v_{11} and v_{21} is at least four in $G - u$. Therefore assume that in the graph $G - u$, $d(v_1, v_{11}) \leq 3$ and $d(v_{11}, v_{21}) \leq 3$. Let P_1 and P_2 be shortest paths in $G - u$ between v_1 and v_{11} and between v_{11} and v_{21}, respectively. Note that due to Lemma 7 these two paths cannot contain any vertex from $V(M)$. By Lemma 4, the areas enclosed by P_1 and (v_1, u, v_{11}), and P_2 and v_{11}, u, v_{21}, respectively, contain an edge strictly inside them. The edge e can be replaced by these two edges to obtain an induced matching of size larger than M, a contradiction to the maximum cardinality of M. This proves our claim. Since there are exactly $2|M|$ vertices in $V(M)$, this shows that the total number of vertices in A_1 is at most $40|M|$.

Next, we bound the size of A_2. Every vertex v in A_2 is adjacent to a vertex $u \in V(M)$ and some boundary vertex $w \notin V(M)$. Vertex w must be adjacent to u, for otherwise there is a path consisting of the vertices (u, v, w) and some subpath on the boundary where w lies which can be added to the region decomposition $\mathcal{R}$, contradicting its maximality. Since there are $O(|M|)$ regions, there are $O(|M|)$ possible boundary vertices adjacent to a vertex in $V(M)$. By Lemma 3, given a vertex $x \in V(M)$ and $y \in V \setminus V(\mathcal{R})$ there can be at most 10 vertices adjacent to both x and y. This shows that A_2 contains $O(|M|)$ vertices. $\square$

It remains to bound the number of vertices in B, that is, the number of vertices outside of regions that are at distance at least two from every vertex in $V(M)$.

Lemma 9. *Given a maximal M-region decomposition with $O(|M|)$ regions, the set B contains $O(|M|)$ vertices.*

Proof. To bound the size of B, observe that $G[B]$ is a graph without edges. Furthermore, observe that $N(B) \subseteq A \cup A'$, where A' is the set of boundary vertices in the M-region decomposition that are different from $V(M)$. By Lemma 8 and since the boundary of each region contains a constant number of vertices, the set $C := A \cup A'$ contains $O(|M|)$ vertices.

First, consider the vertices in B that have degree one. Obviously, there can be at most $|C|$ such vertices due to the Degree One Rule. The remaining vertices are adjacent to at least two vertices in C. We can use an argument similar to the one used in the proof of Lemma 6 (using the Euler formula) to show that there are $O(|C|)$ degree-at-least-two vertices in B. Thus, $|B| = O(|C|) = O(|M|)$. $\square$

Proposition 2. *Given a maximal M-region decomposition with $O(|M|)$ regions, the number of vertices that lie outside of regions is $O(|M|)$.*

Proof. Follows from Lemmas 8 and 9. $\square$

Using Propositions 1 and 2, we can show that, given a reduced plane graph G and a maximum induced matching M of G, there exists an M-region decomposition with $O(|M|)$ regions such that the number of vertices inside and outside of regions is $O(|M|)$. This shows the $O(|M|)$ upper bound on the number of vertices as claimed in Theorem 1.

4   Further Results and Outlook

Zito [23] developed a linear-time dynamic programming algorithm to solve IN-DUCED MATCHING on trees. We generalize this approach to obtain a linear-time algorithm on graphs of bounded treewidth [4]. It is relatively easy to see that a standard dynamic programming approach would result in a running time of $O(9^\omega \cdot n)$, where ω is the width of the given tree decomposition. With an improved dynamic programming algorithm, we obtain a running time of $O(4^\omega \cdot n)$. The description of this algorithm, which is inspired by a similar result for DOM-INATING SET [3], is omitted due to space restrictions.

Theorem 2. *Let $G = (V, E)$ be a graph with a given nice tree decomposition $(\{X_i \mid i \in I\}, T)$. Then, the size of a maximum induced matching of G can be computed in $O(4^\omega \cdot n)$ time, where $n := |I|$ and ω denotes the width of the tree decomposition.*

As our main result, we have shown that INDUCED MATCHING in planar graphs admits a linear problem kernel. The data reduction rules for the planar case are very simple and the kernelization can be done in linear time.

A possible future research topic could be search tree algorithms for planar graphs. For DOMINATING SET in planar graphs, there exists a search tree algorithm [1], and it is open whether a similar result for INDUCED MATCHING on planar graphs is possible. Investigating the parameterized complexity of IN-DUCED MATCHING on other restricted classes of graphs may be of interest. We can show simple problem kernelizations for bounded-degree graphs, graphs of girth at least 6, C_4-free bipartite graphs, and line graphs. A class of major interest are bipartite graphs, where the parameterized complexity of INDUCED MATCHING is open.

Acknowledgements. We thank Jiong Guo and Rolf Niedermeier (University of Jena, Germany) for initiating this research and for several constructive discussions and comments.

References

1. Alber, J., Fan, H., Fellows, M.R., Fernau, H., Niedermeier, R., Rosamond, F.A., Stege, U.: A refined search tree technique for dominating set on planar graphs. Journal of Computer and System Sciences 71(4), 385–405 (2005)
2. Alber, J., Fellows, M.R., Niedermeier, R.: Polynomial-time data reduction for dominating set. Journal of the ACM 51(3), 363–384 (2004)
3. Alber, J., Niedermeier, R.: Improved tree decomposition based algorithms for domination-like problems. In: Rajsbaum, S. (ed.) LATIN 2002. LNCS, vol. 2286, pp. 613–628. Springer, Heidelberg (2002)

4. Bodlaender, H.L.: Treewidth: Characterizations, applications, and computations. In: Fomin, F.V. (ed.) WG 2006. LNCS, vol. 4271, pp. 1–14. Springer, Heidelberg (2006)
5. Cameron, K.: Induced matchings in intersection graphs. Discrete Mathematics 278(1-3), 1–9 (2004)
6. Cameron, K., Sritharan, R., Tang, Y.: Finding a maximum induced matching in weakly chordal graphs. Discrete Mathematics 266(1-3), 133–142 (2003)
7. Chen, J., Fernau, H., Kanj, I.A., Xia, G.: Parametric duality and kernelization: lower bounds and upper bounds on kernel size. SIAM Journal on Computing (to appear)
8. Chlebík, M., Chlebíková, J.: Approximation hardness of dominating set problems. In: Albers, S., Radzik, T. (eds.) ESA 2004. LNCS, vol. 3221, pp. 192–203. Springer, Heidelberg (2004)
9. Downey, R.G., Fellows, M.R.: Parameterized Complexity. Springer, Heidelberg (1999)
10. Duckworth, W., Manlove, D., Zito, M.: On the approximability of the maximum induced matching problem. Journal of Discrete Algorithms 3(1), 79–91 (2005)
11. Fomin, F.V., Thilikos, D.M.: Fast parameterized algorithms for graphs on surfaces: Linear kernel and exponential speed-up. In: Díaz, J., Karhumäki, J., Lepistö, A., Sannella, D. (eds.) ICALP 2004. LNCS, vol. 3142, pp. 581–592. Springer, Heidelberg (2004)
12. Golumbic, M.C., Lewenstein, M.: New results on induced matchings. Discrete Applied Mathematics 101(1-3), 157–165 (2000)
13. Gotthilf, Z., Lewenstein, M.: Tighter approximations for maximum induced matchings in regular graphs. In: Erlebach, T., Persinao, G. (eds.) WAOA 2005. LNCS, vol. 3879, pp. 270–281. Springer, Heidelberg (2006)
14. Guo, J., Niedermeier, R.: Invitation to data reduction and problem kernelization. SIGACT News 38(1), 31–45 (2007)
15. Guo, J., Niedermeier, R.: Linear problem kernels for NP-hard problems on planar graphs. In: Arge, L., Cachin, C., Jurdzinski, T., Tarlecki, A. (eds.) ICALP2007. LNCS, vol. 4596, pp. 375–386. Springer, Heidelberg (2007)
16. Guo, J., Niedermeier, R., Wernicke, S.: Fixed-parameter tractability results for full-degree spanning tree and its dual. In: Bodlaender, H.L., Langston, M.A. (eds.) IWPEC 2006. LNCS, vol. 4169, pp. 203–214. Springer, Heidelberg (2006)
17. Ko, C.W., Shepherd, F.B.: Bipartite domination and simultaneous matroid covers. SIAM Journal on Discrete Mathematics 16(4), 517–523 (2003)
18. Kobler, D., Rotics, U.: Finding maximum induced matchings in subclasses of claw-free and P_5-free graphs, and in graphs with matching and induced matching of equal maximum size. Algorithmica 37(4), 327–346 (2003)
19. Lozin, V.V., Rautenbach, D.: Some results on graphs without long induced paths. Information Processing Letters 88(4), 167–171 (2003)
20. Moser, H., Thilikos, D.M.: Parameterized complexity of finding regular induced subgraphs. In: Proceedings of ACiD'06. Texts in Algorithmics, vol. 7, pp. 107–118. College Publications (2006)
21. Niedermeier, R.: Invitation to Fixed-Parameter Algorithms. Oxford University Press, Oxford, UK (2006)
22. Stockmeyer, L.J., Vazirani, V.V.: NP-completeness of some generalizations of the maximum matching problem. Information Processing Letters 15(1), 14–19 (1982)
23. Zito, M.: Induced matchings in regular graphs and trees. In: Widmayer, P., Neyer, G., Eidenbenz, S. (eds.) WG 1999. LNCS, vol. 1665, pp. 89–100. Springer, Heidelberg (1999)

Removable Edges of a Spanning Tree in 3-Connected 3-Regular Graphs*

Haiyan Kang[1], Jichang Wu[1,**], and Guojun Li[1,2]

[1] School of Mathematics and System Sciences, Shandong University
Jinan, 250100, China
[2] CSBL, Department of Biochemistry and Molecular
Biology Institute of Bioinformatics,
University of Georgia, GA 30602,USA
rdkhy@mail.sdu.edu.cn,jichangwu@126.com,guojun@csbl.bmb.uga.edu

Abstract. Let G be a 3-connected graph and e an edge of G. If, by deleting e from G, the resultant graph $G - e$ is a 3-connected graph or a subdivision of a 3-connected graph, then e is called a *removable edge* of G. In this paper we obtain that there are at least two removable edges in a spanning tree of a 3-connected 3-regular graph. Also we give an $O(n^3)$ time algorithm to find all removable edges in G.

Keywords: 3-connected 3-regular graph, removable edge, edge-vertex-cut fragment.

1 Introduction

The concepts of contractible edges and removable edges of graphs are very important in studying the structures of graphs and in proving some properties of graphs by induction. In 1961, W.T.Tutte [8] gave the structural characterization for 3-connected graphs by using the existence of contractible edges and removable edges. He proved that every 3-connected graph with order at least 5 contains contractible edges. Perhaps, this is the earliest result concerning the concepts of contractible edges and removable edges. In this paper, we shall focus on the study of only removable edges in 3-connected graphs. First of all, we give the definition of a removable edge for a 3-connected graph. Let G be a 3-connected graph and e an edge of G. Consider the graph $G - e$ obtained by deleting the edge e from G. If $G - e$ has vertices of degree 2, we do the following operations on $G - e$. For each vertex x of degree 2 in $G - e$, we delete x from $G - e$ and then connect the two neighbors of x by an edge. If multiple edges occur, we use single edge to replace them. The final resultant graph is denoted by $G \ominus e$. If $G \ominus e$ is still 3-connected, then the edge e is called *removable*; otherwise, it is called *unremovable*. The set of all removable edges of G is denoted by $E_R(G)$; whereas the set of unremovable edges of G is denoted by $E_N(G)$. In [3], Holton

---

* This work was supported from NSFC under Grant No.60673059.
** Corresponding author.

et.al discussed the distribution and number of removable edges in 3-connected graphs. In [6], Su got the lower bound of the number of removable edges of 3-connected graphs and described the structural characterization of graphs which can attain the lower bound. Moreover, Su studied the distribution of removable edges of cycles in 3-connected graphs [7]. Ou et.al got the following result in [5]:

Theorem 1.1. Let G be a 3-connected 3-regular graph. Then any spanning tree of G contains at least one removable edge.

In fact, our main theorem improves the above result.

Let G be a graph. The vertex set and the edge set of G are denoted by $V(G)$ and $E(G)$, respectively. If $x \in V(G)$, we also write $x \in G$. We denote by $|G|$ and $|E(G)|$ the order and the size of G, respectively. For $x \in G$, the set of adjacent vertices of x in G is denoted by $\Gamma_G(x)$ and the degree of x is denoted by $d_G(x)$(or briefly $d(x)$). If the endvertices of an edge of G are x and y, then we write $e = xy$. Let $A, B \subset V(G)$ such that $A \neq \emptyset \neq B$ and $A \cap B = \emptyset$. We define $\langle A, B \rangle = \{xy \in E(G) | x \in A, y \in B\}$. Let F(resp. N) be a nonempty subset of $E(G)$(resp. $V(G)$). The subgraph of G induced by F(resp. N) is denoted by $\langle F \rangle$(resp. $\langle N \rangle$). If H is a subgraph of G, we also say that G contains H. We often identify a subgraph H of G with its vertex set $V(H)$. Let $S \subset V(G)$. $G - S$ denotes the graph obtained by deleting all the vertices in S from G together with all the incident edges. The vertex set S is said to be a vertex cut of G if $G - S$ is disconnected. If $|S| = s$, the vertex set S is said to be a s-vertex cut of G. A l-cycle of G is a cycle with length l in G. The girth of G is the length of a shortest cycle in G and is denoted by $g(G)$. For other graph-theoretic notation not mentioned here, we refer the reader to [1].

2   Preliminary Knowledge

Let G be a 3-connected graph. For $e \in E(G)$ and $S \subset V(G)$ with $|S| = 2$, if $G - e - S$ has exactly two (connected) components, say A and B, such that $|A| \geq 2$ and $|B| \geq 2$, then we say that (e, S) is a *separating pair* and $(e, S; A, B)$ is a *separating group*, in which A and B are called *edge-vertex-cut fragments*. The edge-vertex-cut fragment with the minimum vertices is called an *edge-vertex-cut atom*.

Let $(xy, S; A, B)$ be a separating group of G where $x \in A, y \in B$ and $S = \{a, b\}$. If $|A| = 2$, we take that $A = \{x, z\}$. From the fact that G is 3-connected, we have that $zx, za, zb \in E(G)$ and $\{xa, xb\} \cap E(G) \neq \emptyset$. So we get a 3-cycle $xazx$ or $xbzx$. Then $g(G) = 3$.

Let $E_0 \subset E_N(G)$ and $E_0 \neq \emptyset$, and let $(xy, S; A, B)$ be a separating group where $x \in A, y \in B$. If $xy \in E_0$, we say A and B are E_0-*edge-vertex-cut fragments*. An E_0-edge-vertex-cut fragment is called an E_0-*edge-vertex end-fragment* if it does not contain any other E_0-edge-vertex-cut fragment of G as its proper subset.

In the sequel, we list some known results on removable edges of 3-connected graphs, which will be used in the next section.

Theorem 2.1([3]). Let G be a 3-connected graph with $|G| \geq 6$. An edge e of G is unremovable if and only if there is a separating pair (e, S) or a separating group $(e, S; A, B)$ in G.

Theorem 2.2([3]). Let G be a 3-connected graph with $|G| \geq 6$ and $(e = xy, S)$ a separating pair of G. Then every edge joining S and $\{x, y\}$ is removable.

Theorem 2.3([3]). Let G be a 3-connected graph with $|G| \geq 6$ and C a cycle of G. Suppose that no edges of C are removable. Then there exists a vertex $a \in V(G)$ such that $d(a) \geq 4$.

Theorem 2.4([3]). Let G be a 3-connected with $|G| \geq 6$ and $(xy, S = \{a, b\})$ a separating pair of G. If $ab \in E(G)$, then $ab \in E_R(G)$.

Theorem 2.5([7]). Let G be a 3-connected 3-regular graph with $|G| \geq 6$ and C a 3-cycle of G. then $E(C) \subset E_R(G)$.

3   Main Results

Before we give our main result, we need to show the following lemmas.

Lemma 3.1. Let G be a 3-connected 3-regular graph. $y_1 y_2 ... y_k$ is a path in $\langle E_N(G) \rangle$ where $k \geq 3$. Let $\emptyset \neq D \subset V(G)$. Suppose that there is a separating group $(y_1 y_2, S_1; A_1, B_1)$ of G where $y_1 \in B_1, y_2 \in A_1$ and $D \cap B_1 \neq \emptyset$. We take a separating group $(y_i y_{i+1}, S; A, B)$ where $y_i \in B$, $y_{i+1} \in A$, $D \cap B \neq \emptyset$ such that $|A|$ is as small as possible. If $i \leq k - 2$, we get another separating group $(y_{i+1} y_{i+2}, S'; A', B')$ where $y_{i+1} \in B'$ and $y_{i+2} \in A'$. For those separating groups, we have the following conclusions hold:
$y_{i+1} \in A \cap B', y_i \in B \cap B', y_{i+2} \in A \cap A'. D \cap B = B \cap S' = \{c\}$ where $c \in V(G). |A \cap S'| = |B' \cap S| = 1. S \cap S' = \emptyset = B \cap A'. A' \cap S = \{d\}.$ Then $cd \in E_N(G)$.

Proof. Clearly, $y_{i+1} \in A \cap B'$. It follows from Theorem 2.2 that $y_i \in B \cap B'$ and $y_{i+2} \in A \cap A'$. Let

$$X_1 = (B' \cap S) \cup (S \cap S') \cup (A \cap S'), X_2 = (A \cap S') \cup (S \cap S') \cup (A' \cap S),$$
$$X_3 = (A' \cap S) \cup (S \cap S') \cup (B \cap S'), X_4 = (B \cap S') \cup (S \cap S') \cup (B' \cap S).$$

Since $A \cap A' \neq \emptyset$, we know that X_2 is a vertex cut of $G - y_{i+1} y_{i+2}$. Noticing that G is 3-connected, then $|X_2| \geq 2$. By a similar argument, we can get $|X_4| \geq 2$. From $|X_2| + |X_4| = |S| + |S'| = 4$, we get that $|X_2| = |X_4| = 2$. So $|B' \cap S| = |A \cap S'|$, $|B \cap S'| = |A' \cap S|$.

We claim that $A \cap A' = \{y_{i+2}\}$. Otherwise, $|A \cap A'| \geq 2$. Let $C' = A \cap A', T' = X_2$ and $D' = G - y_{i+1} y_{i+2} - T' - C'$. We get a separating group $(y_{i+1} y_{i+2}, T'; C', D')$. Clearly, $D' \cap D \neq \emptyset$ and $|C'| < |A|$. This contradicts the choice of A.

Since $|A'| \geq 2$ and A' is a connected subgraph of G, we get that $A' \cap S \neq \emptyset$. If $|A' \cap S| = |B \cap S'| = 2$, then $|X_1| = 0$. So $\{y_i, y_{i+2}\}$ is a 2-vertex cut of G. This is a contradiction. Hence, $|A' \cap S| = |B \cap S'| = 1$. From $|S| = |S'| = 2$, we get $|S \cap S'| \leq 1$. Next we discuss the following cases.

Case 1. $|S \cap S'| = 1$. We have that $B' \cap S = \emptyset = A \cap S'$. Let $S \cap S' = \{a\}$, $A' \cap S = \{b\}$ and $B \cap S' = \{c\}$.

We claim that $A \cap B' = \{y_{i+1}\}$. Otherwise, $|A \cap B'| \geq 2$. We can get a 2-vertex cut $\{a, y_{i+1}\}$ of G, which is a contradiction. Since $d(y_{i+1}) = 3$, we have $ay_{i+1} \in E(G)$. Similarly, $ay_{i+2} \in E(G)$.

If $A' \cap B \neq \emptyset$, then $\{a, b, c\}$ is a 3-vertex cut of G. So $\Gamma_G(a) \cap (A' \cap B) \neq \emptyset$. If $A' \cap B = \emptyset$, then $A' = \{y_{i+1}, b\}$. So we get that $ab \in E(G)$. That is to say $\Gamma_G(a) \cap (A' \cap S) \neq \emptyset$. Thus, we deduce that $\Gamma_G(a) \cap ((A' \cap S) \cup (A' \cap B)) \neq \emptyset$.

Since $B \cap B' \neq \emptyset$, we get a 3-vertex cut $\{y_{i+1}, a, c\}$ of G. Then $\Gamma_G(a) \cap (B \cap B') \neq \emptyset$. Then $d(a) \geq 4$. This contradicts the fact that G is 3-regular. Hence, this case is impossible.

Case 2. $|S \cap S'| = 0$. We have that $|B' \cap S| = |A \cap S'| = 1$. Without loss of generality, let $A \cap S' = \{a\}$, $B' \cap S = \{b\}$, $B \cap S' = \{c\}$ and $A' \cap S = \{d\}$. Clearly, $|X_3| = 2$. Since G is 3-connected, we deduce that $B \cap A' = \emptyset$. So $|A'| = 2$. Noticing the minimality of $|A|$ and $|A| \geq 3$, we know that $B' \cap D = \emptyset$. So $B \cap D = B \cap S' = \{c\}$. Hence, we get that $cd \in E(G)$. Let $C_1 = B, T_1 = \{y_{i+1}, b\}$ and $D_1 = G - cd - C_1 - T_1$. Clearly, $(cd, T_1; C_1, D_1)$ is a separating group of G. So $cd \in E_N(G)$. Now, the proof is complete. $\qquad\square$

Lemma 3.2. Let G be a 3-connected 3-regular graph. If $\langle E_N(G) \rangle$ is a tree, then $|\langle E_N(G) \rangle| \leq |G| - 2$.

Proof. By contradiction. We assume that $|\langle E_N(G) \rangle| \geq |G| - 1$. Let x be a leaf of $\langle E_N(G) \rangle$. Since $d_G(x) = 3$ and $|\langle E_N(G) \rangle| \geq |G| - 1$, there exists a vertex $y \in \langle E_N(G) \rangle$ such that $xy \in E_R(G)$. Let P be a path connecting x and y in $\langle E_N(G) \rangle$. Then $P + xy$ is a cycle containing xy in G. So we take a cycle $C = y_1 y_2 ... y_k y_1$ in G such that $y_1 y_k \in E_R(G)$ and $E(C) - \{y_1 y_k\} \subset E_N(G)$.

Let $D = \{y_1\}$. We take a separating group $(y_1 y_2, S_1; A_1, B_1)$ such that $y_1 \in B_1$, $y_2 \in A_1$. Clearly, $D \cap B_1 \neq \emptyset$. We take $i \in \{1, 2, ...k-1\}$ and a separating group $(y_i y_{i+1}, S; A, B)$ where $y_i \in B$, $y_{i+1} \in A$ and $D \cap B \neq \emptyset$ such that $|A|$ is as small as possible.

We claim that $i < k - 1$. Otherwise, $i = k - 1$. So we have that $y_k \in A$. Noticing that $y_1 y_k \in E(G)$, we get that $y_1 \in A \cup S$. This contradicts that $D \cap B \neq \emptyset$.

Now we take another separating group $(y_{i+1} y_{i+2}, S'; A', B')$ where $y_{i+1} \in B'$ and $y_{i+2} \in A'$. This satisfies the condition of Lemma 3.1. Without loss of generality, let $D \cap B = B \cap S' = \{c\}$, $A' \cap S = \{d\}$ and $A \cap S' = \{a\}$. So $c = y_1$. Since $y_1 d(= cd) \in E_N(G)$ and $y_1 y_k \in E_R(G)$, we deduce that $d \neq y_k$ and $y_k \in B' - A$. Since $y_{i+2} \in A \cap A'$, we get that the distance between y_{i+2} and y_k is at least 2. So $i + 2 \leq k - 2$. Noticing that $y_{i+3} \in \Gamma_G(y_{i+2}) = \{y_{i+1}, a, d\}$, we have that $y_{i+3} = a$ or $y_{i+3} = d$. If $y_{i+3} = a$, by Theorem 2.2, $y_{i+2} y_{i+3} \in E_R(G)$. This is a contradiction. So $y_{i+3} = d$. We take a cycle $C' = y_1 y_2 ... y_{i+3} y_1$. Clearly, $E(C') \subset E_N(G)$. However, by Theorem 2.3, we know $\langle E_N(G) \rangle$ does not contain any cycle. So this also leads to a contradiction. Therefore, $|\langle E_N(G) \rangle| \leq |G| - 2$. $\square$

Lemma 3.3. Let G be a 3-connected graph. If there exists a spanning tree T' which contains at most one removable edge, then $g(G) = 3$.

Proof. Let $E_0 = E_N(G) \cap E(T')$ and $F = E(T') - E_0$. Then we have that $|F| \leq 1$. Take a separating group $(uw, S'; A', B')$ where $u \in A'$, $w \in B'$ and $uw \in E_0$. From $|F| \leq 1$, we can get that $(E(A') \cup \langle A', S' \rangle) \cap F = \emptyset$ or $(E(B') \cup \langle B', S' \rangle) \cap F = \emptyset$. Without loss of generality, we assume $(E(A') \cup \langle A', S' \rangle) \cap F = \emptyset$. Since A' is an E_0-edge-vertex-cut fragment, we have an E_0-edge-vertex end-fragment A as its subset. Let $(xy, S; A, B)$ be a separating group where $x \in A$, $y \in B$ and $xy \in E_0$. If $|A| = 2$, the result is easy to be established. Next, we only need to discuss the case $|A| \geq 3$. Since T' is a spanning tree of G, we have that edge $uz \in (E(A) \cup \langle A, S \rangle) \cap E_0$. Take its corresponding separating group $(uz, T; C, D)$ where $u \in C$ and $z \in D$. We assume w.l.o.g. that $u \in A$. Clearly, $u = x$ or $u \neq x$. Let

$$X_1 = (C \cap S) \cup (S \cap T) \cup (A \cap T), X_2 = (A \cap T) \cup (S \cap T) \cup (S \cap D),$$
$$X_3 = (D \cap S) \cup (S \cap T) \cup (B \cap T), X_4 = (B \cap T) \cup (S \cap T) \cup (S \cap C).$$

We will distinguish the following cases to complete the proof.

Case 1. $u = x$. By Theorem 2.2, we know that $z \in A \cap D$ and $y \in B \cap C$. Since $A \cap D \neq \emptyset$, X_2 is 2-vertex cut of $G - xz$. So $|X_2| \geq 2$. By a similar way, we can get $|X_4| \geq 2$. Noticing that $|X_2| + |X_4| = 4$, we have $|X_2| = |X_4| = 2$. So $|D \cap S| = |B \cap T|$.

We claim that $A \cap D = \{z\}$. Otherwise, $|A \cap D| \geq 2$. Let $A_1 = A \cap D$, $S_1 = X_2$ and $B_1 = G - xz - S_1 - A_1$. So $(xz, S_1; A_1, B_1)$ is a separating group of G. Since $xz \in E_0$, A_1 is an E_0-edge-vertex-cut fragment which is a proper subset of A. This contradicts that A is an E_0-edge-vertex end-fragment.

Since $|D| \geq 2$ and D is connected, we have that $|D \cap S| = |B \cap T| \geq 1$. If $|D \cap S| = |B \cap T| = 2$, then $|X_1| = 0$ and hence $\{z, y\}$ is a 2-vertex cut of G. This is a contradiction. So $|D \cap S| = |B \cap T| = 1$. Then $|S \cap T| \leq 1$. If $|S \cap T| = 1$, then $C \cap S = A \cap T = \emptyset$. Since $|A| \geq 3$ and $|A \cap D| = 1$, we get that $|A \cap C| \geq 2$. So $\{x\} \cup (S \cap T)$ is a 2-vertex cut of G. Therefore, $|S \cap T| = 0$. So $|X_3| = 2$. Hence we get that $B \cap D = \emptyset$ and $|D| = 2$. Obviously, $g(G) = 3$.

Case 2. $u \neq x$. Then $uz \in E(A)$ or $uz \in \langle A, S \rangle$. Since $xy \in E_N(G)$, we know that $xy \notin E(T)$. We will discuss the following cases to proceed the proof.

(2.1) $uz \in E(A)$. We get that $u \in A \cap C$ and $z \in A \cap D$.

(2.1.1) $x \in A \cap C$, $y \in B \cap C$. By a similar argument used in Case 1, we have that $A \cap D = \{z\}$ and $|D \cap S| = |B \cap T| = 1$. Then $|S \cap T| \leq 1$.

We claim that $|S \cap T| = 0$. If not, then we have $C \cap S = A \cap T = \emptyset$. Let $A_1 = A \cap C$, $S_1 = \{z\} \cup (S \cap T)$ and $B_1 = G - xy - S_1 - A_1$. Clearly $(xy, S_1; A_1, B_1)$ is a separating group. Since $xy \in E_0$, A_1 is an E_0-edge-vertex-cut fragment, which is a proper subset of A. This contradicts that A is an E_0-edge-vertex end-fragment.

So $|X_3| = 2$ and then $B \cap D = \emptyset$. Therefore $|D| = 2$. It is easy to see that $g(G) = 3$.

(2.1.2) $x \in A \cap T$, $y \in B \cap C$. Since $A \cap D \neq \emptyset$, X_2 is a vertex cut of $G - uz$. So $|X_2| \geq 2$. Similarly, $|X_4| \geq 2$. From $|X_2| + |X_4| = 4$, we have that $|X_2| = |X_4| = 2$. Then, $|C \cap S| = |A \cap T| \geq 1$ and so $|D \cap S| \leq |S| - |C \cap S| = 1$. Hence, $|X_1| \geq 2$. From $|X_1| + |X_3| = 4$, we have $|X_3| \leq 2$. Then $B \cap D = \emptyset$. Since $|X_2| = 2$, similar to Case 1, we know that $A \cap D = \{z\}$. Since $|D| \geq 2$, $|S \cap D| = |D| - |A \cap D| \geq 1$. So we have that $|S \cap D| = 1$ and $|D| = 2$. Obviously, $g(G) = 3$.

(2.1.3) $x \in A \cap C$, $y \in B \cap T$. Since $A \cap D \neq \emptyset$, X_2 is a vertex cut of $G - uz$. So $|X_2| \geq 2$. Then $|X_4| \leq 2$ and $B \cap C = \emptyset$.

We claim that $A \cap T = \emptyset$. Otherwise, $A \cap T \neq \emptyset$. Noticing that $y \in B \cap T$ and $|T| = 2$, we have that $|A \cap T| = 1$. So $|X_1| \geq 1$. Suppose $|X_1| = 1$. Let $A' = A \cap C, S' = X_1 \cup \{z\}$ and $B' = G - xy - S' - A'$. Then $(xy, S'; A', B')$ is a separating group of G. However, $A' \subset A$ and $|A'| < |A|$. This contradicts the fact that A is an E_0-edge-vertex end-fragment. So $|X_1| \geq 2$ and $|X_4| \leq 2$. Therefore, $B \cap D = \emptyset$ and $B = B \cap T = \{y\}$. This contradicts that $|B| \geq 2$.

Since $|X_2| \leq |S| = 2$, we have that $|X_2| = |S|$. So $C \cap S = \emptyset$. Clearly, $C = A \cap C$. We can deduce that C is an E_0-edge-vertex-cut fragment which is a proper subset of A. This is a contradiction. So the subcase is impossible.

By symmetry, the other cases are similarly discussed.

(2.2) $uz \in \langle A, S \rangle$. Then we have that $u \in A \cap C$, $z \in D \cap S$. We will discuss the following cases to complete the proof.

(2.2.1) $x \in A \cap C$, $y \in B \cap C$. Since $B \cap C \neq \emptyset$, we get that X_4 is a vertex cut of $G - xy$. So $|X_4| \geq 2$. Then $|X_2| \leq 2$. We have that $A \cap D = \emptyset$.

We claim that $A \cap T \neq \emptyset$. Otherwise, $A \cap T = \emptyset$. Then $|A| = |A \cap C| \geq 3$. Since X_1 is a vertex cut of $G - xy - uz$, we get that $|X_1| \geq 1$. Noticing that $z \in D \cap S$ and $|S| = 2$, we have $|X_1| = |S \cap (C \cup T)| = 1$. Let $A_1 = A - \{u\}, S_1 = X_1 \cup \{u\}, B_1 = G - xy - S_1 - A_1$. Then $(xy, S_1; A_1, B_1)$ is a separating group of G. Clearly, A_1 is an E_0-edge-vertex-cut fragment which is a proper subset of A. This contradicts that A is an E_0-edge-vertex end-fragment.

Since $A \cap T \neq \emptyset$, we have $|T \cap (S \cup B)| \leq 1$. From $|X_4| \geq 2$, we get $|S \cap C| \geq |A \cap T| \geq 1$. Noticing that $|S| = 2$, we get that $|S \cap D| = 1$ and $|X_3| \leq 2$. Hence, $B \cap D = \emptyset$. Then $D = \{z\}$. This contradicts that $|D| \geq 2$. So this subcase is not possible.

(2.2.2) $x \in A \cap C$, $y \in B \cap T$.

If $A \cap T = \emptyset$, noticing that A is connected, then we have that $A \cap D = \emptyset$. Similar to (2.2.1), we can get an E_0-edge-vertex-cut fragment $A_1 = A - \{u\}$ as a proper subset of A, which contradicts that A is an E_0-edge-vertex end-fragment. So $A \cap T \neq \emptyset$. Since $y \in B \cap T$ and $|T| = 2$, we get that $|B \cap T| = |A \cap T| = 1$ and $S \cap T = \emptyset$.

If $C \cap S = \varnothing$, noticing that C is connected, then $B \cap C = \varnothing$. Clearly, C is an E_0-edge-vertex-cut fragment which is a proper subset of A. This is a contradiction. So $C \cap S \neq \varnothing$. From $|S| = 2$, we know that $|C \cap S| = |D \cap S| = 1$ and $|S \cap T| = 0$.

Since $|X_3| = 2 = |X_2|$, we can deduce that $B \cap D = \varnothing = A \cap D$. Then $D = \{z\}$, which contradict that $|D| \geq 2$. So this subcase is impossible.

(2.2.3) $x \in A \cap T, y \in B \cap C$. Since $B \cap C \neq \varnothing$, we have that $|X_4| \geq 2$. Noticing that $|A \cap T| \geq 1$, we have $|S \cap C| = |X_4| - (|T| - |A \cap T|) \geq 1$. Since $|S| = 2$ and $z \in D \cap S$, we know that $|S \cap C| = |D \cap S| = 1$ and $S \cap T = \varnothing$. From $|X_4| \geq 2$ and $|T| = 2$, we get that $|B \cap T| = |A \cap T| = 1$. Since $|X_2| = |X_3| = 2$, we deduce $A \cap D = \varnothing = B \cap D$. So $D = \{z\}$, which contradicts that $|D| \geq 2$. So this subcase is impossible.

(2.2.4) $x \in A \cap T, y \in B \cap D$. By a similar argument used in (2.2.2), we can deduce that $|C \cap S| = 1 = |D \cap S|$ and $|S \cap T| = 0$. Since X_1 is a vertex cut of $G - uz$, $|X_1| \geq 2$. Similarly, $|X_3| \geq 2$. From $|X_1| + |X_3| = |S| + |T| = 4$, we have that $|X_1| = |X_3| = 2$. So $|A \cap T| = |D \cap S|, |C \cap S| = |B \cap T|$. Then we get $|B \cap T| = |A \cap T| = 1$.

From $|X_2| = |X_4| = 2$, we can deduce that $A \cap D = \varnothing = B \cap C$. Since $|A| \geq 3$, we have $|A \cap C| \geq 2$. Let $A_1 = A \cap C, S_1 = X_1, B_1 = G - uz - S_1 - A_1$. Then $(uz, S_1; A_1, B_1)$ is a separating group of G and A_1 is an E_0-edge-vertex-cut fragment which is a proper subset of A. This is a contradiction. So this subcase is impossible.

(2.2.5) $x \in A \cap D, y \in B \cap T$. Since X_1 is a vertex cut of $G - uz$, we have $|X_1| \geq 2$. Noticing that $|X_1| + |X_3| = 4$ and $|X_3| \geq 2$, we get that $|X_1| = |X_3| = 2$. Then $|B \cap T| = |C \cap S| = |A \cap T| = |D \cap S| = 1$ and $S \cap T = \varnothing$. Since $|X_3| = |X_4| = 2$, we dan deduce that $B \cap D = B \cap C = \varnothing$. So $|B| = |B \cap T| = 1$. This is a contradiction. This subcase is impossible.

(2.2.6) $x \in A \cap D, y \in B \cap D$. By an argument analogous to that used in (2.2.2), we can deduce that $|C \cap S| = 1 = |D \cap S|$ and $|S \cap T| = 0$.

Since X_1 is a vertex cut of $G - uz$, we get $|X_1| \geq 2$. Similarly, we have $|X_3| \geq 2$. From $|X_1| + |X_3| = 4$, we can get $|X_1| = |X_3| = 2$. So we have $|B \cap T| = |C \cap S| = 1 = |A \cap T| = |D \cap S|$. Therefore, $|X_1| = |X_4| = 2$. So $B \cap C = \varnothing$.

We claim that $A \cap C = \{u\}$. Otherwise, $|A \cap C| \geq 2$. Noticing that $|X_1| = 2$, by a similar argument used in (2.2.4), we can get $A_1 = A \cap C$ is an E_0-edge-vertex-cut fragment which is a proper subset of A. This is a contradiction. Thus, $|C| = |A \cap C| + |S \cap C| = 2$ and so $g(G) = 3$.

The proof is now complete. $\square$

Now, we present our main result.

Theorem 3.4. Let G be a 3-connected 3-regular graph and T' a spanning tree of G. Then $|E(T') \cap E_R(G)| \geq 2$.

proof. By Theorem 2.3, we know that $\langle E_N(G)\rangle$ does not contain any cycle. We will distinguish the following cases to proceed the proof.

Case 1. $\langle E_N(G)\rangle$ is connected. It follows from Lemma 3.2 that $|\langle E_N(G)\rangle| \leq |G|-2$. Since T' is a spanning tree of G, it is easy to see that $|E(T')\cap E_R(G)| \geq 2$.

Case 2. $\langle E_N(G)\rangle$ is not connected. If $\langle E_N(G)\rangle$ is not a spanning forest of G, clearly, $|E(T')\cap E_R(G)| \geq 2$. Next we suppose that $\langle E_N(G)\rangle$ is a spanning forest. Let its components be $T_1, T_2, ...T_k$. We claim that $k \geq 3$. By contradiction. We assume that $k = 2$. Let its two components be T_1, T_2. Clearly, $|T_1| \geq 2, |T_2| \geq 2$. Let v be a leaf of T_1 and $\Gamma_G(v) = \{a, b, c\}$ where $a \in T_1$. We claim that $b, c \in T_2$. Otherwise, without loss of generality, $b \in T_1$. Then $bv \in E_R(G)$. So there is a cycle C' which contains bv and $E(C') - bv \subseteq E_N(G)$. Similar to the proof of Lemma 3.2, we can get a contradiction. Let u be a vertex of degree 2 in T_1. If $\Gamma_G(u) \subset V(T_1)$, then there exists a vertex $u_1 \in V(T_1)$ such that $uu_1 \in E_R(G)$. Similar to Lemma 3.2, we can also deduce a contradiction. For T_2, we can discuss it in a similar way. Therefore, $\langle E_R(G)\rangle$ is a bipartite graph. On the other hand, let $T'' = T_1 + T_2 \cup \{uv\}$ where $u \in T_1, v \in T_2$ and $uv \in E_R(G)$. Clearly, T'' is a spanning tree of G and $|E(T'') \cap E_R(G)| = 1$. By Lemma 3.3, $g(G) = 3$. Let C be a 3-cycle of G. Since G is 3-regular, by Theorem 2.5, we know that $E(C) \subset E_R(G)$. That is to say $\langle E_R(G)\rangle$ contains a 3-cycle, which contradicts that $\langle E_R(G)\rangle$ is a bipartite graph. So $k \geq 3$. Now, it is easy to see that $|E(T') \cap E_R(G)| \geq 2$. The proof is now complete. $\square$

Finally, to end the paper, we design a polynomial time algorithm to detect all removable edges in a 3-connected 3-regular graph G. Let $|G| = n$. Clearly, we can observe that

Lemma 3.5. Let G be a 3-connected 3-regular graph and $e = xy$ an edge. If neither x nor y belongs to a triangle of G, then $G \ominus e$ is still 3-regular.

In [2], there is a polynomial time algorithm to find the vertex connectivity of graphs. When the algorithm is used to a 3-regular graph, we can get an $O(n^2)$ algorithm to determine whether it is 3-connected. For convenience, we call the algorithm Fast-Connect. Now, we give our algorithm to determine whether an edge $e = xy$ is removable:

(1) If e is in some triangle of G, according to Theorem 2.5, then e is removable. otherwise, to (2)

(2) If x or y is in some triangle, noticing that the minimum degree of $G \ominus e$ is 2, then we have that e is unremovable. Otherwise, to (3)

(3) Use algorithm Fast-Connect to determine whether $G \ominus e$ is 3-connected. If $G \ominus e$ is 3-connected, then e is removable. Otherwise, e is unremovable.

Next, we simply analyze the complexity of the algorithm. Obviously, it take $O(n^2)$ time to determine if an edge e is removable. So the total time is $O(n^3)$

to detect the number of removable edges of a 3-connected 3-regular graph, since $|E(G)| = 3/2n$.

Acknowledgments. The authors are grateful to the referees for valuable suggestions and comments, which were very helpful for improving the presentation of the paper.

References

1. Bondy, J.A., Murty, U.S.R.: Graph theory with applications. North-Holland, Amersterdam (1976)
2. Gabow, H.N.: Using expander graphs to find vertex connectivity. Journal of ACM 53(5), 800–844 (2006)
3. Holton, D.A., Jackson, B., Saito, A., Wormald, N.C.: Removable edges in 3-connected graphs. Journal of Graph Theory 14(4), 465–473 (1990)
4. Martionov, N.: Uncontractible 4-connected graphs. Journal of Graph theory 6(3), 343–344 (1982)
5. Ou, J.P., Su, J.J.: Distribution of removable edges in 3-connected Graphs. Journal of Guangxi Normal University(Science) 19(1), 25–29 (2001)
6. Su, J.J.: The removable edges in 3-connected graphs. Journal of Combinatorial theory, Series B 75(1), 74–87 (1999)
7. Su, J.J.: Removable edges of cycles in 3-connected graphs. Chinese Science Bulletin 44(9), 921–926 (1999)
8. Tutte, W.T.: A theory of 3-connected graphs. Indagationes Mathematicae 23, 441–455 (1961)

Author Index

Lecture Notes in Computer Science

For information about Vols. 1–4525

please contact your bookseller or Springer